# Modern Digital Control Systems

## ELECTRICAL ENGINEERING AND ELECTRONICS

*A Series of Reference Books and Textbooks*

1. Rational Fault Analysis, *edited by Richard Saeks and S. R. Liberty*
2. Nonparametric Methods in Communications, *edited by P. Papantoni-Kazakos and Dimitri Kazakos*
3. Interactive Pattern Recognition, *Yi-tzuu Chien*
4. Solid-State Electronics, *Lawrence E. Murr*
5. Electronic, Magnetic, and Thermal Properties of Solid Materials, *Klaus Schröder*
6. Magnetic-Bubble Memory Technology, *Hsu Chang*
7. Transformer and Inductor Design Handbook, *Colonel Wm. T. McLyman*
8. Electromagnetics: Classical and Modern Theory and Applications, *Samuel Seely and Alexander D. Poularikas*
9. One-Dimensional Digital Signal Processing, *Chi-Tsong Chen*
10. Interconnected Dynamical Systems, *Raymond A. DeCarlo and Richard Saeks*
11. Modern Digital Control Systems, *Raymond G. Jacquot*
12. Hybrid Circuit Design and Manufacture, *Roydn D. Jones*
13. Magnetic Core Selection for Transformers and Inductors: A User's Guide to Practice and Specification, *Colonel Wm. T. McLyman*
14. Static and Rotating Electromagnetic Devices, *Richard H. Engelmann*
15. Energy-Efficient Electric Motors: Selection and Application, *John C. Andreas*
16. Electromagnetic Compossibility, *Heinz M. Schlicke*
17. Electronics: Models, Analysis, and Systems, *James G. Gottling*
18. Digital Filter Design Handbook, *Fred J. Taylor*
19. Multivariable Control: An Introduction, *P. K. Sinha*
20. Flexible Circuits: Design and Applications, *Steve Gurley, with contributions by Carl A. Edstrom, Jr., Ray D. Greenway, and William P. Kelly*
21. Circuit Interruption: Theory and Techniques, *Thomas E. Browne, Jr.*
22. Switch Mode Power Conversion: Basic Theory and Design, *K. Kit Sum*
23. Pattern Recognition: Applications to Large Data-Set Problems, *Sing-Tze Bow*
24. Custom-Specific Integrated Circuits: Design and Fabrication, *Stanley L. Hurst*
25. Digital Circuits: Logic and Design, *Ronald C. Emery*
26. Large-Scale Control Systems: Theories and Techniques, *Magdi S. Mahmoud, Mohamed F. Hassan, and Mohamed G. Darwish*
27. Microprocessor Software Project Management, *Eli T. Fathi and Cedric V. W. Armstrong (Sponsored by Ontario Centre for Microelectronics)*
28. Low Frequency Electromagnetic Design, *Michael P. Perry*
29. Multidimensional Systems: Techniques and Applications, *edited by Spyros G. Tzafestas*
30. AC Motors for High-Performance Applications: Analysis and Control, *Sakae Yamamura*

31. Ceramic Motors for Electronics: Processing, Properties, and Applications, *edited by Relva C. Buchanan*
32. Microcomputer Bus Structures and Bus Interface Design, *Arthur L. Dexter*
33. End User's Guide to Innovative Flexible Circuit Packaging, *Jay J. Miniet*
34. Reliability Engineering for Electronic Design, *Norman B. Fuqua*
35. Design Fundamentals for Low-Voltage Distribution and Control, *Frank W. Kussy and Jack L. Warren*
36. Encapsulation of Electronic Devices and Components, *Edward R. Salmon*
37. Protective Relaying: Principles and Applications, *J. Lewis Blackburn*
38. Testing Active and Passive Electronic Components, *Richard F. Powell*
39. Adaptive Control Systems: Techniques and Applications, *V. V. Chalam*
40. Computer-Aided Analysis of Power Electronic Systems, *Venkatachari Rajagopalan*
41. Integrated Circuit Quality and Reliability, *Eugene R. Hnatek*
42. Systolic Signal Processing Systems, *edited by Earl E. Swartzlander, Jr.*
43. Adaptive Digital Filters and Signal Analysis, *Maurice G. Bellanger*
44. Electronic Ceramics: Properties, Configuration, and Applications, *edited by Lionel M. Levinson*
45. Computer Systems Engineering Management, *Robert S. Alford*
46. Systems Modeling and Computer Simulation, *edited by Naim A. Kheir*
47. Rigid-Flex Printed Wiring Design for Production Readiness, *Walter S. Rigling*
48. Analog Methods for Computer-Aided Circuit Analysis and Diagnosis, *edited by Takao Ozawa*
49. Transformer and Inductor Design Handbook: Second Edition, Revised and Expanded, *Colonel Wm. T. McLyman*
50. Power System Grounding and Transients: An Introduction, *A. P. Sakis Meliopoulos*
51. Signal Processing Handbook, *edited by C. H. Chen*
52. Electronic Product Design for Automated Manufacturing, *H. Richard Stillwell*
53. Dynamic Models and Discrete Event Simulation, *William Delaney and Erminia Vaccari*
54. FET Technology and Application: An Introduction, *Edwin S. Oxner*
55. Digital Speech Processing, Synthesis, and Recognition, *Sadaoki Furui*
56. VLSI RISC Architecture and Organization, *Stephen B. Furber*
57. Surface Mount and Related Technologies, *Gerald Ginsberg*
58. Uninterruptible Power Supplies: Power Conditioners for Critical Equipment, *David C. Griffith*
59. Polyphase Induction Motors: Analysis, Design, and Application, *Paul L. Cochran*

60. Battery Technology Handbook, *edited by H. A. Kiehne*
61. Network Modeling, Simulation, and Analysis, *edited by Ricardo F. Garzia and Mario R. Garzia*
62. Linear Circuits, Systems, and Signal Processing: Advanced Theory and Applications, *edited by Nobuo Nagai*
63. High-Voltage Engineering: Theory and Practice, *edited by M. Khalifa*
64. Large-Scale Systems Control and Decision Making, *edited by Hiroyuki Tamura and Tsuneo Yoshikawa*
65. Industrial Power Distribution and Illuminating Systems, *Kao Chen*
66. Distributed Computer Control for Industrial Automation, *Dobrivoje Popovic and Vijay P. Bhatkar*
67. Computer-Aided Analysis of Active Circuits, *Adrian Ioinovici*
68. Designing with Analog Switches, *Steve Moore*
69. Contamination Effects on Electronic Products, *Carl J. Tautscher*
70. Computer-Operated Systems Control, *Magdi S. Mahmoud*
71. Integrated Microwave Circuits, *edited by Yoshihiro Konishi*
72. Ceramic Materials for Electronics: Processing, Properties, and Applications, Second Edition, Revised and Expanded, *edited by Relva C. Buchanan*
73. Electromagnetic Compatibility: Principles and Applications, *David A. Weston*
74. Intelligent Robotic Systems, *edited by Spyros G. Tzafestas*
75. Switching Phenomena in High-Voltage Circuit Breakers, *edited by Kunio Nakanishi*
76. Advances in Speech Signal Processing, *edited by Sadaoki Furui and M. Mohan Sondhi*
77. Pattern Recognition and Image Preprocessing, *Sing-Tze Bow*
78. Energy-Efficient Electric Motors: Selection and Application, Second Edition, *John C. Andreas*
79. Stochastic Large-Scale Engineering Systems, *edited by Spyros G. Tzafestas and Keigo Watanabe*
80. Two-Dimensional Digital Filters, *Wu-Sheng Lu and Andreas Antoniou*
81. Computer-Aided Analysis and Design of Switch-Mode Power Supplies, *Yim-Shu Lee*
82. Placement and Routing of Electronic Modules, *edited by Michael Pecht*
83. Applied Control: Current Trends and Modern Methodologies, *edited by Spyros G. Tzafestas*
84. Algorithms for Computer-Aided Design of Multivariable Control Systems, *Stanoje Bingulac and Hugh F. VanLandingham*
85. Symmetrical Components for Power Systems Engineering, *J. Lewis Blackburn*
86. Advanced Digital Signal Processing: Theory and Applications, *Glenn Zelniker and Fred J. Taylor*
87. Neural Networks and Simulation Methods, *Jian-Kang Wu*
88. Power Distribution Engineering: Fundamentals and Applications, *James J. Burke*

89. Modern Digital Control Systems, Second Edition, *Raymond G. Jacquot*
90. Adaptive IIR Filtering in Signal Processing and Control, *Philip Regalia*

*Additional Volumes in Preparation*

Integrated Circuit Quality and Reliability, Second Edition, Revised and Expanded, *Eugene R. Hnatek*

Handbook of Electric Motors, *edited by Richard Engelmann and William H. Middendorf*

# Modern Digital Control Systems

## Second Edition

**Raymond G. Jacquot**
*University of Wyoming*
*Laramie, Wyoming*

**Materials formerly available with this book on CD-ROM or floppy disk are now available for download from our website www.crcpress.com. Go to the Downloads & Updates tab on the book's page.**

CRC Press
Taylor & Francis Group
Boca Raton London New York

CRC Press is an imprint of the
Taylor & Francis Group, an **informa** business

CRC Press
Taylor & Francis Group
6000 Broken Sound Parkway NW, Suite 300
Boca Raton, FL 33487-2742



**Visit the Taylor & Francis Web site at**
**http://www.taylorandfrancis.com**

**and the CRC Press Web site at**
**http://www.crcpress.com**

To the teachers of teachers:
Raymond Cohen, Edward M. Lonsdale, and Robert L. Sutherland

# Preface to the Second Edition

This book is written for students who have had one course in conventional feedback control systems. The essentials needed to begin this course are use of the root locus, the Nyquist and Bode techniques for single-loop control system design, and an introduction to system representation using state variables. It is assumed that students will be well grounded in the use of the Laplace transform and $s$-domain concepts and will possess a background in elementary matrix theory and linear algebra. For study of the final chapters, it is also assumed that the student is versed in probability theory and possesses introductory knowledge in the area of continuous-time stochastic processes. The material in this book should also be useful to practicing engineers who have studied the concepts of continuous-time control of dynamic systems and wish to extend their knowledge to the domain of computer-controlled systems.

In the author's experience, it seemed that learning discrete-data concepts has, in the past, been unnecessarily complicated by the impulse sampling model, which bears no resemblance to what happens in most digital control systems except those involving data representation and transmission using pulse-width or pulse-amplitude modulation. After considerable study, the author decided that the subject of discrete-time control systems could be introduced with reasonable rigor without the complication of modulated impulse functions. This approach works well for single-rate sampling but is troublesome for the multirate problem. Several example physical problems, which provide an interesting common thread throughout the book, tie the theoretical developments together.

This second edition differs from the first in that the chapter on the impulse sampling model has been deleted because the author felt that it was not necessary for complete coverage. The material from the former Chapter 3 has been split and amended to form what are now Chapters 3 and 4, which include a detailed exposition of the Ragazzini direct design method. The former Chapter 4, now Chapter 5, on digital filtering has been revised considerably, including all new examples and use of the Butterworth prototype filters. Chapter 8 has been revised extensively to include concepts that are useful in the computer-aided control system design (CACSD) environment. The problem of transfer function identification is addressed in the new Chapter 13, in which a least-squares approach is taken.

Also included with this edition is an integrated software package, *Digital Control Designer*, which can be useful in teaching and learning of the material through Chapter 9, including the solution of the problems. Several versions of this software have been employed at the University of Wyoming over the past decade and found to relieve the computational tedium associated with higher-order problems and allow students to concentrate on design concepts rather than getting the correct numerical answers to low-order problems.

The author has taught several courses from the preliminary versions of the second edition materials and found it was possible to cover the materials of Chapters 1 through 6, all of Chapter 8, and the first half of Chapter 9 in a three-semester-hour course to beginning graduate students. The materials in the latter portions of the book have been employed as supplementary materials in an estimation theory course. For a number of years the author has used an associated set of laboratory exercises based on a personal computer with analog-to-digital and digital-to-analog converters and with an analog computer for modeling simple continuous-time plant models.

The material contained herein is largely the product of concepts developed by the author during a sabbatical leave to the Department of Aeronautics and Astronautics at Stanford University. The author is particularly indebted to Professors Arthur E. Bryson, Jr., and J. David Powell for the valuable discussions that stimulated this book.

The author would like to thank his colleagues Dr. Jerry Cupal and Dr. Stanislaw Legowski, who were valuable sources of information on the state-of-the-art microcontroller technology. Dr. Jerry Hamann was also very helpful with his information on modern digital signal processor (DSP) hardware.

Many students who studied early versions of the manuscript made suggestions that resulted in a vastly improved written product. One student, Robert Thurston, was particularly diligent in this effort and pointed out many errors in the written English and some conceptual errors in the early chapters of the manuscript. Thanks are still due to those students who were helpful with the first edition in that it served as a skeleton on which this second edition is constructed.

The author is much indebted to Shelley Straley for her patience and accuracy in the word processing of the manuscript. Jean Richardson is also to be thanked for production of the high-quality illustrations, which was no mean feat.

Last, but not least, the author would like to thank the administration of the College of Engineering at the University of Wyoming for nurturing an academic atmosphere in which efforts of this type are still valued.

Raymond G. Jacquot

# Preface to the First Edition

This book is written for students who have had one course in conventional feedback control systems with or without an introduction to the modern concepts of state variables. Sufficient material is provided for students lacking background in modern concepts. It is assumed, however, that students will be well grounded in the use of the Laplace transform and $s$-domain concepts and will also possess a background in elementary matrix theory and linear algebra. For study of the final chapters, it is also assumed that the student is versed in the theory of probability and possesses introductory knowledge in the area of continuous-time stochastic processes. This material should also be useful to practicing engineers who have studied the concepts of continuous-time control of dynamic systems.

In the author's experience, it seems that learning discrete data concepts was unnecessarily complicated by introduction of the artifice of impulse sampling. After considerable study of the subject, the author decided that those subjects necessary for discrete-time control systems and discrete-time signal processing could be introduced without the complication of modulated impulse functions. A chapter on impulse sampling is provided for completeness but is not a prerequisite to any other material in the text.

Several example problems with a physical basis provide interesting applications of the theory developed. These examples present a recurring theme in the text and tie the theoretical developments together. These examples are introduced in Chapter 3 and are continuously used through the final chapter.

The author has taught a course from preliminary versions of this material and most recently found it possible to cover all the material in Chapters 1 through 7 with fundamental points taken from Chapters 8 and 9 in a semester. The material in the latter portions of the text has been used as supplementary material in an estimation theory course. The author's course meets for three lectures weekly with one laboratory exercise every two weeks. The importance of these laboratory exercises cannot be overemphasized in the understanding of the theoretical developments they provide.

The material contained herein is largely the product of ideas developed by the author during a sabbatical leave to the Department of Aeronautics and Astronautics at Stanford University. The author is particularly indebted to those valuable discussions with Professors Arthur E. Bryson, Jr., and J. David Powell that stimulated the effort in this writing.

The author is particularly grateful to his colleague Dr. John Steadman, who continually provided encouragement and advice in the development of this work. He was also responsible for the development of the minicomputer system which inspired the laboratory exercises that are so valuable in the teaching of this material.

A course built around this material is no substitute for knowledge of digital hardware and software design. Knowledge of the hardware and software of mini- and microcomputers is highly desirable because these are the devices which are the heart of any digital control system.

The author was considerably aided in the development of the manuscript by the careful attention to detail by students Ken Jensen, Kim Sturm, Mizan Rahman, Mike Petrea, Mark Hepworth, Barry Eklund, Paul Nelson, and Mike Mundt. Without these diligent students, no effort to write text material would have been successful.

Raymond G. Jacquot

# Contents

# Modern Digital Control Systems

# 1 Introduction to Digital Control

## 1.1 THE BASIC IDEA OF SYSTEM CONTROL

The topic with which we deal in this book is that of the control of a dynamic system, the plant, by employing feedback which incorporates a digital computer in the control loop. A dynamic system is one that is described by differential equations, and as such the output variables will not exactly track the reference input variables.

Generally, this plant has continuous-in-time inputs and outputs. In general, the plant has $r$ input variables (the control efforts) and $m$ output variables. There may also be internal variables that are not outputs. Which variables are outputs is a matter for the system designer to decide. The control problem is one of manipulating the input variables in an attempt to influence the output variables in a desired fashion, for example, to achieve certain values or certain rates.

If we know the system model and the initial conditions with reasonable accuracy, we can manipulate the input variables so as to drive the outputs in the desired manner. This is what we would call *open-loop control* since it is accomplished without knowledge of the current outputs. This situation is depicted in Fig. 1.1. This type of control also assumes that the system operates in the absence of disturbances that would cause the system outputs to vary from those predicted by a model of the plant, which is seldom the case.

Because of uncertainties in the system model and initial conditions and because of disturbances there is a better way to accomplish the control task. This technique involves using sensors to measure the behavior of

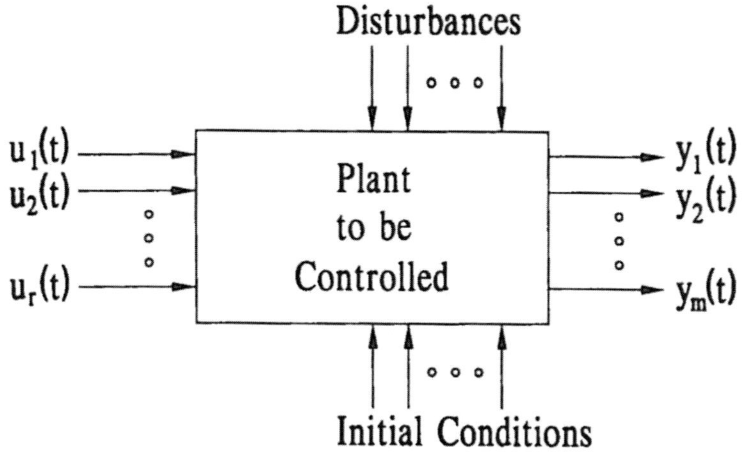


**Figure 1.1.** Open-loop-controlled plant.

some subset of the output variables (those we want to control) $y_1(t), \ldots, y_k(t)$ $(k \leq m)$ and, after measurement, comparing them with what we would like each of them to be at time $t$ and calling the difference between the desired value $r_i(t)$ and the measured value $w_i(t)$ the *error*. Using the errors in each of the variables, we can generate the control efforts so as to drive the errors toward zero. This situation, depicted in Fig. 1.2, is referred to as *feedback control*. Due to friction, inertia, and other dynamic

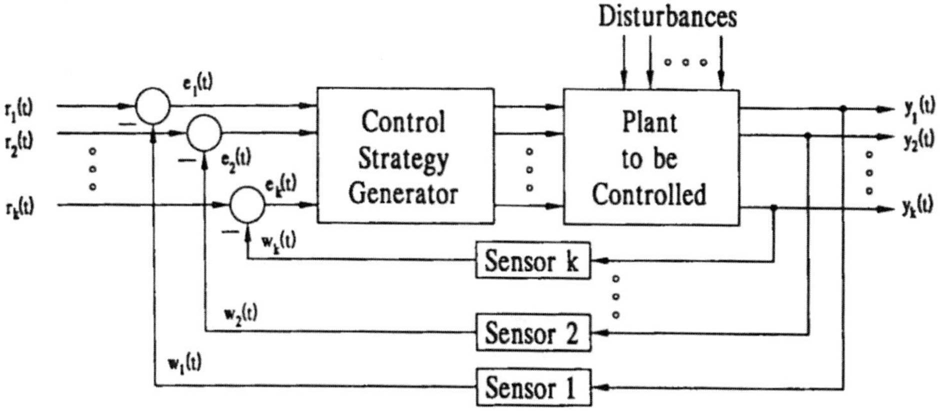


**Figure 1.2.** Feedback-controlled multivariable system with sensors.

properties, it is impossible to drive a system instantaneously to zero error. Thus the system outputs $y_i(t)$ will "lag" the desired outputs $r_i(t)$, and sometimes they will overshoot or oscillate about a zero error condition. Often, the dynamic character of available sensors makes the measurements not exactly true representations of the outputs.

For five decades the sensors and the synthesis of the control strategy generator have been the center of considerable engineering activity, and in general these devices are electrical or electromechanical in nature—more specifically, filters, amplifiers, power amplifiers, motors, and actuators. In all these cases the associated signals are continuous functions of the temporal variable $t$. We commonly refer to such systems as continuous-time control systems. As systems have become more and more complex with a multitude of control variables and output variables, the hardware synthesis job becomes a difficult task. Typical systems are airplane autopilot controls, chemical-processing controls, nuclear power plant controls, and a host of others generally involving large-scale systems.

With the advent of the digital computer, engineers began to explore the possibilities of having the computer keep track of the various signals and make logical decisions about control signals based on the measured signals and the desired values of the outputs. We know, however, that a digital computer is capable only of dealing with numbers and not signals, so if a digital computer is to accomplish the task, the sensor signals must be converted to numbers while the output control decisions must be output in the form of continuous-time signals. We discuss these conversion processes in Section 1.5. As an example, let us consider a single-loop position servomechanism in continuous form as shown in Fig. 1.3. The reference signal is in the form of a voltage, as is the feedback signal, both generated by mechanically driven potentiometers.

## 1.2 THE COMPUTER AS A CONTROL ELEMENT

Let us now investigate how this relatively simple task, outlined in Section 1.1, might be accomplished by employing a digital computer to generate the signal to the power amplifier. We must first postulate the existence of two devices. The first of these devices is the *analog-to-digital* (A/D) *converter*, which will sample the output signal periodically and convert these samples to digital words to be processed by the digital computer, which generates a control strategy in the form of a number. The second device is a *digital-to-analog* (D/A) *converter*, which converts the numerical control strategy generated by the digital computer from a digital word to an analog

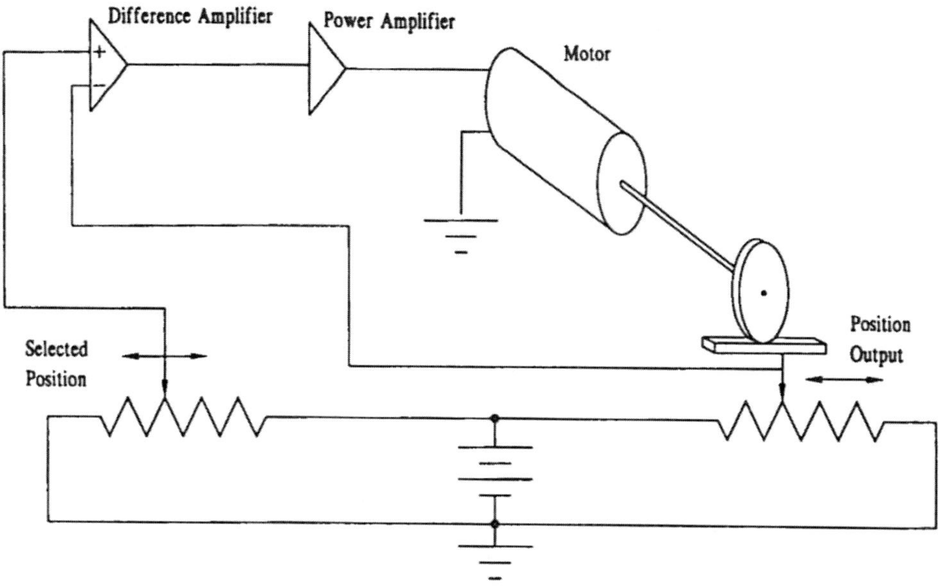


**Figure 1.3.** Position servomechanism with continuous signals.

signal. The position servomechanism is shown in Fig. 1.4 controlled by a digital computer.

This system was not chosen because it was realistic but rather to show the principle of digital control of a single-loop system. Generally, the A/D and D/A converters operate periodically, and hence the closer together in time the samples are taken and the more often the output of the D/A converter is updated, the closer the digital control system will approach the continuous-time system. As we shall learn later, it is not always desirable to have the system approach the continuous system in that there are desirable attributes to a discrete-time system. The limitation on the rate at which sampling can be done is that the sampled signal must be processed by the computer before the computer can drive the D/A converter with the new control strategy. Any algorithm to provide the control signal takes time to execute and hence limits the rate at which control effort update can occur. The general computer-controlled multivariable system is shown in Fig. 1.5.

The engineering problems involved with the design and construction of digital control systems concern the design of faster and more accurate A/D and D/A converters and the synthesis of control algorithms to be executed by the digital computer. Other problems of engineering interest

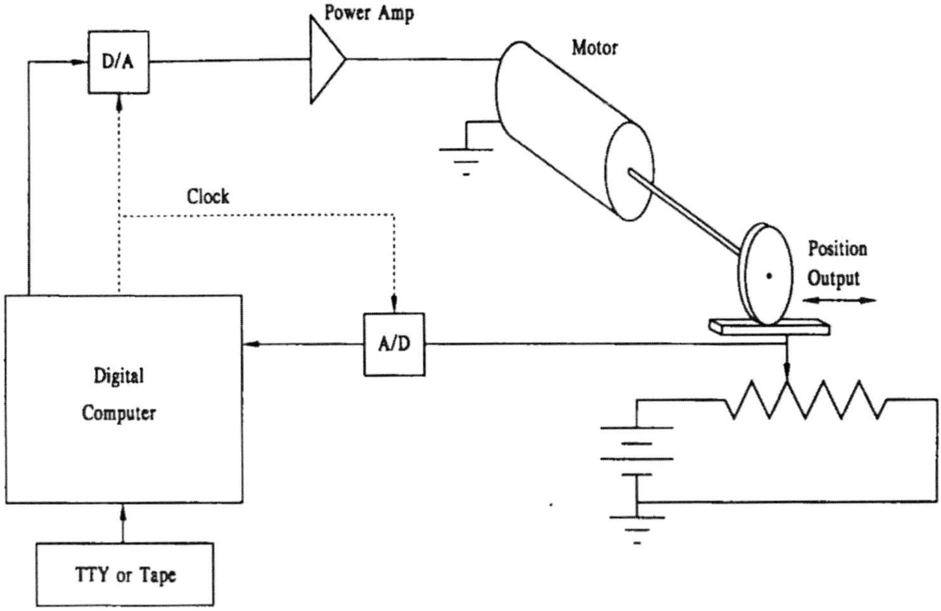


**Figure 1.4.** Digitally controlled positioning system.

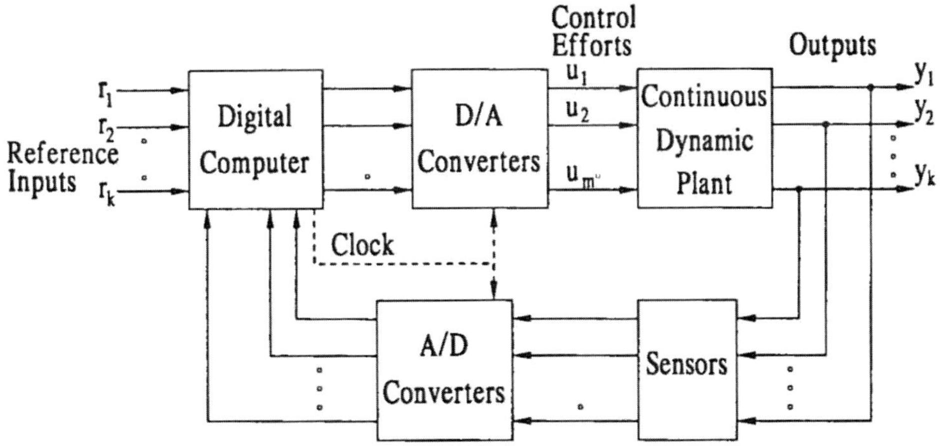


**Figure 1.5.** Digitally controlled multivariable system.

are those of quantization and of finite-word-length representation of numbers which in general require infinite-word-length representation.

Techniques have been established for designing the digital computer algorithm to obtain the best system performance given the dynamics of the plant to be controlled. Generally, the more system variables measured on which to base the control decisions, the better will be the degree of control.

## 1.3 SINGLE-LOOP DIGITAL CONTROL SYSTEM

Often, the problem is control of a single variable of a system which may have a multitude of other variables that are not necessarily to be controlled. There are several configurations of a single-loop control system, two of which are shown in Fig. 1.6a and b. In both cases a single continuous-time variable $y(t)$ is being controlled to follow some reference signal $r(t)$ which might be zero or constant, as in the case of a regulator.

The information leaving the digital computer in both cases is a sequence of numbers written periodically to the D/A converter which represent the control strategy as generated by the computer. The input to the digital computer is a periodic sequence of numbers that represent the samples

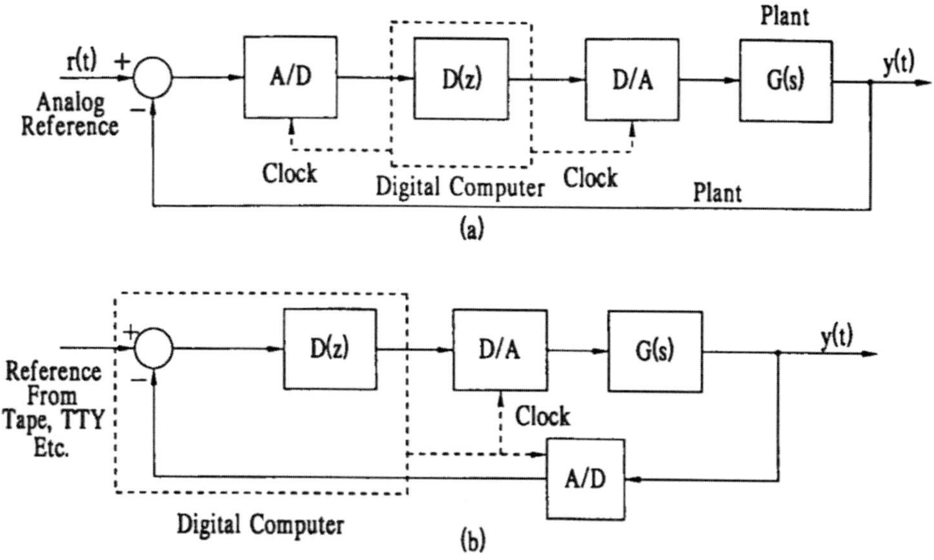


**Figure 1.6.** Several configurations of a digital control system.

taken periodically from the continuous signal that is input to the A/D converter. The purpose of developing a digital control theory is to be able to design desirable algorithms by which the digital computer converts the input sequence into the output sequence, which is the numerical control strategy. The design process is one of selecting the algorithm that reflects the function $D(z)$, which we shall discuss at length in Chapters 3 and 4.

## 1.4 WHY DIGITAL CONTROL INSTEAD OF ANALOG?

Since almost all control functions can be achieved with analog (continuous-time) hardware, one is tempted to ask why we might wish to study digital control theory. Engineers have long been interested in the possibility of incorporating digital computers into the control loop because of the ability of a digital computer to process immense quantities of information and base control strategies logically on that information. If we follow the historical development of digital computers, we find that they were initially very complex large devices which generally cost too much for application in control systems of the moderate degree of complexity found in industry. At this stage cost limitations relegated digital control to only the largest control systems, such as those for oil refineries or large chemical-processing plants.

With the introduction of the minicomputer in the mid-1960s the possibilities for digital control were greatly expanded because of reductions in both size and cost. This allowed application of computers to control smaller and less costly systems. The advent of the microprocessor unit in the early 1970s similarly expanded the horizons, since capabilities that formerly cost thousands of dollars could be purchased for, at most, a few hundreds of dollars. These prices make digital control hardware competitive with analog control hardware for even the simplest single-loop control applications: for example, applications from the automotive and appliance industries. We now see that microprocessor units complete with system controller, arithmetic unit, clock, limited read-only memory (ROM), random-access memory (RAM), and input–output ports on a single very large scale integrated (VLSI), 40-pin integrated circuit chip. It is acknowledged that other hardware is required for A/D and D/A conversion, but inexpensive hardware is also available to accomplish these tasks. Some of these devices now appear as part of the central processing unit (CPU) chip. In Section 1.6 we discuss briefly the current state of microprocessor hardware.

A cost which is not easily estimated is that of software development, which is necessary in control applications, but it is safe to say that the more complex the control task, the more complex the software required, regardless of the digital system employed. Some of these difficulties are now being alleviated by development of high-level-language cross-assemblers for such languages as Pascal and C. The availability of hardware in LSI form has made digital hardware attractive from space, weight, power consumption, and reliability points of view.

Computational speed is affected directly by hardware speed, and hence considerable effort has been expended to increase component speed. This speed has increased exponentially in the past two decades and is illustrated by increasing clock speeds and instruction execution speeds, as typified by the data of Fig. 1.7, which are for the Intel family of microprocessors.

We have discussed the development of VLSI circuit density, which is, in itself, a measure of the progress in the field of digital hardware. It is interesting to note that the MC68020 Motorola microprocessor has in excess of 200,000 transistors on a single integrated circuit chip, and projections are for 50 million transistors per die by the year 2000 (Gelsinger et al., 1989).

Any engineer who has casually witnessed the development of the electronic hand calculator can attest to the rate of change of hardware in

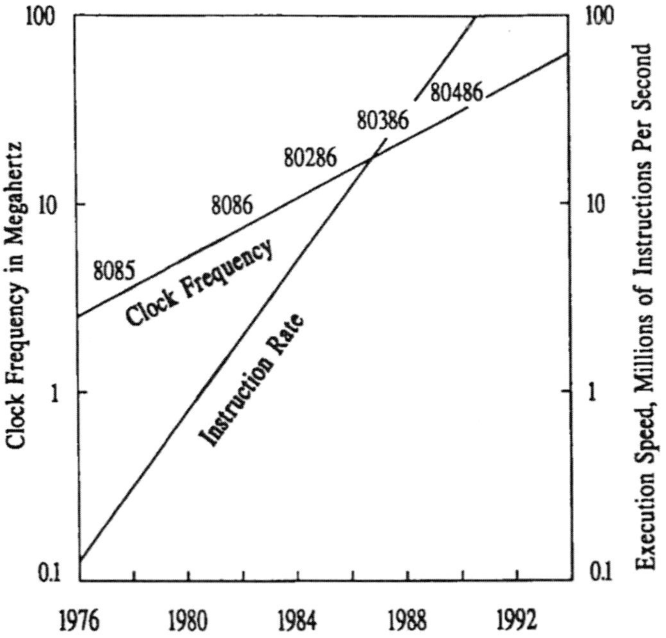


**Figure 1.7.** Progress in microprocessor speed over several decades.

microprocessor-based systems. Currently, microprocessors are employed in the control of many home appliances, automotive fuel metering and spark advance system, inventory control systems, and a host of home entertainment products. It is clear by observation of product applications that the microprocessor will continue to play a larger role in future control and communication applications, particularly those of cellular telephony and computer networking.

## 1.5 DATA CONVERTERS

In Sections 1.2 and 1.3 we discussed the general structure of a digital control system, so it is now appropriate to discuss the interface hardware, which allows the digital computer to take in data from the measured variables of the plant to be controlled and for it to issue control signals for the control of such plants. Let us initiate this discussion with the discussion of the solid-state switch, which is an integral part of the digital-to-analog converters we discuss.

### Solid-State Switch

An important component in both types of digital-to-analog converters which we shall discuss is the solid-state switch shown in Fig. 1.8. The functioning of the switch is controlled by the logical input. If the input is a logical "1" voltage, the switch is closed and the voltage labeled $V_a$ appears at the output terminal, while if the input is a logical "0" voltage, $V_b$ appears at the output. The operation of the switch is undefined for logical input levels other than the acceptable levels for logical "0" and logical "1", which for TTL logic are 0 and 5 V, respectively.

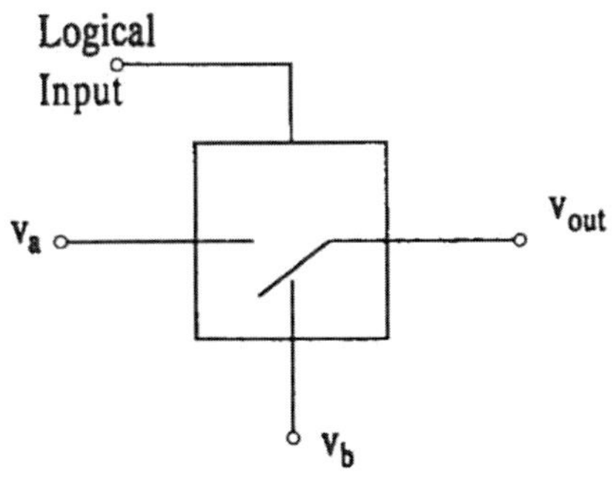


**Figure 1.8.** Solid-state switch.

### Summing Digital-to-Analog Converter

The device that we consider next, illustrated in Fig. 1.9, is chosen to be compatible with transistor–transistor logic (TTL) levels of 0 V for logical "0" and 5 V for logical "1." We consider only the 4-bit version of the device, as it illustrates the general principles involved. We are interested in converting a 4-bit positive binary input to a positive voltage between 0 and 10 V.

Note that the final inverter stage has a gain of negative unity and is present to give a positive analog voltage output. Note that the gain from each of the inputs to the summing amplifier varies from that for adjacent inputs by a factor of 2. If $B_i$ represents a binary digit (0 or 1), the output voltage is

$$v_o = 5\left(B_3 + \frac{B_2}{2} + \frac{B_1}{4} + \frac{B_0}{8}\right) \quad \text{volts} \qquad (1.5.1)$$

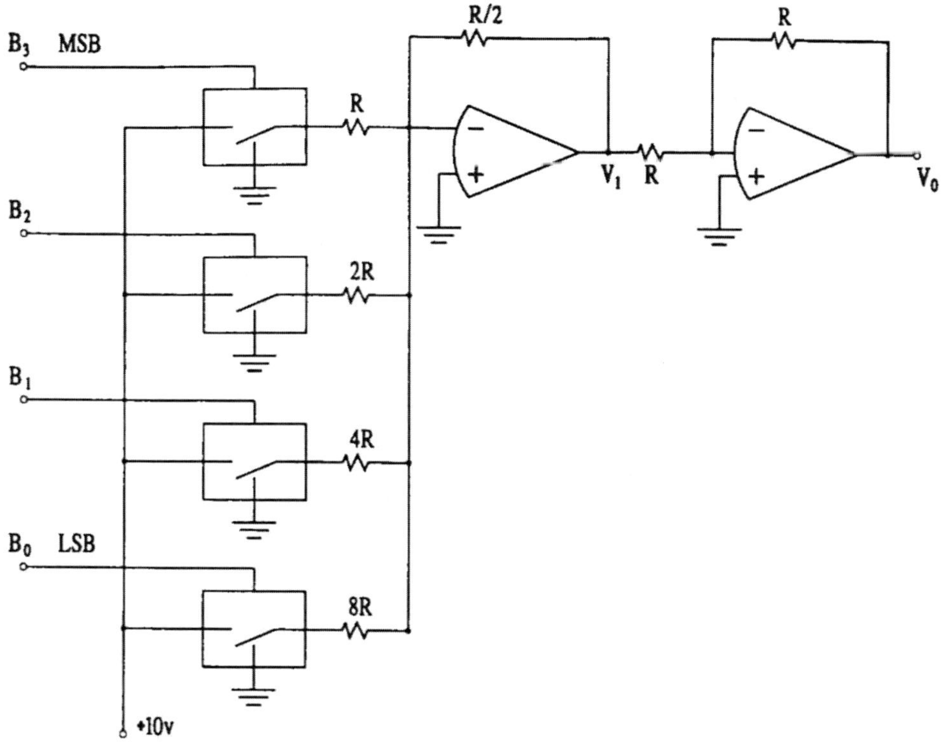


**Figure 1.9.** Summing digital-to-analog converter.

When the binary number at the input is 0000, the output voltage is 0 V. When the input number is as large as it possibly can be (1111), the output is 75/8 = 9.375 V. The steps in the output voltage are 0.625 V for variations in the least significant bit (LSB). In other words, with this 4-bit device the highest resolution in the output voltage is 0.625 V. This is a useful device concept that could be extended to a larger number of bits by including more resistors and switches at the input to the summing amplifier. The implementation of this concept in monolithic silicon circuitry is made difficult by the wide range of resistors $R/2$, $R$, $2R$, . . . , the ratios of which must be controlled accurately in the manufacturing process. As we shall see, an alternative circuit alleviates the problem of a wide range of resistors with little or no complication in the circuitry.

### Ladder Digital-to-Analog Converter

The solid-state switches defined earlier will now be employed with some simple circuitry to illustrate the type of digital-to-analog converter available from a large number of manufacturers. The circuitry for the 4-bit version of the device for an output range of 0 to 10 V is shown in Fig. 1.10.

For arbitrary inputs the output may be calculated by the method of superposition for each bit of input, although the calculations are quite tedious for the lower bits. The input–output relation is

$$v_o = 5\left(B_3 + \frac{B_2}{2} + \frac{B_1}{4} + \frac{B_0}{8}\right) \qquad \text{volts} \qquad (1.5.2)$$

This design eliminates the need to have a wide range of resistances by having only two values, and only the ratio of these must be controlled in the manufacturing process.

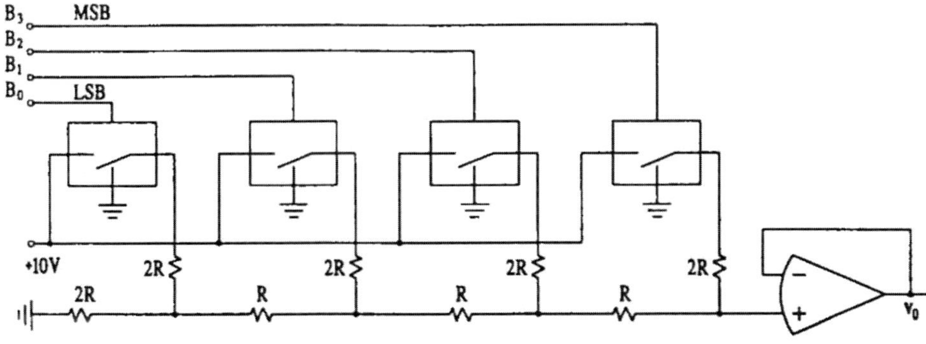


**Figure 1.10.** *R*–2*R* ladder D/A converter.

### Sample-and-Hold Circuit

We need to explore circuitry that will give us an instantaneous (or nearly so) snapshot of some voltage that we wish to convert to a corresponding binary member. We accomplish our task by storing a "sample" of the input voltage on a capacitor which will then be converted to a binary number. The circuit to accomplish this task is shown in Fig. 1.11. The voltage $v_1(t)$ is the voltage we would like to convert to a binary number. In fact, $v_2(t)$ is actually the voltage that will be converted to a binary number. The terminal labeled $v_g(t)$ is driven by the pulse train shown in Fig. 1.12a.

The pulse train $v_g(t)$ causes the field-effect transistor (FET) to attain a low resistance when the pulse is high and a very high resistance (open circuit) when the voltage $v_g$ is zero. Assuming that the input impedance of the operational amplifier is very high when the transistor acts as a short, the capacitor will change to the input voltage with a time constant of

$$\tau = R_{FET}C \tag{1.5.3}$$

The duration of the pulses is chosen to be roughly four times this time constant. After the capacitor is charged, the voltage $v_g$ goes low, thus rendering the FET essentially an open circuit, and then the voltage on the capacitor does not change except for the small charge loss due to the finite (but high) input impedance of the operational amplifier. After the voltage $v_2(t)$ has settled, the analog-to-digital conversion process may take place, so a slightly delayed version of $v_g(t)$ can be used to drive the start-conversion pin on the analog-to-digital converter.

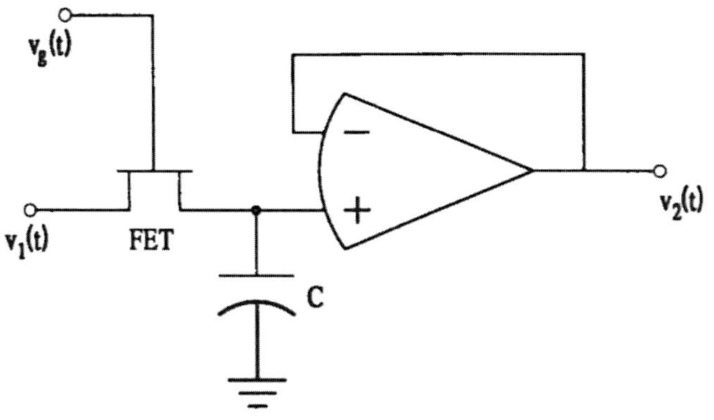


**Figure 1.11.** Sample-and-hold circuit.

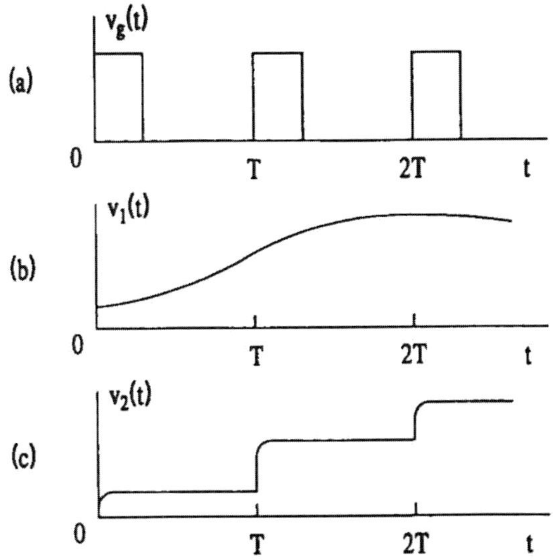


**Figure 1.12.** Typical voltages for the sample-and-hold circuit.

### Successive Approximation Analog-to-Digital Converters

The successive approximation analog-to-digital converter is perhaps the most popular design available in integrated circuit form at a reasonable cost. The conceptual hardware for a 4-bit 0- to 10-V version of such a device is illustrated in Fig. 1.13. The control logic shown is set up to control each bit of the register separately. Assume that the register bits are all initially zeros and hence no voltage is output from the D/A converter. When a start-conversion signal is present, the most significant bit of the register is driven to a high state and the output of the D/A converter is then half of full scale (5 V). If this voltage is larger than the input voltage, the comparator will change state, indicating that the D/A output is higher than the held input voltage, which will, in turn cause the logic to reset the most significant bit to zero and then set the next most significant bit to a high state. The process is then repeated by the logic, and the bit MSB-1 is left set or reset to zero. The process thus continues to work its way through the bits downward to the least significant bit (LSB), at which time the register contains the closest 4-bit digital representation, which is less than the actual value. At this time a conversion-complete signal is issued and the device waits for another start-conversion signal.

Assume that a voltage of 4 V is the input to a 4-bit device with a dynamic range of 0 to 10 V. The highest bit is then set high, the output

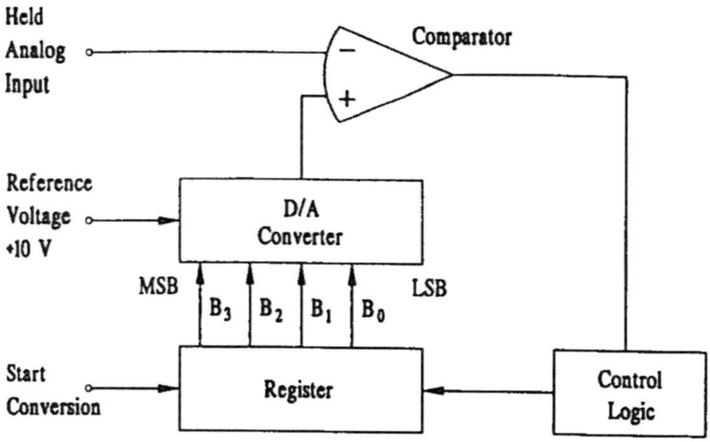


**Figure 1.13.** Successive approximation analog-to-digital converter.

from the D/A converter will be 5 V, and the comparator will change state, causing the most significant bit to be reset to zero. Following this, the next most significant bit is set high, the output of the D/A converter is then 2.5 V, and the comparator output does not change, hence the second most significant bit will remain high. Next, bit 1 is set high such that the register now reads 0110 and the output of the D/A converter is 3.75 V. Again the comparator output does not change, leaving bits 1 and 2 both high. The least significant bit is then set high and the output of the D/A converter will be 4.375 V, which is more than the input voltage, and as a result the least significant bit is reset to zero and the digital representation is now 0110, which has a D/A output of 3.75 V, which is as close as we can get to 4 V with 4 bits. This illustrates the need for more bits in the representation of a range of voltages, hence the need for 10- and 12-bit converters, which are now available in integrated circuit form at low cost.

### Tracking Analog-to-Digital Converter

There are many schemes for analog-to-digital conversion and many involve the use of a digital-to-analog converter in a feedback loop with some technique for comparing the converted value with the input value. A popular scheme known as the tracking A/D is illustrated in Fig. 1.14. This device involves an operational amplifier used as a comparator, a high-frequency clock, a binary up-down counter, and a digital-to-analog converter. The output voltage of the D/A is compared to the input voltage

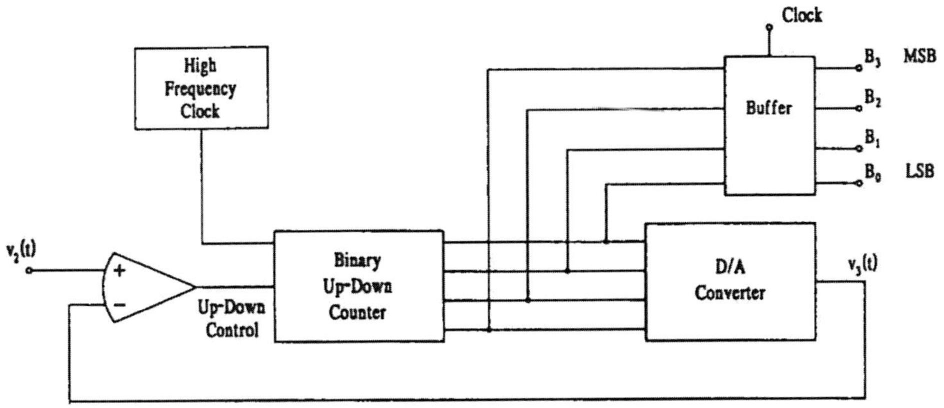


**Figure 1.14.** Tracking analog-to-digital converter.

$v_2(t)$, and the direction of count for the counter is controlled by the polarity of the difference in the D/A output and input voltage.

The converter runs in a continuous fashion, attempting to track the input voltage $v_2(t)$, and a "snapshot" of the input is given at the desired time by clocking the buffer on the output lines. This scheme works well as long as the clock frequency is much greater than the highest frequency in the input signal.

### Flash Analog-to-Digital Converter

There is a scheme which is faster than converters that involve integrators or counters, such as the tracking converter. That is the class of converters known as flash converters. A typical 4-bit converter operating in the 0- to 10-V range is shown in Fig. 1.15. The comparators are configured such that their outputs are either logic level "1" or logic level "0." For a given positive input voltage, all the comparators with reference inputs lower than the value of input voltage $v_2(t)$ will have outputs of logic level "0." The outputs of the comparators, $A_0$ to $A_{14}$, form a 15-bit binary word which must then be decoded using combinational logic to a 4-bit binary word. Large-scale monolithic circuit fabrication techniques have made the implementation of this complicated circuitry (all comparators, resistors, and logic) on a single silicon chip possible, although the largest versions available at this writing are 9-bit devices that are relatively expensive compared to slower-tracking devices. These devices are commonly found in oscilloscopes and other high-speed digital instruments.

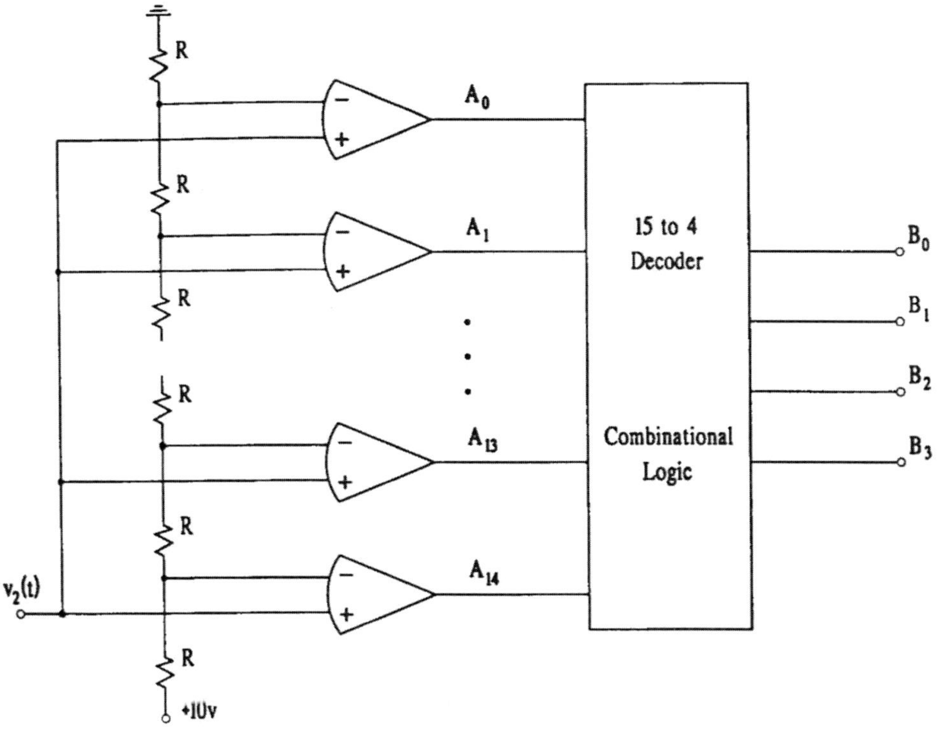


**Figure 1.15.** Flash analog-to-digital converter.

## 1.6 THE STATE OF MICROPROCESSOR TECHNOLOGY

In the past decade we have witnessed a change in microprocessor data path and register width from 8 bits to 32 bits. This is accomplished because of the ability to construct VLSI circuitry with a much higher density of circuit components. This increase in data path width thus assures that data can be transferred from one location to another in fewer clock cycles, thus resulting in faster read/write and register-to-register transfer operations and an accompanying increase in overall processor speed. For arithmetic-intensive computing, hardware multipliers have been incorporated into the hardware. Extensive software support is available, including floating-point and trigonometric function routines.

### Microcontrollers

With the growth in the use of computer power in the automotive and home appliance industries, specialized lines of microprocessors have been developed for these applications. Such processors are most commonly of the 8-, 16-, and 32-bit varieties. These processors often incorporate a multichannel multiplexed analog-to-digital converter as part of the integrated circuitry. The purpose of these analog input channels is to gather sensor data about the process being controlled, whether the fuel injection process for an engine, an antilock brake system, or the wash–rinse–dry cycle for a dishwasher. Outputs for the control of process variables (e.g., flow rates, heater controls, and pump operation) are usually derived at a processor output port.

Popular 8-bit microcontrollers currently in use are the Motorola MC68HC11 and the Intel 8051. The 8-bit devices are adequate for the control precision required in the home appliance and automotive industries; however, more bits are required for precise control tasks—hence the popularity of the Motorola 68HC16 and Intel 80196 processors. Some of the features of these two devices are shown in Table 1.1. The two devices have different attractive features and hence are not directly comparable but rather, represent the capabilities available in single-chip microcontrollers.

### Digital Signal Processors

Also useful in digital control applications requiring high speed is the family of special-architecture microprocessors designed for the implementation of digital signal processing algorithms in both the time and frequency domains. Previous generations of these processors incorporated signal-conversion hardware (A/D and D/A); however, current versions have many high-speed serial communication ports which handle data input and output. These multiple communication ports allow these processors to be used in parallel and exchange numerical data rapidly between processors. These

**Table 1.1.** Feature of Two Typical 16-Bit Microcontrollers

| | Max.clock rate | EPROM | Masked ROM | RAM | A/D ch. | A/D bits | Timer/ counters | PWM outputs | PLL |
|---|---|---|---|---|---|---|---|---|---|
| Intel 8XC196KD | 20 | 32K or | 32K | 1K | 8 | 8/10 | 2 | 3 | |
| Motorola M68HC16Z2 | 16 | 0 | 8K | 2K | 8 | 8/10 | 2 | 2 | 1 |

are particularly useful in robot control, vision systems, and in graphics accelerators for workstation applications.

The architecture of this class of devices is designed to perform the multiply and accumulate operation very efficiently. This operation, which is common to signal-processing operations in both the frequency and time domains, is of the form

$$\text{newsum} = \text{oldsum} + \text{coef} * \text{variable} \tag{1.6.1}$$

The current versions of the DSP hardware support floating-point operations in hardware along with other operations formerly requiring support software (slow) or an added coprocessor (expensive). Compared with the microcontroller class of hardware, this hardware is both more capable and more expensive.

## 1.7 EXAMPLE OF A MICROPROCESSOR-BASED THERMAL CONTROLLER

As an example of how a digital processor might be used to control a physical process, consider the bacterial growth chamber of Fig. 1.16, in which it is desired to control the temperature to be somewhat higher than room temperature. The chamber is subject to changing environmental temperature $T_e(t)$.

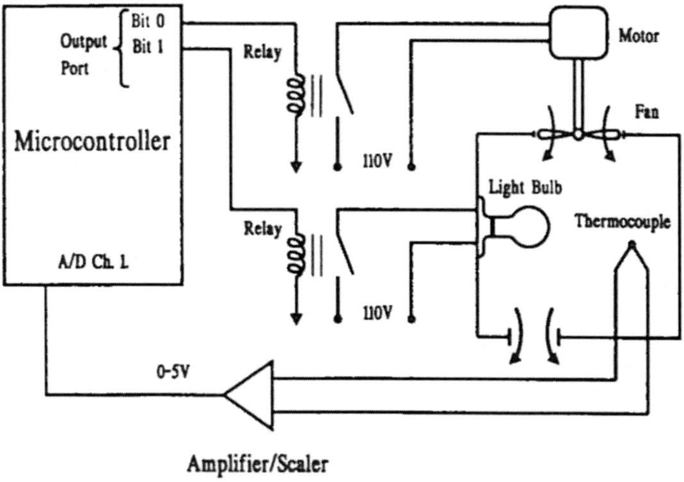


**Figure 1.16.** Bacterial growth chamber control problem.

The temperature in the chamber is measured by a thermocouple and some associated amplifying and scaling circuitry. The voltage output from the amplifier is then converted by the on-board analog-to-digital converter of the microcontroller. If the temperature exceeds some predetermined value, bit 0 of the output port will be driven high and the fan will be turned on to bring in cool external air. When the temperature gets below a predetermined value, bit 1 of the output port will turn on the light to heat the air in the chamber. The predetermined values for light on and fan on need not be the same. It is even conceivable that both the fan and light could be on at the same time.

This example has many of the features of the types of systems in which we are interested. These features are:

1. The output variable (temperature) is measured periodically and converted to a binary number.
2. The resulting number is used to decide what control strategy will be used (fan on, light on, both on, both off).
3. There are actuators (the light and fan) to provide the controlling action.

The on–off nature of the actuation in this problem will make it such that the precision of the control will not be too great, as the temperature will tend to "hunt" because there is no way to decrease the heat addition as the system nears the temperature desired.

## 1.8 SUMMARY

In this chapter we have explored the possibilities and reasons for the application of the digital computer to accomplish feedback control of physical systems. We have attempted to justify the use of digital control elements from the points of view of economics, reliability, and ability to handle complex control tasks with great speed. In addition, we have discussed the elements of data conversion hardware, which comprises the digital computer and the "real" world. It has been found that estimation and control tasks can be accomplished in real time at high speeds such that application to the control of fast systems such as high-performance aircraft is now possible.

## REFERENCES

Altman, L., 1976. Advances in designs and new processes yield surprising performance, *Electronics* *49*(7): 73–81.

Bibbero, R. J., 1977. *Microprocessors in Instruments and Control*, Wiley, New York.

Bose, B. K., 1992. "Recent advances in power electronics," *IEEE Trans. on Power Electronics, 7*(2): 2–16.

Faggin, F., 1978. How VLSI impacts computer architecture, *IEEE Spectrum, 15*(5): 28–31.

Gelsinger, P. P., P. A. Gargini, G. H. Parker, and A. Y. C. Yu, 1989. Microprocessors circa 2000, *IEEE Spectrum, 26*(10): 43–47.

Pickett, G., J. Marley, and J. Gragg, 1975. Automotive electronics, II: The microprocessor is in, *IEEE Spectrum, 14*(11): 37–45.

Queyssac, D., 1979. Projecting VLSI's impact on microprocessors, *IEEE Spectrum, 16*(5): 38–41.

Sheingold, D. H., Ed., 1986. *Analog–Digital Conversion Handbook*, Prentice Hall, Englewood Cliffs, NJ.

# 2
# Linear Difference Equations and the *z*-Transform

## 2.1 INTRODUCTION

As we discussed in Chapter 1, the element in the closed loop of a digital control system that makes it different from a continuous-time system is the digital computer. The purpose of this chapter is to delineate the theory necessary to design algorithms that will be programmed into the digital computer to provide the control efforts at the outputs of the D/A converter based on the input to the A/D converter. Digital computers are general-purpose devices designed with the flexibility to do many tasks other than the control function. In fact, sometimes one computer can accomplish a variety of tasks by time sharing, such as control of several systems, data acquisition, and so on.

We restrict our discussion in this book to examination of techniques by which the digital computer can be applied to the control of dynamic continuous-time systems. Examples of such systems are the positioning of a radar antenna, an autopilot for controlling the motions of an airplane in one of the phases of a routine flight, the control of chemical-production processes, or the control of the head of an automatic machine tool.

## 2.2 SCALAR DIFFERENCE EQUATIONS

For purposes of simplicity, we shall first be concerned with sequences that are scalar real values associated with the temporal index $k$. These scalars will be defined for integer values of $k$ ($k = 0, 1, 2, \ldots$). Generally, we are concerned with the generation of a sequence $u(k)$ (the controls) given a sequence $y(k)$ (the sampled data measurement sequence).* We generally would like to define the $k$th control effort in terms of the $k$th measurement or sample, a finite number of previous measurements or samples, and a finite number of previous control efforts, or, written mathematically,

$$u(k) = f[y(k), y(k-1), \ldots, y(k-m), u(k-1), u(k-2), \ldots, u(k-n)] \tag{2.2.1}$$

Of course there are an infinite number of ways the $n + m + 1$ values on the right side can be combined to form $u(k)$, but for the majority of this book we shall be interested only in the case where the right side involves a linear combination of the measurements and past controls, or

$$u(k) = b_{n-1}u(k-1) + \cdots + b_0u(k-n) + a_my(k) + a_{m-1}y(k-1) + \cdots + a_0y(k-m) \tag{2.2.2}$$

This is a linear difference equation, and if the parameters $a_i$ and $b_j$ are independent of $k$, it will be said to be a time-invariant or constant-coefficient difference equation. The algorithm of the form of (2.2.2) is satisfactory to perform many control tasks, along with a host of other tasks, including real-time digital filtering. What we need is a rational approach to selection of the $a_i$ and $b_j$ such that the algorithm has the correct dynamic properties to perform a desired task.

**Example 2.1.** Let us consider a problem that carries over into control theory from numerical analysis. That is the approximate numerical integration of an arbitrary function $y(t)$ based on equally spaced samples of that function as shown in Fig. 2.1. Let $x(k)$ be the value of the integral at time $kT$, while $x(k-1)$ is the value of the integral at the previous time. If we approximate the integration with a simple sum and employ a *rectangular* approximation, we can write

$$x(k) = x(k-1) + y(k-1)T \tag{2.2.3}$$

We see here that we have an algorithm for successively approximating the integral of a continuous-time function $y(t)$, and the algorithm takes the form of a homogeneous linear difference equation with constant coefficients. It is interesting to note that there are other approximations for the integral, yielding different difference equations.

*In all of the work that follows, we employ several forms of notation interchangeably. A sequence of numbers may be designated in three ways: $u_k = u(k) = u(kT)$.

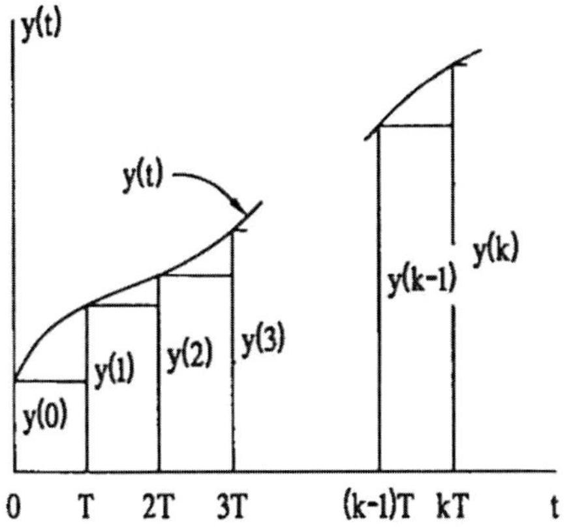


**Figure 2.1.** Approximate rectangular numerical integration.

**Example 2.2.** We can use the field of numerical analysis as the source of another example. Consider the solution of a linear constant-coefficient first-order differential equation

$$\frac{dx}{dt} + x = 0 \qquad \text{(a)}$$

Let us use the central difference approximation where the first derivative is approximated by the average of the first derivatives at half a time interval before and after time $kT$, or

$$\frac{dx(kT)}{dt} \simeq \frac{1}{2}\left[\dot{x}\left(\left(k + \frac{1}{2}\right)T\right) + \dot{x}\left(\left(k - \frac{1}{2}\right)T\right)\right]$$

or approximating the derivatives

$$\frac{dx(kT)}{dt} \simeq \frac{1}{2}\left[\frac{x((k+1)T) - x(kT)}{T} + \frac{x(kT) - x(k-1)T)}{T}\right]$$

and simplifying,

$$\frac{dx(kT)}{dt} \simeq \frac{x((k+1)T) - x((k-1)T)}{2T}$$

Let us simplify the notation by letting $x(kT) = x(k)$, and so on, so that we can write the differential equation approximately as

$$\frac{x(k+1) - x(k-1)}{2T} + x(k) = 0$$

or solving for the most advanced (in time) response,

$$x(k + 1) = 2Tx(k) + x(k - 1)$$

This is a linear homogeneous difference equation with constant coefficients that could be evaluated recursively to generate an approximate solution of the original differential equation.

In continuous-time control systems described by linear constant-coefficient differential equations, the Laplace transformation serves as a useful tool for the solution of these equations or exploration of the dynamic stability of such systems, and the analyses of such systems is most often carried out in the complex frequency (or frequencies) domain.

It has been found that linear constant-coefficient difference equations can be solved by employing the $z$-transform, which is the topic of the next section. The $z$-transform has properties that make it as useful in the solution of linear constant-coefficient difference equations as the Laplace transform for the solution of ordinary differential equations. Similarly, the stability and qualitative response character can be explored in the $z$-domain.

## 2.3 z-TRANSFORM OF SIMPLE SEQUENCES

In the solution of linear constant-coefficient difference equations, we are concerned with sequences of numbers $u(0), u(1), u(2), \ldots$ and $y(0), y(1), y(2), \ldots$ which we shall denote as $u(k)$ and $y(k)$, respectively. We shall define the *one-sided z-transform* of a sequence $f(k)$ by the infinite series in the complex variable $z$:

$$F(z) = \mathfrak{Z}[f(k)] = f(0) + f(1)z^{-1} + f(2)z^{-2} + \cdots \tag{2.3.1}$$

or

$$F(z) = \mathfrak{Z}[f(k)] = \sum_{k=0}^{\infty} f(k)z^{-k} \tag{2.3.2}$$

We tacitly assume that there is some region of the complex $z$-plane where the series of (2.3.2) converges in order that the $z$-transform be well defined. The region of convergence clearly depends on the nature of the coefficients of the series, which are the values of the sequence $f(k)$, and thus the region of convergence will potentially be different for different sequences.

As is common with the Laplace transform, let us construct a table of $z$-transforms of some simple sampled functions which are commonly encountered in discrete-time system analysis. The three functions to be initially transformed are shown in Fig. 2.2a–c.

### Unit-Step Sequence

Consider the unit-step sequence of Fig. 2.2a. The function is defined as

$$u(kT) = \begin{cases} 0 & k < 0 \\ 1 & k \geq 0 \end{cases} \tag{2.3.3}$$

By application of the definition of the $z$-transform (2.3.2) and that of the function (2.3.3) which defines the samples, we get

$$U(z) = \mathcal{Z}[u(kT)] = \sum_{k=0}^{\infty} z^{-k} = 1 + \frac{1}{z} + \frac{1}{z^2} + \cdots \tag{2.3.4}$$

It is clear that this series converges for $|z| > 1$, and a glance at a set of

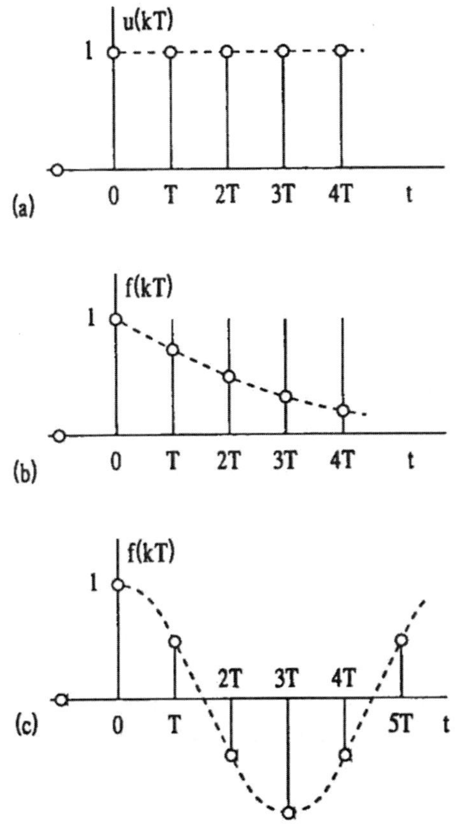


**Figure 2.2.** Typical sequences of sampled data to be $z$-transformed: (a) discrete-step function; (b) discrete exponential; (c) sampled cosine.

ordinary math tables will give the limiting form of such a convergent geometric series as

$$\mathfrak{Z}[u(kT)] = \frac{z}{z-1} \qquad \text{for } |z| > 1 \tag{2.3.5}$$

which can easily be verified by long division. The requirement that $|z| > 1$ defines what is known as the region of convergence, which in this case is the area of the complex $z$-plane exterior to the unit circle.

### Exponential Function

Consider now the sampled exponential function illustrated in Fig. 2.2b. The sequence is defined by

$$f(k) = f(kT) = \begin{cases} 0 & k < 0 \\ e^{-akT} & k \geq 0 \end{cases} \tag{2.3.6}$$

Substitution of (2.3.6) into (2.3.3) yields

$$\mathfrak{Z}[e^{-akT}] = \sum_{k=0}^{\infty} (e^{-aT}z^{-1})^k \tag{2.3.7}$$

and from the previous transform the limit of the series is

$$\mathfrak{Z}[e^{-akT}] = \frac{z}{z - e^{-aT}} \qquad \text{for } |z| > e^{-aT} \tag{2.3.8}$$

### General Exponential Function

Similar to the development in the last transform, we can examine the sequence

$$f(k) = r^k \tag{2.3.9}$$

Using the definition of the $z$-transform, we get

$$\mathfrak{Z}[r^k] = \sum_{k=0}^{\infty} (r^{-1}z)^{-k} \tag{2.3.10}$$

Clearly, this series converges as long as $|z| > |r|$, so the region of convergence is the exterior of a circle of radius $|r|$. The limiting form of the series is

$$\mathfrak{Z}[r^k] = \frac{z}{z - r} \qquad \text{for } |z| < |r| \tag{2.3.11}$$

Similarly, we can show that

$$\mathscr{Z}[r^{-k}] = \frac{z}{z - r^{-1}} \tag{2.3.12}$$

### Cosine Function

Consider the sampled cosine function of radian frequency $\Omega$ which is shown in Fig. 2.2c. The sequence is defined by

$$f(k) = f(kT) = \begin{cases} 0 & k < 0 \\ \cos k\Omega T & k \geq 0 \end{cases} \tag{2.3.13}$$

The cosine function can be rewritten using the Euler identity as

$$\cos k\Omega T = \frac{1}{2}(e^{jk\Omega T} + e^{-jk\Omega T}) \tag{2.3.14}$$

Since the $z$-transform of a sum is the sum of individual $z$-transforms, the result of (2.3.8) can be used to give

$$\mathscr{Z}[\cos k\Omega T] = \frac{1}{2}\left(\frac{z}{z - e^{j\Omega T}} + \frac{z}{z - e^{-j\Omega T}}\right) \tag{2.3.15}$$

and finding a common denominator yields

$$\mathscr{Z}[\cos k\Omega T] = \frac{z^2 - z\cos \Omega T}{z^2 - z \cdot 2\cos \Omega T + 1} \tag{2.3.16}$$

The region of convergence is the region of the $z$-plane exterior to the unit circle. The sampled sine function will be left as an exercise for the reader but is given in Table 2.1.

### Discrete Impulse Function

Consider a sequence $\delta(k)$ that is defined by

$$\delta(k) = \begin{cases} 1 & k = 0 \\ 0 & k \neq 0 \end{cases} \tag{2.3.17}$$

Using the definition of (2.3.2) the result is

$$\mathscr{Z}[\delta(k)] = 1 \tag{2.3.18}$$

In a similar fashion we can show that a delayed impulse function defined by

$$\delta(k - n) = \begin{cases} 0 & k \neq n \\ 1 & k = n > 0 \end{cases} \tag{2.3.19}$$

has a z-transform

$$\mathfrak{Z}[\delta(k-n)] = z^{-n} \tag{2.3.20}$$

**Ramp Function**

The sampled ramp function is defined by

$$f(k) = kT \qquad k = 0, 1, 2, \ldots \tag{2.3.21}$$

and application of definition (2.3.2) yields

$$\mathfrak{Z}[kT] = T \sum_{k=0}^{\infty} kz^{-k} \tag{2.3.22}$$

Again consulting a table of mathematical functions, the limit of the series is

$$\mathfrak{Z}[kT] = \frac{Tz}{(z-1)^2} \qquad \text{for } |z| > 1 \tag{2.3.23}$$

We have just derived a number of transform pairs from which we can derive further transforms to form a complete table of transform pairs similar to that employed in Laplace transform theory. The results just given are summarized in Table 2.1; a more extensive table may be found in Appendix A. With the aid of this table and the idea of partial fraction expansion, it is possible to generate a large number of transform pairs. Some theorems involving some properties of the z-transform developed in the next section and will aid in application of the transform in the solution of difference equations and in the development of additional transform pairs. Again we should emphasize that we have considered only the one-sided z-transform, and in this book we will always assume that all sequences are identically zero for all negative values of $k$.

## 2.4 USEFUL THEOREMS ASSOCIATED WITH THE z-TRANSFORM

It has been our experience that certain theorems were quite useful in the theory of the Laplace transform, and hence a few necessary and very useful theorems associated with the z-transforms will now be developed.

**Theorem 2.1.** Linearity. We shall show that the z-transform is a linear transformation which implies that

$$\mathfrak{Z}[\alpha f(k)] = \alpha\mathfrak{Z}[f(k)] = \alpha F(z) \tag{2.4.1}$$

**Table 2.1.** Short Table of z-Transforms of Sampled Continuous Functions

| $f(t),\ t \geq 0$ | $L[f(t)]$ | $F(z) = \mathfrak{Z}[f(kT)] = \mathfrak{Z}[f(k)]$ |
|---|---|---|
| $u(t)$ | $\frac{1}{s}$ | $\frac{z}{z-1}$ |
| $t$ | $\frac{1}{s^2}$ | $\frac{Tz}{(z-1)^2}$ |
| $e^{-at}$ | $\frac{1}{s+a}$ | $\frac{z}{z-e^{-aT}}$ |
| $\cos \Omega t$ | $\frac{s}{s^2+\Omega^2}$ | $\frac{z^2-(\cos \Omega T)z}{z^2-(2\cos \Omega T)z+1}$ |
| $\sin \Omega t$ | $\frac{\Omega}{s^2+\Omega^2}$ | $\frac{(\sin \Omega T)z}{z^2-(2\cos \Omega T)z+1}$ |

and

$$\mathfrak{Z}[\alpha f(k) + \beta g(k)] = \alpha F(z) + \beta G(z) \tag{2.4.2}$$

The proof of (2.4.1) follows easily from the definition, or

$$\mathfrak{Z}[\alpha f(k)] = \sum_{k=0}^{\infty} \alpha f(k) z^{-k}$$

The factor $\alpha$ may be factored to give

$$\mathfrak{Z}[\alpha f(k)] = \alpha \sum_{k=0}^{\infty} f(k) z^{-k}$$

or

$$\mathfrak{Z}[\alpha f(k)] = \alpha F(z) \tag{2.4.3}$$

and (2.4.1) is proven. Now consider (2.4.2) and the definition

$$\mathfrak{Z}[\alpha f(k) + \beta g(k)] = \sum_{k=0}^{\infty} [\alpha f(k) + \beta g(k)] z^{-k}$$

$$\mathfrak{Z}[\alpha f(k) + \beta g(k)] = \sum_{k=0}^{\infty} \alpha f(k) z^{-k} + \sum_{k=0}^{\infty} \beta g(k) z^{-k}$$

and employing (2.4.3), we get

$$\mathfrak{Z}[\alpha f(k) + \beta g(k)] = \alpha F(z) + \beta G(z) \tag{2.4.4}$$

and the theorem is proven.

**Theorem 2.2.** Delay Theorem. The $z$-transform of a delayed sequence shifted one step to the right is given by

$$\mathfrak{Z}[f(k-1)] = z^{-1}F(z) \tag{2.4.5}$$

The proof again comes from the definition, or

$$\mathfrak{Z}[f(k-1)] = \sum_{k=0}^{\infty} f(k-1)z^{-k}$$

and rewriting gives

$$\mathfrak{Z}[f(k-1)] = f(-1)z^{-0} + \sum_{k=1}^{\infty} f(k-1)z^{-k} \tag{2.4.6}$$

We have stipulated that the sequence was zero for negative values of $k$; hence the first term vanishes. We may also change the subscript by letting $j = k - 1$; then

$$\mathfrak{Z}[f(k-1)] = \sum_{j=0}^{\infty} f(j)z^{-(j+1)}$$

or factoring $z^{-1}$, we have

$$\mathfrak{Z}[f(k-1)] = z^{-1}\sum_{j=0}^{\infty} f(j)z^{-j} \tag{2.4.7}$$

We note from the original definition (2.3.2) that the sum is identically the transform of $f(k)$ or $F(z)$, so

$$\mathfrak{Z}[f(k-1)] = z^{-1}F(z) \tag{2.4.8}$$

We can also generalize this for a delay of $n$ steps in time, or

$$\mathfrak{Z}[f(k-n)] = z^{-n}F(z) \tag{2.4.9}$$

**Theorem 2.3.** Advance Theorem. The $z$-transform of a sequence that has been shifted one step to the left is

$$\mathfrak{Z}[f(k+1)] = zF(z) - zf(0) \tag{2.4.10}$$

The proof of this theorem again stems from the definition (2.3.2), which yields

$$\mathfrak{Z}[f(k+1)] = \sum_{k=0}^{\infty} f(k+1)z^{-k} = z \sum_{k=0}^{\infty} f(k+1)z^{-(k+1)} \tag{2.4.11}$$

Now note that we can add and subtract a term to yield

$$\mathfrak{Z}[f(k+1)] = z\left\{\left[f(0) + \sum_{k=0}^{\infty} f(k+1)z^{-(k+1)}\right] - f(0)\right\} \tag{2.4.12}$$

and changing the index to let $k + 1 = m$ and incorporating one of the $f(0)$ terms under the summation, we get

$$\mathfrak{Z}[f(k+1)] = z \sum_{m=0}^{\infty} f(m)z^{-m} - zf(0) \tag{2.4.13}$$

or recognizing the definition of the transform of $f(k)$, we get

$$\mathfrak{Z}[f(k+1)] = zF(z) - zf(0) \tag{2.4.14}$$

and we have completed the proof.

We could carry this process out successively to show that signals advanced by $n$ steps in time have the transform

$$\mathfrak{Z}[f(k+n)] = z^n F(z) - z^n f(0) - \cdots - zf(n-1) \tag{2.4.15}$$

The terms $f(0), f(1), \ldots, f(n-1)$ will hereafter be referred to as the starting conditions.

This concludes the derivation of the initial theorems, which in conjunction with the derived transforms will be useful in the solution of linear constant-coefficient difference equation.

## 2.5 INVERSION OF THE z-TRANSFORM

We have discussed how to obtain the z-transform of a sequence that may or may not be the samples of some continuous-time function. Generally, the z-transforms of interest here will be the ratios of polynomials in the complex variable $z$ with the numerator polynomial being of no higher order than the denominator. The question to be answered in this and the next section of this chapter is that of finding the sequence associated with a given polynomial ratio. There are several ways to answer this question; the first stems from the original definition of the z-transform as given by expression (2.3.2). If we know the z-domain function as a polynomial ratio, we can first develop this into a ratio of polynomials in $z^{-1}$ which can then be divided by long division to give the series in $z^{-1}$ that fits the definition of (2.3.2) and the coefficients are the temporal sequence $f(k)$. An example will serve to illustrate this technique.

**Example 2.3.** Find the inverse sequence for the following function:

$$F(z) = \frac{z^2 + z}{z^2 - 3z + 4}$$

Multiplication by $z^{-2}$ in numerator and denominator gives

$$F(z) = \frac{1 + z^{-1}}{1 - 3z^{-1} + 4z^{-2}}$$

Now carry out formal long division to yield

$$
\begin{array}{r l}
 & 1 + 4z^{-1} + 8z^{-2} \\
\cline{2-2}
1 - 3z^{-1} + 4z^{-2} & \big)\, 1 + z^{-1} \\
 & \underline{1 - 3z^{-1} + 4z^{-2}} \\
 & \quad 4z^{-1} - 4z^{-2} \\
 & \quad \underline{4z^{-1} - 12z^{-2} + 16z^{-3}} \\
 & \qquad 8z^{-2} - 16z^{-3} \\
 & \qquad \underline{8z^{-2} - 24z^{-3} + 32z^{-4}} \\
 & \qquad\quad 8z^{-3} - 32z^{-4} \\
 & \qquad\quad \vdots
\end{array}
$$

Now upon examination of the coefficients of the infinite series answer, the sequence is

$$
\begin{aligned}
f(0) &= 1 \\
f(1) &= 4 \\
f(2) &= 8
\end{aligned}
$$

and so on. This process could be carried out any desired number of steps by the digital computer. The disadvantage of this technique is that it does not give a closed form of the resulting sequence but rather, numerical values of the sequence. Often, we need a closed-form result to infer general qualitative properties about the sequence $f(k)$.

Another method employing the complex inversion integral is discussed in Appendix A; however, for most engineering work the method of partial fraction expansion and a good table of $z$-transforms is sufficient. This method is the subject of the next section.

Any engineer who has worked with the Laplace transform is familiar with the use of a table of Laplace transform pairs and how the table may be used to find inverse transforms. A table of $z$-transforms such as that given in Appendix A may be employed to give the appropriate sequence $f(k)$ associated with a function $F(z)$. It is clear from the long-division method of inversion that for any set of coefficients in polynomials of the numerator and denominator a unique set of samples will result. It is then appropriate to employ the table of $z$-transform pairs to invert the transforms.

There is, of course, a limitation to the size of any $z$-transform table and hence the order of polynomial denominators which may be considered to give closed-form representations of the associated sequence. To alleviate this problem, we shall discuss the extension of the $z$-transform table by the method of partial fraction expansion.

## 2.6 METHOD OF PARTIAL FRACTION EXPANSION

In Section 2.5 the long-division method of $z$-transform inversion was examined and reference was made to the complex inversion integral. The long-division method leaves much to be desired because closed forms do not result and the inversion integral requires some knowledge of complex variable theory. The method of partial fraction expansion which is somewhat similar in nature to that used in the inversion of high-order Laplace transforms was also mentioned. The technique of partial fraction expansion is developed so that a relatively small table of transforms with a few standard forms will be usable in the inversion process.

We shall assume that the polynomial ratios involved have all been rationalized such that only positive powers of the complex variable $z$ are present in the denominator. If they are not of this form, they must be made so before the following results are applicable. We shall also assume that the roots of the denominator polynomial have been found by whatever means necessary. If there is a constant term in the numerator, then for equality in the partial fraction we must make the expansion have a constant term which is the transform of a discrete impulse at $k = 0$. For example, if

$$F(z) = \frac{1}{(z - 1)(z - 0.8)}$$

the correct form of the expansion is

$$F(z) = A + \frac{Bz}{z - 1} + \frac{Cz}{z - 0.8}$$

If the numerator of the function has a constant term and a term in $z^{-1}$, the expansion should also include a term in $z^{-1}$, and so on.

We shall divide the discussion here into two cases: that concerned with real distinct denominator roots and that associated with distinct complex roots.

### Case I. Distinct Real Roots

We shall assume that the z-domain fraction can be written as

$$F(z) = \frac{N(z)}{(z - p_1)(z - p_2) \cdots (z - p_n)} \tag{2.6.1}$$

where the $p_i$ are real and no two of them are the same. If we examine the transforms of Table 2.1 or the transform table of Appendix A, we find that the form of the answers for application of the table should be

$$\begin{aligned} F(z) &= \frac{N(z)}{(z - p_1) \cdots (z - p_n)} \\ &= A_0 + \frac{A_1 z}{z - p_1} + \frac{A_2 z}{z - p_2} + \cdots \end{aligned} \tag{2.6.2}$$

where the constant term is appended to ensure equality in the case where there is a constant term in the numerator of (2.6.1). If we now multiply both sides of (2.6.2) by $(z - p_1)/z$, we get

$$\begin{aligned} &\frac{N(z)}{z(z - p_2) \cdots (z - p_n)} \\ &\qquad = \frac{A_0(z - p_1)}{z} + A_1 + \frac{A_2(z - p_1)}{z - p_2} + \cdots \end{aligned} \tag{2.6.3}$$

If this relation is to hold for all $z$, it must hold at $z = p_1$. If we let $z = p_1$, we get

$$A_1 = \frac{N(p_1)}{p_1(p_1 - p_2) \cdots (p_1 - p_n)} \tag{2.6.4}$$

and we can generalize the process to yield all the coefficients, or

$$A_i = \frac{z - p_i}{z} F(z)\Big|_{z = p_i} \qquad i = 1, 2, \ldots, n \tag{2.6.5}$$

The coefficient $A_0$ can thus be given by evaluating (2.6.2) at $z = 0$.

**Example 2.4.** Consider the following $z$-domain function:

$$F(z) = \frac{z^2 + z}{(z - 0.6)(z - 0.8)(z - 1)}$$

Find the partial fraction expansion and invert the resulting transform. The expansion will be of the form

$$F(z) = \frac{A_1 z}{z - 0.6} + \frac{A_2 z}{z - 0.8} + \frac{A_3 z}{z - 1}$$

where the coefficients are given by (2.6.5). The constant term in the expansion has been omitted bcause there is not a constant term in the numerator polynomial of the original function, or

$$A_1 = \left.\frac{z + 1}{(z - 0.8)(z - 1)}\right|_{z=0.6} = \frac{1.6}{(-0.2)(-0.4)} = 20$$

$$A_2 = \left.\frac{z + 1}{(z - 0.6)(z - 1)}\right|_{z=0.8} = \frac{1.8}{(0.2)(-0.2)} = -45$$

and

$$A_3 = \left.\frac{z + 1}{(z - 0.6)(z - 0.8)}\right|_{z=1} = \frac{2}{(0.4)(0.2)} = 25$$

So upon inversion of the transform,

$$f(k) = 20(0.6)^k - 45(0.8)^k + 25$$

In the case that the poles are a complex-conjugate pair, one can proceed as above with the added complication of complex numbers or consider that case separately, as in the next section.

### Case II. Some Distinct Complex Roots

We shall assume here that all the roots are distinct and that there exists at least one pair of complex-conjugate roots. One could factor them and use the method of the preceding section and obtain complex coefficients in the partial fraction expansion, then combine the resulting terms using Euler's identities to yield exponentially multiplied sinusoidal sequences. A quick look at the transforms of damped sinusoidal sequences gives a hint as to what we must do as an alternative. From Appendix A we find the following $z$-transform pairs:

$$\mathcal{Z}[e^{-\alpha kT} \cos \Omega kT] = \frac{z^2 - ze^{-\alpha T} \cos \Omega T}{z^2 - z2e^{-\alpha T} \cos \Omega T + e^{-2\alpha T}} \quad (2.6.6)$$

and

$$\mathfrak{Z}[e^{-\alpha kT}\sin \Omega kT] = \frac{ze^{-\alpha T}\sin \Omega T}{z^2 - z2e^{-\alpha T}\cos \Omega T + e^{-2\alpha T}} \tag{2.6.7}$$

Now if we examine the pole locations for these transforms, we find in both cases that the roots in polar form are

$$z_{1,2} = e^{-\alpha T}(\cos \Omega T \pm j \sin \Omega T) \tag{2.6.8}$$

or

$$z_{1,2} = e^{-\alpha T}e^{\pm j\Omega T} = Re^{\pm j\theta} \tag{2.6.9}$$

where $R = e^{-\alpha T}$ and $\theta = \Omega T$. Let us now consider functions of the form

$$F(z) = \frac{N(z)}{P(z)(z^2 - 2R\beta z + R^2)} \qquad 0 < \beta < 1 \tag{2.6.10}$$

where $R = e^{-2\alpha T}$ and $\beta = \cos \Omega T$ as in relations (2.6.6) and (2.6.7). The limits on the value of $\beta$ imply that the poles associated with the quadratic term are complex and hence associated with damped or unstable oscillatory sequences. The partial fraction expansion should be chosen to be of the form

$$F(z) = \frac{A(z^2 - zR\beta)}{z^2 - 2R\beta z + R^2} + \frac{Bz^2R\sin \Omega T}{z^2 - 2R\beta z + R^2} + Q(z) \tag{2.6.11}$$

The two terms with the quadratic denominator are the $z$-transforms, respectively, of $R^k \cos \Omega kT$ and $R^k \sin \Omega kT$ and $Q(z)$ represents the remaining terms in the partial fraction expansion. A good rule of thumb is first to evaluate the coefficients in the whole expansion associated with $Q(z)$ and then evaluate $A$ and $B$ by brute force, usually by finding a common denominator and then equating the coefficients of the numerator polynomials. An example at this point will be useful in illustrating the method.

**Example 2.5.** Find the inverse of the following function using the method of partial fraction expansion

$$F(z) = \frac{z^2 + z}{(z^2 - 1.13z + 0.64)(z - 0.8)}$$

Let us first identify $R$, $\beta$, and $\Omega T$. By inspection,

$$R = \sqrt{0.64} = 0.8$$

and

$$\beta = \frac{1.13}{2R} = \frac{1.13}{1.6} = 0.7063$$

Since $\beta = \cos \Omega T$, then

$$\Omega T = \cos^{-1} \beta = \cos^{-1}(0.7063) = 0.7865 \text{ rad}$$

so

$$\sin \Omega T = 0.7079$$

Thus the correct form for the partial fraction expansion

$$F(z) = \frac{A(z^2 - z(0.8)(0.7063))}{z^2 - 1.13z + 0.64} + \frac{Bz(0.8)(0.7079)}{z^2 - 1.13z + 0.64} + \frac{Cz}{z - 0.8}$$

Multiplying out the factors above gives

$$F(z) = \frac{A(z^2 - z0.565)}{z^2 - 1.13z + 0.64} + \frac{Bz(0.5663)}{z^2 - 1.13z + 0.64} + \frac{Cz}{z - 0.8}$$

We may evaluate $C$ by the usual Heaviside method to give

$$C = \frac{0.8 + 1}{(0.8)^2 - 1.13(0.8) + 0.64} = \frac{1.8}{0.376} = 4.787$$

Now find a common denominator and equate the numerators to give

$$z^2 + z = A(z^2 - 0.565z)(z - 0.8) + Bz(0.5663)(z - 0.8) + 4.787z(z^2 - 1.13z + 0.64)$$

or

$$z^2 + z = (A + 4.787)z^3 + [(0.565 - 0.8)A + 0.5663B + 4.787(-1.13)]z^2 + [(0.565)(0.8)A - 0.8B + (4.787)(0.64)]z$$

Equating the coefficients of line powers of $z$, we get

$$z^3: \quad 0 = A + 4.787$$

$$z^2: \quad 1 = -1.365A + 0.5663B - 5.409$$

$$z^1: \quad 1 = 0.452A - 0.8B + 3.064$$

The first equation implies that

$$A = -4.787$$

The second equation implies that

$$B = \frac{1 + 1.365(-4.787) + 5.409}{0.5663} = \frac{-0.1253}{0.5663} = 0.222$$

Now that the coefficients have been found the inverse transform may be written as

$$f(k) = -4.787(0.8)^k \cos(0.7865k)$$
$$= 0.222(0.8)^k \sin(0.7865k) + 4.787(0.8)^k$$

The reader is encouraged to go through the numerical details of the foregoing developments so as to understand the details of this expansion.

If the partial fraction expansion is made for $F(z)/z$, the expansion can be made as it is in the Laplace transform case, and then the resultant terms multiplied by $z$ will have inverses that are in the $z$-transform tables. There are a number of computational aids now available to alleviate the tedium of these calculations. For the case of repeated roots the reader is referred to any text on Laplace transforms, as the analysis is very much the same.

An alternative to the methods of long division or partial fraction expansion using tables is that of the complex inversion integral. This technique employs the theory of residues from complex variable theory and is no saving in labor over the method of partial fraction expansion; in fact, some of the operations are exactly the same.

## 2.7 SOLVING LINEAR DIFFERENCE EQUATIONS WITH THE z-TRANSFORM

By employing the delay and advance theorems (Theorems 2.2 and 2.3) and the transforms of known functions, we are now prepared to solve linear constant-coefficient difference equations, but first let us write down the results of these valuable theorems:

$$\text{Theorem 2.2:} \quad \mathfrak{Z}[f(k-n)] = z^{-n}F(z) \tag{2.7.1}$$

$$\text{Theorem 2.3:} \quad \mathfrak{Z}[f(k+n)] = z^{n}F(z) - z^{n}f(0) - \cdots - zf(n-1) \tag{2.7.2}$$

The technique is similar to that of using Laplace transforms and is best illustrated by some example problems, which follow.

**Example 2.6.** Consider the homogeneous first-order difference equation

$$x(k+1) - 0.8x(k) = 0$$

with initial value $x(0) = 1$. Now take the $z$-transform to yield

$$zX(z) - zx(0) - 0.8X(z) = 0$$

Solving for $X(z)$ yields

$$X(z) = \frac{z}{z - 0.8}$$

and this is of the same form as the transform developed in expression (2.3.11), which implies that the solution is

$$x(k) = 0.8^k$$

**Example 2.7.** Consider the same example as before with starting condition $x(0) = 2$ and an inhomogeneous term on the right side, or

$$x(k + 1) - 0.8x(k) = 1$$

Taking the $z$-transform yields

$$zX(z) - 2z - 0.8X(z) = \frac{z}{z - 1}$$

Solving for $X(z)$ yields

$$X(z) = \frac{2z}{z - 0.8} + \frac{z}{(z - 1)(z - 0.8)}$$

We can readily invert the first term as in Example 2.6, but we must now expand the second term as

$$\frac{z}{(z - 1)(z - 0.8)} = \frac{Az}{z - 1} + \frac{Bz}{z - 0.8}$$

Solving for $A$ and $B$ yields

$$A = \frac{z - 1}{z} \frac{z}{(z - 1)(z - 0.8)}\bigg|_{z=1} = 5$$

and

$$B = \frac{z - 0.8}{z} \frac{z}{(z - 1)(z - 0.8)} = -5$$

So the $z$-domain solution is

$$X(z) = \frac{2z}{z - 0.8} + \frac{5z}{z - 1} - \frac{5z}{z - 0.8}$$

and the total solution is

$$x(k) = -3(0.8)^k + 5$$

We can check that the initial value is indeed satisfied, and by substitution into the difference equation we see that it is identically satisfied.

As a third example, let us consider a problem where complex roots are involved, or the following difference equation.

**Example 2.8.** Solve the following difference equation

$$x(k + 2) - x(k + 1) + 0.24x(k) = \sin(k)$$

with starting conditions $x(0) = x(1) = 0$. Note first that for the sampled sinusoid on the right side that $\Omega T = 1$. Transform the left side using the advance theorem and the right side using the last entry in Table 2.1 to give

$$z^2X(z) - z^2x(0) - zx(1) - zX(z) + zx(0) + 0.24X(z) = \frac{\sin(1)z}{z^2 - 2\cos(1)z + 1}$$

Note that $\sin(1) = 0.8415$ and $\cos(1) = 0.54$ and that $x(0) = x(1) = 0$, so

$$(z^2 - z + 0.24)X(z) = \frac{0.8415z}{z^2 - 1.08z + 1}$$

so

$$X(z) = \frac{0.8415z}{(z - 0.4)(z - 0.6)(z^2 - 1.08z + 1)} \qquad \text{(a)}$$

Note now that the expansion must contain both the transforms of $\sin(k)$ and $\cos(k)$, so let us make the expansion in the form

$$X(z) = 0.8415\left[\frac{Az}{z - 0.4} + \frac{Bz}{z - 0.6} + \frac{C(0.8415)z}{z^2 - 1.08z + 1} + \frac{D(z^2 - 0.54z)}{z^2 - 1.08z + 1}\right] \qquad \text{(b)}$$

Coefficients $A$ and $B$ are given by the usual Heaviside method to be

$$A = \frac{1}{(0.4 - 0.6)(0.16 - 0.432 + 1)} = -6.868$$

and

$$B = \frac{1}{(0.6 - 0.4)(0.36 - 0.648 + 1)} = 7.0225$$

To evaluate the remaining coefficients, find a common denominator in (b) and equate the numerator with that of (a) to yield

$$z = Az(z - 0.6)(z^2 - 1.08z + 1) + Bz(z - 0.4)(z^2 - 1.08z + 1) + C(0.8415)z(z^2 - z + 0.24) + D(z^2 - 0.54z)(z^2 - z + 0.24)$$

Equating the coefficients of like powers of $z$, we get

$$z^4: \quad 0 = A + B + D$$
$$z^3: \quad 0 = -1.68A - 1.48B + 0.8415C - 1.54D$$

Thus the first of these equations gives

$$D = -A - B = 6.868 - 7.0225 = -0.1545$$

while the second gives

$$0.8415C = 1.68A + 1.48B + 1.54D$$

or

$$C = \frac{1.68A + 1.48B + 1.54D}{0.8415}$$

and plugging the values of $A$, $B$ and $D$ gives

$$C = 1.6471$$

Thus the inverse transform is the sequence $x(k)$, or

$$x(k) = 0.8415[-6.868(0.4)^k + 7.0225(0.6)^k + 1.6471 \sin(k) - 0.1545 \cos(k)]$$

At this point it might be useful to see if the starting conditions are met, which in fact, they are.

## 2.8 z-DOMAIN TRANSFER FUNCTION AND IMPULSE RESPONSE SEQUENCE

Consider now a general discrete-time system described by the linear constant-coefficient difference equation

$$y(k + n) + a_{n-1}y(k + n - 1) + \cdots + a_1y(k + 1) + a_0y(k) = d_{n-1}u(k + n) + d_{n-1}u(k + n - 1) + \cdots + d_0u(k) \quad (2.8.1)$$

If we now take the a transform employing the advance theorem (Theorem 2.3) and ignoring the starting conditions $y(0), \ldots, y(n - 1)$ and $u(0), \ldots, u(n - 1)$, we get

$$(z^n + a_{n-1}z^{n-1} + \cdots a_1z + a_0)Y(z) = (d_nz^n + \cdots + d_1z + d_0)U(z) \qquad (2.8.2)$$

We can solve for the ratio of output $Y(z)$ to the input $U(z)$ to give

$$H(z) = \frac{Y(z)}{U(z)} = \frac{d_n z^n + \cdots + d_1 z + d_0}{z^n + a_{n-1}z^{n-1} + \cdots + a_1 z + a_0} \tag{2.8.3}$$

which is the $z$-domain transfer function.

In a similar fashion the single-input/single-output system described by a set of simultaneous difference equations could be reduced to a transfer function by first $z$-transforming the equations and then solving via Cramer's rule for the desired output/input ratio. Note that transfer functions ignore starting conditions and hence are not useful in solving problems that involve nonzero starting conditions.

**Example 2.9.** Find the transfer function associated with the second-order difference equation

$$y(k+2) - 2y(k+1) + y(k) = u(k+1) - u(k)$$

where $u(k)$ is a general sequence and not the unit-step sequence. Taking the $z$-transform and applying the advance theorem, we get

$$(z^2 - 2z + 1)Y(z) = (z - 1)U(z)$$

so the transfer function is

$$H(z) = \frac{Y(z)}{U(z)} = \frac{z - 1}{z^2 - 2z + 1}$$

**Example 2.10.** Find the transfer function $Y(z)/U(z)$ associated with the simultaneous difference equations

$$y(k+1) - 2y(k) + x(k) = u(k)$$

and

$$x(k+1) - y(k) = 3u(k)$$

Taking the $z$-transform of these equations while ignoring starting conditions $x(0)$ and $y(0)$, we get the linear $z$-domain equations

$$(z-2)Y(z) + X(z) = U(z)$$
$$-Y(z) + zX(z) = 3U(z)$$

Now using Cramer's rule to solve for the output variable, we get

$$Y(z) = \frac{\begin{vmatrix} U(z) & 1 \\ 3U(z) & z \end{vmatrix}}{\begin{vmatrix} z-2 & 1 \\ -1 & z \end{vmatrix}}$$

on carrying out the multiplication and combining like terms,

$$Y(z) = \frac{(z - 3)U(z)}{z^2 - 2z + 1}$$

Solving for the output/input ratio gives

$$H(z) = \frac{Y(z)}{U(z)} = \frac{z - 3}{z^2 - 2z + 1}$$

Consider a single-input/single-output system with input sequence $u(k)$ and output sequence $y(k)$, so the $z$-domain transfer function is defined as

$$H(z) = \frac{\mathfrak{Z}[y(k)]}{\mathfrak{Z}[u(k)]} = \frac{Y(z)}{U(z)} \tag{2.8.4}$$

For the type of system of interest in this context, $H(z)$ will be the ratio of two polynomials in the complex variable $z$. For realizable systems the denominator polynomial will be at least as high in degree as the numerator.

Let us now specify the input sequence $u(k)$ to be a unit discrete impulse at $k = 0$, or

$$u(k) = \delta(k) \tag{2.8.5}$$

and the $z$-transform of the input sequence was given in Section 2.3 to be

$$\mathfrak{Z}[u(k)] = \mathfrak{Z}[\delta(k)] = 1 \tag{2.8.6}$$

Since the $z$-transform of the output sequence is given by the product of the transfer function and the input function, the output in this case is

$$Y(z) = H(z) \tag{2.8.7}$$

Upon inverting the $z$-transform the result is

$$y(k) = h(k) \tag{2.8.8}$$

Thus the inverse transform of the transfer function is commonly referred to as the *impulse response sequence*. We discuss the role of the impulse response sequence further in Section 2.12.

**Example 2.11.** Find the impulse response sequence for a discrete-time system in which the transfer function was found to be

$$H(z) = \frac{z - 3}{z^2 - z + 1}$$

First, it is appropriate to check the pole locations, which are

$$z_{1,2} = \tfrac{1}{2} \pm \sqrt{\tfrac{1}{4} - 1} = 0.5 \pm j0.866$$

These are complex roots on the unit circle, and as such the time-domain sequence $h(k)$ must contain undamped sines and cosines. Also, we note that the numerator polynomial has a constant term and thus we must include a constant term in the partial fraction expansion. The denominator is of the form $z^2 - (2\cos \Omega T)z + 1$; thus

$$2\cos \Omega T = 1$$

and then

$$\Omega T = \cos^{-1}(\tfrac{1}{2}) = 1.047 \text{ rad}$$

so

$$\sin \Omega T = 0.866$$

The partial fraction expansion must be

$$\frac{z-3}{z^2 - z + 1} = A + \frac{B(z^2 - 0.5z)}{z^2 - z + 1} + \frac{C(0.866)z}{z^2 - z + 1}$$

Evaluating at $z = 0$ gives

$$A = \frac{-3}{1} = -3$$

Finding a common denominator, the equated numerators are

$$z - 3 = A(z^2 - z + 1) + B(z^2 - 0.5z) + C(0.866)z$$

Equating coefficients of like powers of $z$ gives

$$z^2\!: \quad 0 = A + B$$
$$z^1\!: \quad 1 = -A - 0.5B + 0.866C$$

and we do not need the final equation since we have already evaluated $A$. Thus

$$B = -A = 3$$

and

$$C = \frac{1 + A + 0.5B}{0.866} = -0.577$$

Thus the impulse response sequence $h(k)$ is

$$h(k) = -3\delta(k) + 3\cos(1.047k) - 0.577\sin(1.047k)$$

A plot of this impulse response is shown in Fig. 2.3.

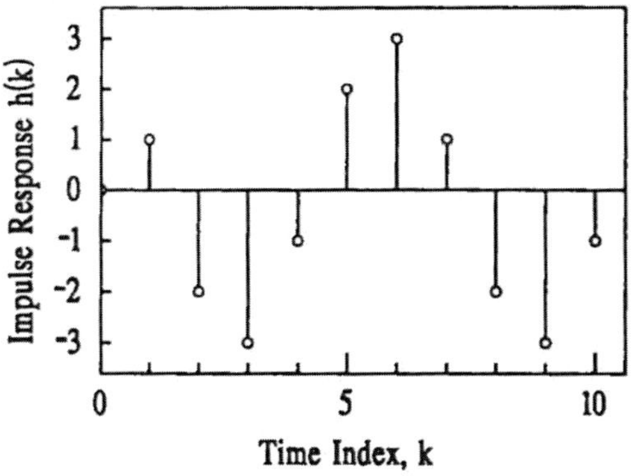


**Figure 2.3.** Impulse response sequence.

## 2.9 RELATION BETWEEN z-PLANE POLE LOCATIONS AND THE NATURE OF TEMPORAL SEQUENCES

We have seen that Laplace transforms of continuous-time functions which have poles in the right half of the $s$-plane imply that the associated time-domain functions increase without bound as time becomes large. We shall now discuss a similar property for $z$-transforms of discrete-time sequences.

First let us consider a $z$-domain function of the form

$$F(z) = \frac{Az}{z - p} \tag{2.9.1}$$

where $p$ is a real number or a real pole of the function $F(z)$; the associated time-domain sequence is

$$f(k) = Ap^k \qquad k > 0 \tag{2.9.2}$$

Clearly, if $p < -1$, the solution will oscillate and increase in magnitude for large $k$. If $-1 < p < 0$, the solution will decay in an oscillatory fashion, and if $0 < p < 1$, it will decay in an exponential manner as $k$ becomes large. Also, if $p > 1$, the sequence will grow with an exponential nature.

Now let us consider a function with a quadratic denominator of the form

$$F(z) = \frac{N(z)}{z^2 - bz + c} \tag{2.9.3}$$

The denominator has complex roots $p$ and $p^*$, and hence a partial fraction expansion of the form of (2.9.4) can be made to yield

$$F(z) = \frac{Az}{z - p} + \frac{A^*}{z - p^*} \tag{2.9.4}$$

where the asterisk denotes the complex conjugate. We can write the complex pole in polar form as

$$p = Re^{j\theta} \tag{2.9.5}$$

and the conjugate pole as

$$p^* = Re^{-j\theta} \tag{2.9.6}$$

The coefficients $A$ and $A^*$ are also complex conjugates, and we can let

$$A = \alpha + j\beta \tag{2.9.7}$$

If we substitute (2.9.5), (2.9.6), and (2.9.7) into (2.9.4) and invert the $z$-transforms, we get

$$f(k) = (\alpha + j\beta)R^k e^{jk\theta} + (\alpha - j\beta)R^k e^{-jk\theta} \tag{2.9.8}$$

We may rewrite this as

$$f(k) = R^k[\alpha(e^{jk\theta} + e^{-jk\theta}) + j\beta(e^{jk\theta} - e^{-jk\theta})] \tag{2.9.9}$$

and upon employing the Euler identities we get

$$f(k) = R^k(2\alpha \cos k\theta - 2\beta \sin k\theta) \tag{2.9.10}$$

If we now examine this, we see that the sine and cosine functions are bounded by plus and minus unity, and hence the $R^k$ factor determines the asymptotic nature of the discrete-time sequence. If $R$ is greater than unity, the sequence will be unbounded for large $k$, and if $R$ is less than unity (but still positive), the sequence will converge to zero for large $k$.

Since $R$ is the radial distance from the origin to the complex poles, it is clear that the poles which lie within the unit circle of the $z$-plane contribute to decreasing sequences, while poles on the unit circle contribute to an oscillatory sequence, and those that lie outside the unit circle are associated with sequences that become unbounded. The region of boundedness is shown in Fig. 2.4. We have seen that various pole locations in the $z$-plane yield time-domain sequences with quite different characters. A summary of these locations and the type of sequences they yield is given in Fig. 2.5.

We know that the poles of a transfer function will contribute to the response sequence for a discrete-time system, and hence if one of those transfer function poles lies exterior to the unit circle, the output sequence will be unbounded and the system will be said to be unstable. The unit

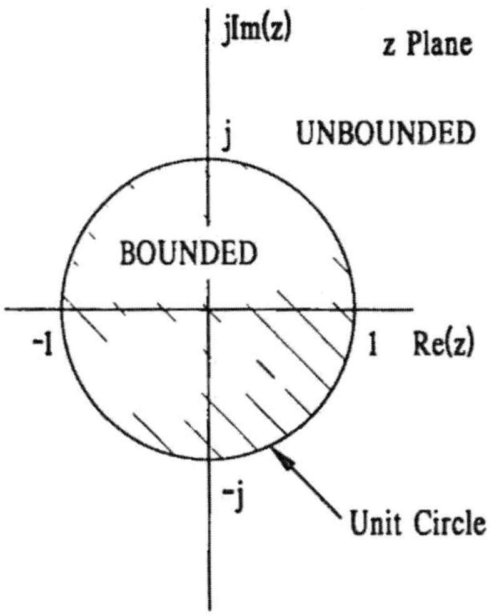


**Figure 2.4.** Regions of poles for bounded and unbounded sequences in the *z*-plane.

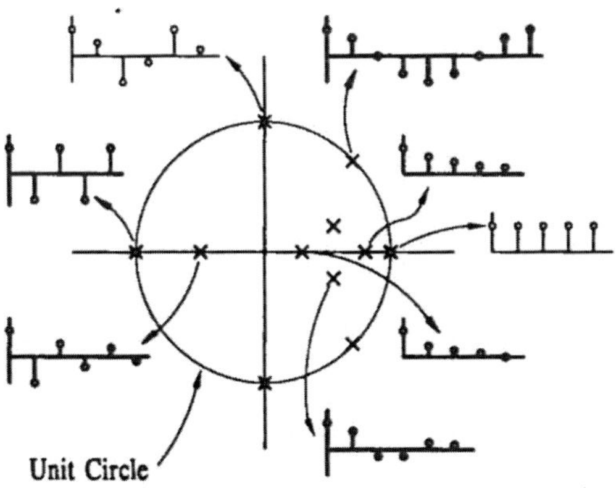


**Figure 2.5.** Pole locations and associated time-domain sequences.

circle is thus the stability boundary, as was the imaginary axis on the $s$-plane in analysis of the continuous-time systems.

## 2.10 FREQUENCY RESPONSE OF DISCRETE-DATA SYSTEMS

In the design of control systems, it is often important to know how a system will respond to a sinusoidal driving function, and we find that the frequency response design methods (Nyquist, Bode, and Nichols) are still useful tools in the design of digital control systems. It is for that reason that we examine the steady-state sinusoidal response of a discrete-time system to a sampled sinusoid.

In order for the system to possess a steady-state response to a sinusoidal input, it must be stable [i.e., all the poles of the associated transfer function $H(z)$ must lie within the unit circle of the complex $z$-plane]. The system transfer function of interest is

$$H(z) = \frac{Y(z)}{U(z)} \tag{2.10.1}$$

Let the input sequence to the system be a cosine sequence of radian frequency $\omega$, or

$$u(k) = A\cos(\omega kT) \tag{2.10.2}$$

This may be rewritten by employing Euler's identity or

$$u(k) = \frac{A}{2}\left(e^{j\omega kT} + e^{-j\omega kT}\right) \tag{2.10.3}$$

This form may now be transformed by employing the transform of the exponential sequence so that

$$U(z) = \frac{A}{2}\left(\frac{z}{z - e^{j\omega T}} + \frac{z}{z - e^{-j\omega T}}\right) \tag{2.10.4}$$

The transfer function may be written with a factored denominator

$$H(z) = \frac{N(z)}{(z - p_1)(z - p_2)\cdots(z - p_n)} \tag{2.10.5}$$

where for stability we have assumed that $|p_i| < 1$ for all $i$. The $z$-domain output is then

$$\begin{aligned} Y(z) = {} & \frac{A}{2}\left(\frac{z}{z - e^{j\omega T}} + \frac{z}{z - e^{-j\omega T}}\right) \\ & \times \left[\frac{N(z)}{(z - p_1)(z - p_2)\cdots(z - p_n)}\right] \end{aligned} \tag{2.10.6}$$

A partial fraction expansion is then of the form

$$Y(z) = \frac{Bz}{z - e^{j\omega T}} + \frac{Cz}{z - e^{-j\omega T}} + \sum_{i=1}^{n} \frac{D_i z}{z - p_i} \tag{2.10.7}$$

Each term of the summation yields a time-domain sequence of the form $D_i p_i^k$, which if $|p_i| < 1$ vanishes as $k$ gets large and hence does not contribute to the steady-state response. The coefficient $B$ may be evaluated by the Heaviside method to give

$$B = \lim_{z \to e^{j\omega T}} \left[\frac{A}{2} + \frac{C(z - e^{j\omega T})}{z - e^{-j\omega T}}\right] H(z) \tag{2.10.8}$$

$$= \frac{A}{2} H(e^{j\omega T}) \tag{2.10.9}$$

Similarly, we can see that $C$ is given by

$$C = \frac{A}{2} H(e^{-j\omega T}) \tag{2.10.10}$$

Thus the transform of the steady-state response may be written as

$$Y(z) = \frac{A}{2}\left[\frac{H(e^{j\omega T})z}{z - e^{j\omega T}} + \frac{H(e^{-j\omega T})z}{z - e^{-j\omega T}}\right] \tag{2.10.11}$$

Now let us examine the nature of $H(e^{j\omega T})$. Since $H(z)$ is a ratio of polynomials in $z$, $H(e^{j\omega T})$ is a complex number that may be written in polar form as

$$H(e^{j\omega T}) = |H(e^{j\omega T})| e^{j\phi} \tag{2.10.12}$$

where $\phi$ is the angle of $H(e^{j\omega T})$. Also, since $H(z)$ is a ratio of polynomials in $z$, the function $H(e^{-j\omega T})$ has the same magnitude and the conjugate angle, hence

$$H(e^{-j\omega T}) = |H(e^{j\omega T})| e^{-j\phi} \tag{2.10.13}$$

Using relations (2.10.11) and (2.10.12), relation (2.10.10) may be written as

$$Y(z) = \frac{A}{2} |H(e^{j\omega T})| \left(\frac{e^{j\phi} z}{z - e^{j\omega T}} + \frac{e^{-j\phi} z}{z - e^{-j\omega T}}\right) \tag{2.10.14}$$

Inverting the transforms, the result is

$$y(k) = \frac{A}{2} |H(e^{j\omega T})| (e^{j\phi} e^{j\omega T} + e^{-j\phi} e^{-j\omega T}) \tag{2.10.15}$$

Combining the exponentials gives

$$y(k) = A|H(e^{j\omega T})|\tfrac{1}{2}[e^{j(\omega T+\phi)} + e^{-j(\omega T+\phi)}] \qquad (2.10.16)$$

Again using the Euler identity $\cos x = (e^{jx} + e^{-jx})/2$, this yields the steady-state response

$$y(k) = A|H(e^{j\omega T})| \cos(\omega kT + \phi) \qquad (2.10.17)$$

We see that the steady-state response is indeed sinusoidal and shifted in phase from the driving sinusoidal sequence. The amplitude of the output is attenuated (or amplified) by a factor of $|H(e^{j\omega T})|$, which thus will be referred to as the gain associated with $H(z)$ at frequency $\omega$. The complex function of $\omega$, $H(e^{j\omega T})$, is referred to as the *frequency response function.* The frequency response function is related to the transfer function as

$$H(z)\Big|_{z=e^{j\omega T}} = H(\cos \omega T + j \sin \omega T) \qquad (2.10.18)$$

As in the continuous-time case, we are usually interested in the magnitude and phase character of this function as a function of radian frequency $\omega$. It is important to note that the dc gain corresponds to $\omega = 0$, which yields

$$\text{dc gain} = H(e^{j\omega T})|_{\omega=0} = H(z)|_{z=1} = H(1) \qquad (2.10.19)$$

Several important properties of $H(\cos \omega T + j \sin \omega T)$ should now be discussed. Since both $\cos \omega T$ and $\sin \omega T$ are periodic in $\omega$ (for fixed $T$), then $H(e^{j\omega T})$ will also be periodic in $\omega$ and will repeat every $\omega_s = 2\pi/T$ rad/s. Since $e^{-j\omega T}$ is the complex conjugate of $e^{j\omega T}$, we can write for negative frequencies

$$H(e^{-j\omega T}) = H^*(e^{j\omega T}) \qquad (2.10.20)$$

which implies that

$$|H(e^{-j\omega T})| = |H(e^{j\omega T})| \qquad (2.10.21)$$

and

$$\angle H(e^{-j\omega T}) = -\angle H(e^{j\omega T}) \qquad (2.10.22)$$

Equation (2.10.21) along with the periodic condition for $H(e^{j\omega T})$ indicates that the magnitude function will be "folded" about frequency $\omega_s/2$, or

$$|H(e^{j\omega T})| = |H(e^{j(\omega_s-\omega)T})| \qquad (2.10.23)$$

and for the phase shift

$$\angle H(e^{j\omega T}) = -\angle H(e^{j(\omega_s-\omega)T}) \qquad (2.10.24)$$

**Example 2.12.** Consider a discrete-time system with a transfer function

$$H(z) = \frac{(e^{-\alpha T}\sin \Omega T)z}{z^2 - (2e^{-\alpha T}\cos \Omega T) - z + e^{-2\alpha T}}$$

To find the frequency response we simply let $z = \cos \omega T + j \sin \omega T$, or

$$H(e^{j\omega T}) = \frac{(e^{-\alpha T}\sin \Omega T)(\cos \omega T + j\sin \omega T)}{(\cos \omega T + j\sin \omega T)^2 + (2e^{\ \alpha T}\cos \Omega T)(\cos \omega T + j\sin \omega T) + e^{-2\alpha T}}$$

Although the evaluation of this expression by hand for a range of frequencies $\omega$ is a tedious task, a simple computer program would suffice to give the gain and phase shift over that range. This has been done for a range of dimensionless frequency for one value of $\alpha$ and several values of sampling interval and results are presented in Fig. 2.6. We can even see the folding of the frequency response for $\Omega T = \pi/2$ at a dimensionless frequency of 2.

## 2.11 RELATIONSHIP BETWEEN *s*- AND *z*-DOMAIN POLES OF SAMPLED FUNCTIONS

Let us consider a simple continuous-time exponential function

$$f(t) = e^{-at} \qquad t > 0 \tag{2.11.1}$$

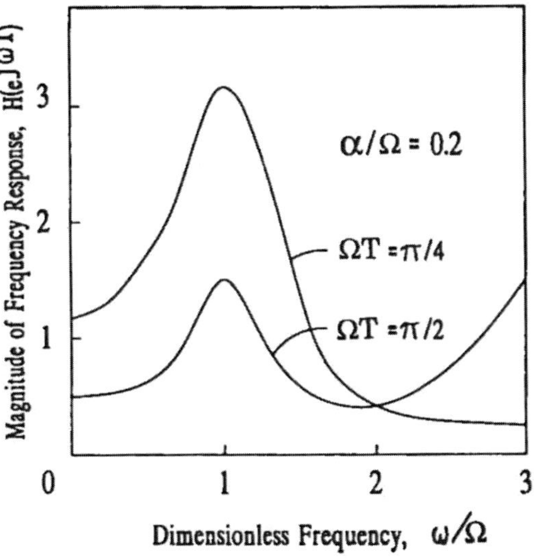


**Figure 2.6.** Frequency response of a discrete-time transfer function.

which has Laplace transform

$$F(s) = \mathcal{L}[f(t)] = \frac{1}{s + a} \tag{2.11.2}$$

This transform has a pole at $s = -a$ in the complex $s$-plane. Now let us sample this function with sampling interval of $T$ seconds, which will give a sequence

$$f(k) = e^{-akT} \tag{2.11.3}$$

which upon examination of Table 2.1 gives a $z$-transform of

$$F(z) = \frac{z}{z - e^{-aT}} \tag{2.11.4}$$

This function has a pole at

$$z = e^{-aT} \tag{2.11.5}$$

Before we rush to any hasty conclusions, let us examine a damped or undamped sine function. We shall consider only the damped sine function of the form

$$f(t) = e^{-\alpha t} \sin \Omega t \tag{2.11.6}$$

We know that this function has a Laplace transform of

$$F(s) = \frac{\Omega}{(s + \alpha)^2 + \Omega^2} \tag{2.11.7}$$

The poles of this function are located at

$$s_{1,2} = -\alpha \pm j\Omega \tag{2.11.8}$$

Now if we take the $z$-transform of the sampled version of (2.11.6), we get

$$F(z) = \frac{ze^{-\alpha T} \sin \Omega T}{z^2 - z \cdot 2e^{-\alpha T} \cos \Omega T + e^{-2\alpha T}} \tag{2.11.9}$$

The denominator polynomial of this function is

$$P(z) = z^2 - z \cdot 2e^{-\alpha T} \cos \Omega T + e^{-2\alpha T} \tag{2.11.10}$$

The locations of the $z$-domain poles are found by equating $P(z)$ to zero and solving for $z$. The pole locations given by the quadratic formula are

$$z_{1,2} = e^{-\alpha T} \cos \Omega T \pm \sqrt{e^{-2\alpha T} \cos^2 \Omega T - e^{-2\alpha T}} \tag{2.11.11}$$

or since $\cos \Omega T$ is bounded above by unity,

$$z_{1,2} = e^{-\alpha T}(\cos \Omega T \pm j\sqrt{1 - \cos^2 \Omega T}) \tag{2.11.12}$$

$$= e^{-\alpha T}(\cos \Omega T \pm j \sin \Omega T) \tag{2.11.13}$$

and employing the Euler identity gives

$$z_{1,2} = e^{-\alpha T \pm j\Omega T} \tag{2.11.14}$$

or the pole locations are at

$$z_{1,2} = e^{(-\alpha \pm j\Omega)T} \tag{2.11.15}$$

If one is curious about the location of the poles, return to relationship (2.11.13). If we look for the polar form of the poles, we can see that the magnitude of the distance of the root from the origin of the *z*-plane is

$$|z_{1,2}| = e^{\alpha T} \tag{2.11.16}$$

and the angle from the positive real axis is

$$\angle z_{1,2} = \tan^{-1}\left(\pm \frac{\sin \Omega T}{\cos \Omega T}\right) \tag{2.11.17}$$

or the poles are located at an angle of

$$\angle z_{1,2} = \pm\, \Omega T \tag{2.11.18}$$

This situation is illustrated in Fig. 2.7.

Note that the pole of the exponential and damped sinusoidal function are mapped exponentially to the *z*-plane by the operation of sampling, suggesting the relation

$$z = e^{sT} \tag{2.11.19}$$

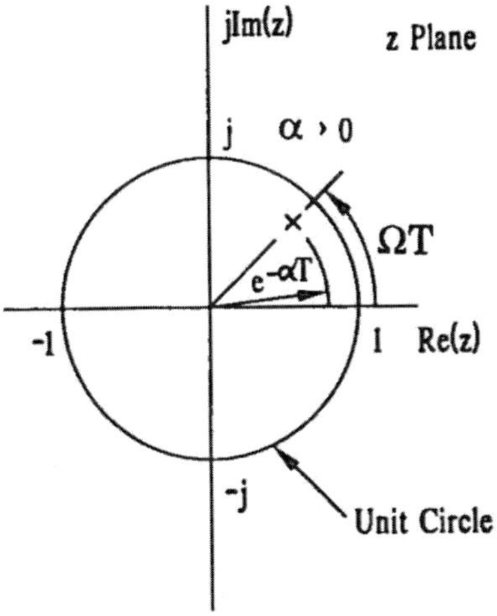


**Figure 2.7.** Complex roots in the *z*-plane.

as the relationship between $s$-domain poles and the poles of the sampled function in the $z$-plane. Note that the $z$-domain pole locations are functions of the sampling interval $T$ *and* the original $s$-domain locations. This sort of relationship does not necessarily hold between the zeros of the $s$-domain and $z$-domain functions. The mapping given in (2.11.19) also holds for functions that are polynomial in form in the time domain and have poles at the origin of the $s$-plane and at $z = +1$ in the $z$-plane.

Let us return to expression (2.11.19) and substitute $s = \sigma + j\omega$ into it to yield

$$z = e^{\sigma T}e^{j\omega T} = e^{\sigma T}(\cos \omega T + j \sin \omega T) \tag{2.11.20}$$

or

$$z = e^{\sigma T} \cos \omega T + je^{\sigma T} \sin \omega T \tag{2.11.21}$$

Note now that the magnitude of the $z$-domain pole location is

$$|z| = e^{\sigma T} \tag{2.11.22}$$

and the angle of $z$ is given by

$$\angle z = \omega T \tag{2.11.23}$$

From (2.11.22) it is clear that lines of constant $\sigma$ (vertical lines in the $s$-plane) are lines of constant radial distance or concentric circles in the $z$-plane. Also from (2.11.23) it is clear that lines of constant $\omega$ (horizontal lines in the $s$-plane) are lines of constant angle or radial lines in the $z$-plane.

From (2.11.23) we can see that an angle of $\omega T = 3\pi/4$ gives the same angular location as $\omega T = 11\pi/4$, and this indicates that there are infinitely many horizontal lines in the $s$-plane with the same image in the $z$-plane. For values of $-\pi/T < \omega < \pi/T$ there are unique images in the $z$-plane. Note here that $2\pi/T = \omega_s$, the radian sampling frequency, so the range of the primary strip in the $s$-plane is $-\omega_s/2 < \omega < \omega_s/2$. Note also that the image of the $j\omega$-axis in the $s$-plane between $-\omega_s/2$ and $\omega_s/2$ is the unit circle.

The left half of the primary strip is mapped to the inside of the unit circle and the right half is mapped to the outside of the unit circle. These ideas are illustrated by Fig. 2.8a and b. In a similar fashion the next strip where $\omega_s/2 < \omega < 3\omega_s/2$ is mapped onto the entire $z$-plane.

The facts exposed above are quite useful in the digital filter synthesis problem, since they give us some indication of where $z$-domain poles must be located to have similar dynamical character as some given $s$-domain poles. We discuss this at length in Chapter 5.

If one desires the correct $z$-plane pole locations for a sampled function, the sampling frequency $\omega_s$ must be chosen such that the poles of the

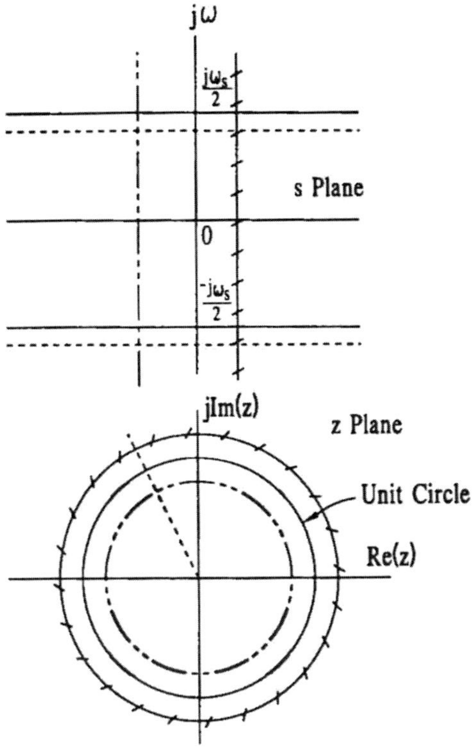


**Figure 2.8.** (a) $s$-Plane loci; (b) corresponding $z$-plane loci.

$s$-domain function lie within the primary strip of the $s$-plane. This amounts to a crude statement of the sampling theorem, which we discuss in more detail later.

## 2.12 DISCRETE-TIME CONVOLUTION THEOREM (THEOREM 2.4)

In Section 2.8 we discussed the response of a discrete-time system to a unit discrete impulse and we showed that the response was the inverse transform of the transfer function. We would now like to consider the response to some arbitrary input sequence $u(k)$. The arbitrary input could be thought of as a sequence of weighted discrete impulses.

$$u(k) = u(0)\delta(k) + u(1)\delta(k - 1) + u(2)\delta(k - 2) + \cdots \quad (2.12.1)$$

The response due to the first impulse is $u(0)h(k)$ and that due to the second is $u(1)h(k - 1)$, so we can write the response at some time $k$ as the sum of all the responses due to the previous impulses plus the one at time $k$, or

$$y(k) = u(0)h(k) + u(1)h(k - 1) + u(2)h(k - 2) + \cdots + u(k)h(0) \qquad (2.12.2)$$

This can be written as

$$y(k) = \sum_{i=0}^{k} u(i)h(k - i) \qquad k = 0, 1, 2, \ldots \qquad (2.12.3)$$

or if we change the subscript by letting $k - i = j$, we get

$$y(k) = \sum_{j=0}^{k} u(k - j)h(j) \qquad k = 0, 1, 2, \ldots \qquad (2.12.4)$$

Expressions (2.12.3) and (2.12.4) represent discrete-time convolutions. In short, the response sequence could have been calculated in the $z$-domain by

$$y(k) = \mathfrak{Z}^{-1}[Y(z)] = \mathfrak{Z}^{-1}[H(z)U(z)] \qquad (2.12.5)$$

or equivalently, by either expression (2.12.3) or (2.12.4). In fact, convolution in the discrete-time domain is equivalent to multiplication in the $z$-domain. In discrete-time system studies, the discrete-time convolution can be carried out by digital computer to calculate system response to a known input sequence. This works well for small values of $k$ but becomes inefficient for simulations of long-time duration. The trade-off point is a function of the complexity of the system, the availability of digital computer time, and the amount of time available for the inversion of $z$-transforms.

## 2.13 FINAL VALUE THEOREM (THEOREM 2.5)

It is often useful in the analysis of steady-state errors in control systems to be able to examine the asymptotic behavior of such systems without actually having to evaluate the high-order $z$-transforms associated with transfer functions. Sequences that have a limiting value for large $k$ must be stable functions, or all the poles must lie inside the unit circle except for one at $z = 1$ if the steady-state value is to be nonzero. In other words, a step function must be present in the function. The function should be expandable by partial fractions into the form

$$F(z) = \frac{Az}{z - 1} + \sum_{i=1}^{n} \frac{B_i z}{z - p_i} \qquad |p_i| < 1 \qquad (2.13.1)$$

where $|p_i| < 1$ is the condition of stable transients in the sequence. If we invert the transforms, we get

$$f(k) = A + \sum_{i=1}^{n} B_i p_i^k \qquad (2.13.2)$$

and as $k$ gets large, the limiting value will be $A$ as the terms contributed by the sum approach zero. If we evaluate $A$ by the usual method of partial fractions, we simply multiply (2.13.1) by $(z - 1)/z$ and let $z$ approach unity:

$$A = \lim_{z \to 1} \left(\frac{z - 1}{z}\right) F(z) \qquad (2.13.3)$$

The conclusion is the formal mathematical statement of the final value theorem:

$$\lim_{k \to \infty} f(k) = \lim_{z \to 1} \left(\frac{z - 1}{z}\right) F(z) \qquad (2.13.4)$$

and the proof is complete.

**Example 2.13.** For a discrete-data system with transfer function

$$H(z) = \frac{Y(z)}{U(z)} = \frac{z + 1}{z^2 - 1.4z + 0.48}$$

and a unit-step input for which the $z$-transform is

$$U(z) = \frac{z}{z - 1}$$

find the final value of the response sequence $y(k)$. The response in the $z$-domain is

$$Y(z) = \frac{z(z + 1)}{(z^2 - 1.4z + 0.48)(z - 1)}$$

and employing the final value theorem after checking that the poles are inside the unit circle:

$$y(\infty) = \frac{1 + 1}{1 - 1.4 + 0.48} = \frac{2}{0.08} = 25$$

This answer could also be verified by examining the dc gain of the system, or

$$H(1) = \frac{2}{0.08} = 25$$

Multiplication of this by the unit-step input also gives the correct steady-state output.

## 2.14 BACKDOOR APPROACH TO THE SAMPLING THEOREM

Let us consider in this section the properties of sampled sinusoids. We know that the $z$-transforms of a sampled cosine and sine waves are, respectively,

$$\mathfrak{Z}[\cos k\Omega T] = \frac{z^2 - (\cos \Omega T)z}{z^2 - 2(\cos \Omega T)z + 1} \tag{2.14.1}$$

and

$$\mathfrak{Z}[\sin k\Omega T] = \frac{(\sin \Omega T)z}{z^2 - 2(\cos \Omega T)z + 1} \tag{2.14.2}$$

Note that the denominators of these two transforms are identical and hence the transforms possess identical poles. These poles are complex and lie on the unit circle at respective angles of $\pm\Omega T$ radians from the positive real axis. Any sampled sinusoid of radian frequency $\Omega$ would, in fact, have the same poles. As long as $\Omega T$ is less than $\pi$ radians, the roots are clearly distinguishable. When $\Omega T$ becomes greater than $\pi$ radians, say, $\pi + \alpha$, the pole at an angle of $-(\pi + \alpha)$ is indistinguishable from one at an angle of $+(\pi - \alpha)$ associated with $\Omega T < \pi$. The conclusion of this argument is that if $\Omega > \pi/T$, the frequency of the sinusoid is too high relative to the sampling frequency and the sampled sinusoid cannot, by its samples, be distinguished from one at some lower frequency. Another way of saying this is that if $T > \pi/\Omega$, this indicates that the sampling interval $T$ was chosen too large relative to the period of the sinusoid to permit us to distinguish this sinusoid from one of a longer period. These points are best illustrated by the examination of Fig. 2.9, wherein two sinusoids of quite different frequencies with identical samples are shown. Clearly, there are an infinite number of sinuoids which will have the samples shown in Fig. 2.9.

If the samples are identical, the $z$-transforms of both sequences will be the same, and there is no way to infer anything about the continuous-time functions from whence they were taken. If one can be assured that the original time-domain functions are bandlimited to a region lower than half the sampling frequency, the sampling theorem assures that they can be reconstructed from the samples since no ambiguity of the $z$-transform will exist. If the samples are identical, then by Eq. (2.3.2) the $z$-transforms will be the same, and there is no way to infer anything about the continuous-time functions from which they were taken.

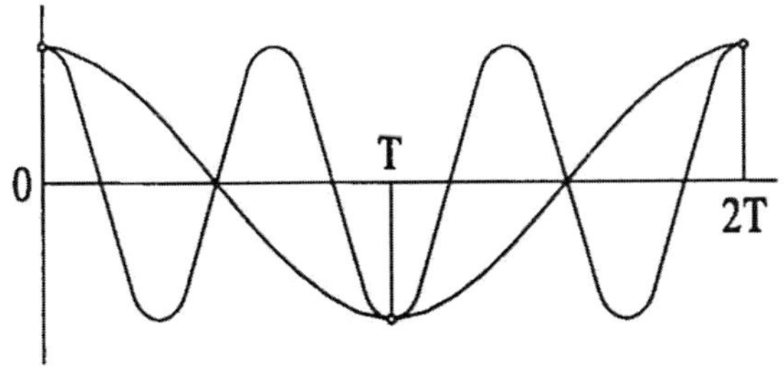


**Figure 2.9.** Sampled cosine waves of different frequencies with identical samples.

## 2.15 SUMMARY

In this chapter we have examined the fundamental mathematics necessary to discuss the discrete-time control problem on a rational basis. The $z$-transform was developed as a tool for the solution of linear difference equations with constant coefficients. Whenever possible the parallel structure with the Laplace transform is pointed out. The concept of the discrete-time transfer function was developed, and from this the frequency response function resulted for sinusoidal excitation. The significance of the location of the $z$-domain poles of sampled functions and how these locations reflect the nature of the sequence itself were pointed out. The relationship between $s$-domain poles of continuous-time functions and the $z$-domain poles of their sampled versions is examined and general conclusions are drawn. The sampling theorem is introduced from a $z$-transform uniqueness point of view, and the necessary condition for uniqueness of $z$-transform is developed.

## PROBLEMS

2.1. Find the $z$-transform of the two sequences shown below by using step sequences to turn them on and off.

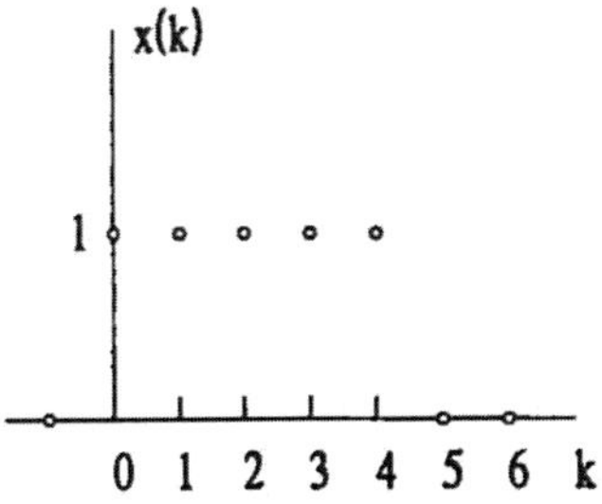


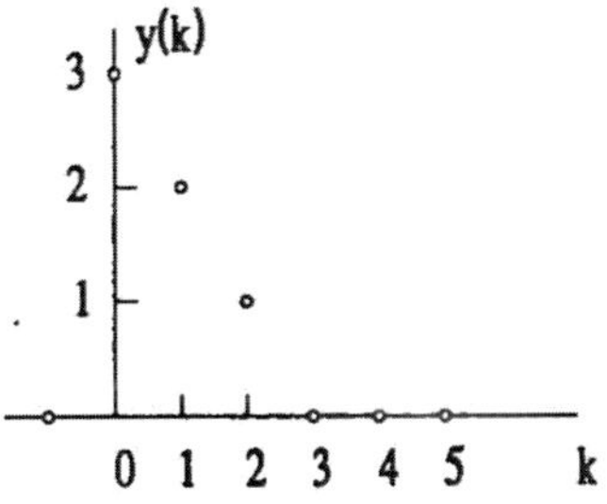

2.2. Show that the inverse transform of $A/z^2$ is the sequence $y(k) = \{0, 0, A, 0, 0, \ldots\}$.

2.3. Find the $z$-transform of the following time-domain sequences: (a) $f(k) = 2k$ $(k \geq 0)$; (b) $f(k) = \sin(kT)$ $(k \geq 0)$. (You may use only results that have been proven from first principles, i.e., the definition.)

2.4. Find the $z$-transform of the following sequence. This transform should not be in infinite series form

$$x(k) = A \sin(kT + \phi) \qquad \alpha \neq 0$$

(*Hint:* You may want to resolve this into a combination of sine and cosine functions and use their associated transforms.)

2.5. Given the linear difference equation $x(k + 1) = \alpha x(k)$ $[x(0) = 1]$, calculate the first five steps $x(1), x(2), \ldots, x(5)$ for the initial condition given and the following values of $\alpha$: (a) $\alpha = 0.8$; (b) $\alpha = 1.2$; (c) $\alpha = -0.8$; (d) Can you draw any conclusion about the value of $\alpha$ and stability? If so, what?

2.6. Find the closed-form solution to the following difference equations using the $z$-transform method.

(a) $y(k + 1) - 0.8y(k) = 3k$, $y(0) = 2$

(b) $y(k + 1) - 0.25y(k) = u(k)$, $y(0) = 0$ where $u(k)$ is a unit-step sequence.

2.7. Solve the following difference equations using the $z$-transform.

(a) $w(k + 2) - 1.3w(k) = 0$, $w(0) = 1$, $w(1) = 3$

(b) $w(k + 1) + 2w(k) = 5k$, $w(0) = -1$

2.8. A good example of a discrete-time system is the process of paying interest on a savings account in a bank. Let $b(k)$ be the balance at the beginning of the $k$th month prior to when any deposit might have been made. Let the total deposits made during the $k$th month be denoted by $d(k)$ and let the monthly rate of interest be denoted by $r(0 < r < 1)$.

(a) Derive an expression for the balance at the beginning of the $(k + 1)$st month given that this balance is the sum of the previous balance, the interest accrued on that balance, and the deposits made during the $k$th month.

(b) Let us assume that the initial balance $b(0)$ is zero and that the rate of interest is ½% per month and that a \$20.00 deposit is made each month. Solve the difference equation derived in part (a) via the $z$-transformation and find the balance at the end of 10 years (120 months).

2.9. Consider the difference equation

$$x(k+2) - 0.4x(k+1) + 0.2x(k) = u(k)$$

with starting conditions $x(0) = x(1) = 0$. Assume that $u(k)$ is the unit-step sequence and find the solution $x(k)$ by employing the $z$-transform.

2.10. Use partial fractions to find the inverse $z$-transform of the following $z$-domain functions:

(a) $F(z) = \dfrac{z}{(z-1)(z-0.125)}$

(b) $F(z) = \dfrac{z-2}{z(z-0.4)(z+0.25)}$ (Note that you will need both a constant term and a term in $1/z$ for this problem.)

(c) $F(z) = \dfrac{z}{(z-1)(z-0.5)^2}$

2.11. Find the $z$-domain transfer function $Y(z)/X(z)$ and the impulse response sequence associated with the following difference equations.

(a) $y(k+2) - 1.4y(k+1) + 0.4y(k) = x(k+1)$

(b) $y(k) = \frac{1}{3}[x(k-2) + x(k-1) + x(k)]$
which is a smoothing algorithm for data processing.

2.12. Find the $z$-domain transfer functions associated with the following impulse response sequences. The transfer functions should be in polynomial ratio form.

(a) $h(k) = (0.4)^k$

(b) $h(k) = (0.25)^k + (-0.2)^k$

(c) $h(k) = ku(k)$

(d) $h(k) = (0.8)^k \sin(0.6k)$

2.13. For the two sequences $x(k)$ and $y(k)$ of Problem 2.1 perform the convolution of one of them into the other graphically/numerically.

2.14. A discrete-time system has a $z$-domain transfer function

$$H(z) = \frac{z+1}{z^2 - 1.4z + 0.48} = \frac{Y(z)}{U(z)}$$

Find the response sequence $y(k)$ if the driving function $u(k)$ is a unit-step sequence with $z$-transform

$$U(z) = \frac{z}{z-1}$$

Assume that starting conditions $y(0)$ and $y(1)$ are zero.

2.15. Find the impulse response function for the transfer function given in Problem 2.14.

2.16. Find the difference equation associated with the following transfer functions, where $H(z) = Y(z)/X(z)$.

(a) $$H(z) = \frac{z^2}{(z - 1)(z + 0.12)}$$

(b) $$H(z) = \frac{z(z - 1)}{z^2 - 0.5z + 0.97}$$

2.17. (a) For the ladder network shown below using Kirchhoff's current law, show that the appropriate difference equation is

$$v_{k+2} - 3v_{k+1} + v_k = 0 \qquad k = 0, 1, 2, \ldots, 4$$

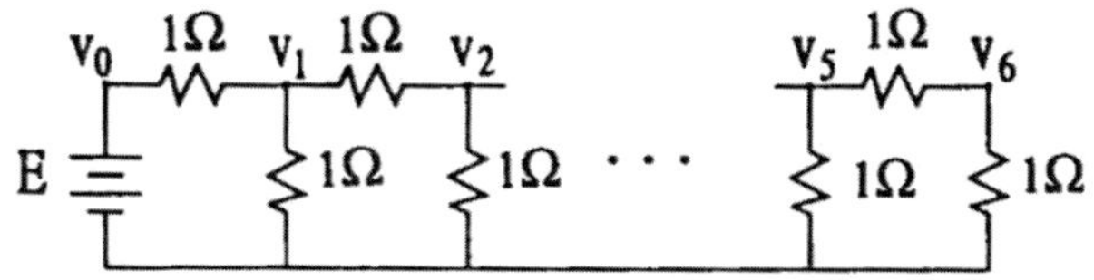


(b) Find a boundary condition at the left end of the network.

(c) Find a boundary condition at the right end of the network. [You should not solve the difference equation to get the answers to parts (b) and (c).]

2.18. Find the final value of the other time-domain sequence $f(k)$ associated with the following $z$-domain function. (First you must check that all the poles are in the unit circle except for the one at $z = 1$.)

$$F(z) = \frac{3z^3 - z^2}{(z - 1)(z^2 - 1.4z + 0.45)}$$

2.19. Find the $z$-domain transfer function $H(z) = Y(z)/U(z)$ associated with the following set of difference equations.

$$y(k + 1) - 0.5y(k) + 0.2x(k) = u(k)$$
$$x(k + 2) + 0.2x(k + 1) + x(k) - 0.3y(k) = 0$$

2.20. A first-order system has a transfer function

$$H(z) = \frac{0.2z}{z - 0.6} = \frac{X(z)}{U(z)}$$

Find the steady-state output if $u(k) = 2\sin(kT)$ by employing the frequency response method for $T = 0.5$ s.

2.21. For the discrete-time transfer function

$$H(z) = \frac{z + 1}{z^2 - 1.4z + 0.48} = \frac{Y(z)}{U(z)}$$

Find the steady-state output sinusoid for driving sinusoid

$$u(k) = 3\cos(0.9k)$$

This implies that $\omega T = 0.9$ rad.

## REFERENCES


Lathi, B. P., 1992. *Linear Systems and Signals*, Berkeley-Cambridge, Carmichael, CA.

Soliman, S. S., and M. D. Srinath, 1990. *Continuous and Discrete Signals and Systems*, Prentice Hall, Englewood Cliffs, NJ.

Strum, R. D., and D. E. Kirk, 1988. *First Principles of Discrete Systems and Digital Signal Processing*, Addision-Wesley, Reading, MA.

Strum, R. D., and D. E. Kirk, 1994. *Contemporary Linear Systems Using MATLAB*, PWS-Kent, Boston.

# 3
# Elementary Digital Control System Design Using Transform Techniques

## 3.1 INTRODUCTION

The conventional techniques for design of continuous-time control systems have been well developed in the past five decades. These techniques are built around the Laplace transform representation of the differential equations describing the system dynamics. The analysis for closed-loop stability can be accomplished by examining the location of the closed-loop characteristic roots (eigenvalues) in the complex $s$-plane by the Routh–Hurwitz criterion (Routh, 1877). Another tool that gives more information for design purposes is the root-locus technique developed by Evans (1948), which gives the location of the closed-loop characteristic roots as a loop gain parameter is varied. Nyquist (1932) and Bode (1945) have shown how the closed-loop stability can be examined by examining the open-loop frequency response (phase *and* magnitude).

In Chapter 2 we developed the mathematics of discrete-time systems, but the systems examined there were essentially open loop in nature. In this chapter we explore techniques for digital control system design that are similar to those conventionally employed in continuous-time control system design. The cornerstone of this chapter is the $z$-transform and the $z$-domain representation of the discrete-time dynamics of the plant and proportional controllers which are inserted into the control loop to improve the system dynamics and error character.

## 3.2 ANALOG-TO-DIGITAL AND DIGITAL-TO-ANALOG CONVERTERS

As discussed in Chapter 1, special electronic devices are necessary in order that the digital processor used to provide control strategy be able to converse with the continuous-time system to be controlled. This conversation must be bidirectional in order that the control be of the feedback type. In other words, the digital processor will base the control strategy on measurements made of the variables to be controlled. The two devices discussed here are those that allow the digital processor to communicate with the outside continuous-time world.

### Analog-to-Digital Converter

The analog-to-digital (A/D) converter is the device that allows the computer to obtain the measured data about the system state. This device may involve one or more data channels, perhaps as many as 16. We shall consider only a single channel, but the ideas are easily extended to a multichannel converter by use of time-division multiplexing, whereby the numbers representing converted variables at a given time are transmitted to the digital processor serially. Here we use the symbol given in Fig. 3.1 to represent the analog-to-digital converter, which samples the input periodically every $T$ seconds. Henceforth in this book the analog-to-digital converter and an ideal sampler will be treated as the same device. Multichannel analog-to-digital converters on a single LSI integrated circuit chip are available for a few dollars.

### Digital-to-Analog Converter

The digital-to-analog converter is the device that converts the numerical content of some register of the digital processor to an analog voltage and holds that voltage constant until the content of the register is updated, and then the output of the digital-to-analog converter is updated and held again. We shall model the D/A converter as a zero-order hold because the device output has no slope information. In short, a zero-order hold (ZOH) is a

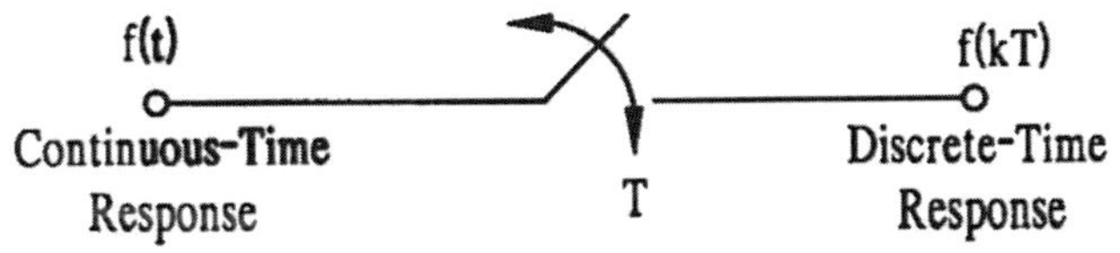


**Figure 3.1.** Symbol for analog-to-digital converter.

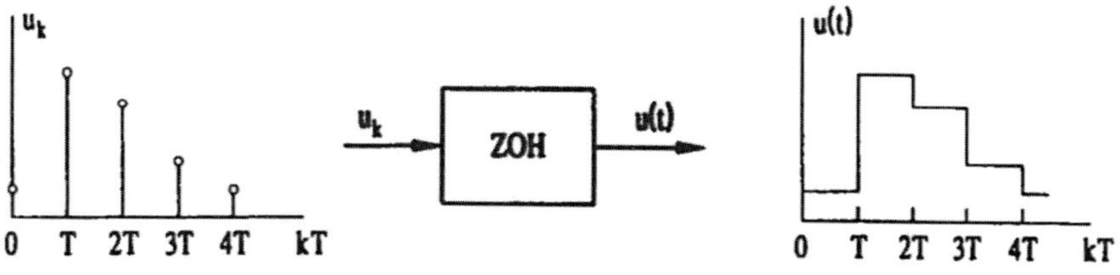


**Figure 3.2.** Zero-order-hold operation.

device such that given a uniformly spaced sequence of numbers $u_0$, $u_1$, $u_2$, . . . , a stairstep output $u(t)$ is produced such that

$$u(t) = u_k \qquad kT \le t < (k + 1)\, T$$

A typical situation of this type is illustrated in Fig. 3.2. Digital-to-analog converters are also available in inexpensive integrated circuit form.

Having discussed the two devices with which the digital processor converses with the continuous-time world around it, we shall now continue to develop the digital control problem.

## 3.3 CONTINUOUS-TIME PLANT DRIVEN BY A ZERO-ORDER HOLD WITH SAMPLED OUTPUT

In a large number of physical control systems, the plant to be controlled by the digital computer is one that is described by ordinary differential equations. Typical plants are aircraft, missiles, and spacecraft in which the dynamics are governed by Newton's second law. Others are chemical or nuclear reactors, milling machines, pipeline systems, and speed or positioning systems which are parts of larger systems. To analyze the digital control of any such system, a discrete-time description of the plant to be controlled must first be obtained.

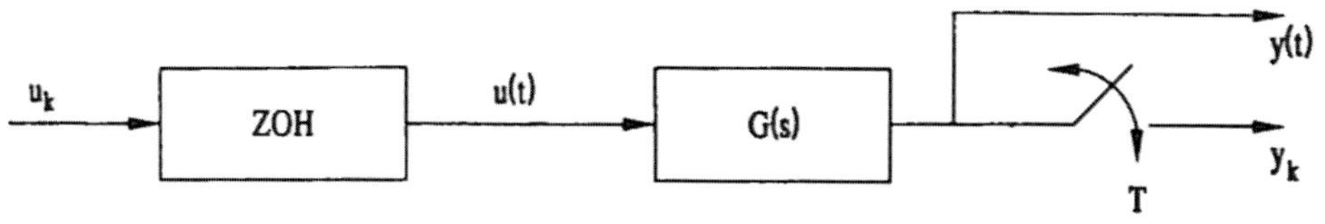


**Figure 3.3.** Continuous-time plant to be digitally controlled.

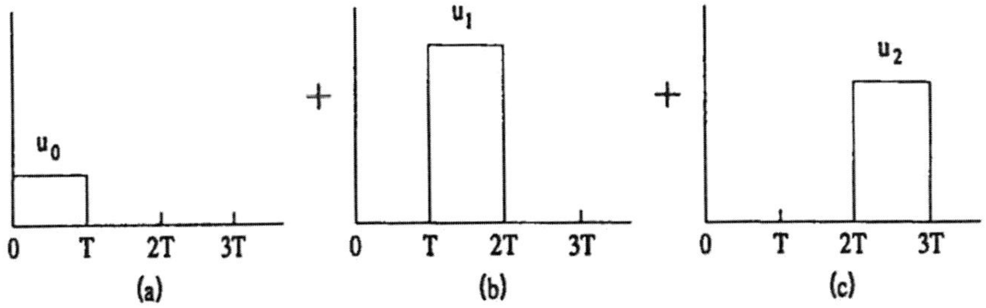


**Figure 3.4.** Gate function decomposition of the zero-order-hold output.

The continuous-time single-input/single-output plant represented by transfer function $G(s)$ is shown in Fig. 3.3 driven by a zero-order hold (a model of the D/A converter) and followed by an output sampler (A/D converter). These devices function as outlined in Section 3.2. The operation of the zero-order hold is shown in Fig. 3.2, whereby a discrete-time signal is converted to a stairstep function according to expression (3.2.1). The signal at the output of the zero-order hold could be decomposed into a series of gate functions as shown in Fig. 3.4. The $u_0$ gate function of Fig. 3.4a could be further decomposed into step functions as shown in Fig. 3.5.

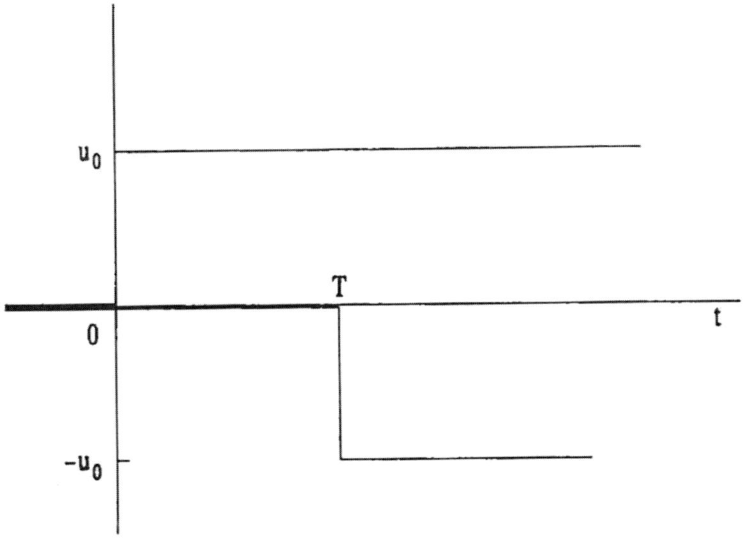


**Figure 3.5.** Step function decomposition of a gate function.

The continuous-time response to the first step of magnitude $u_0$ is

$$y(t) = u_0\mathcal{L}^{-1}\left[\frac{G(s)}{s}\right] \tag{3.3.1}$$

and the $z$-transform of the sampled sequence $y_k$ is

$$Y(z) = u_0\mathcal{Z}\mathcal{L}^{-1}\left[\frac{G(s)}{s}\right] \tag{3.3.2}$$

The response due to the negative-going step is exactly the same except that it is negative and delayed by one sample period so that the total $z$-domain response to the $u_0$ *gate function* may be calculated by superposition to be

$$Y(z) = u_0\left\{\mathcal{Z}\mathcal{L}^{-1}\left[\frac{G(s)}{s}\right] - z^{-1}\,\mathcal{Z}\mathcal{L}^{-1}\left[\frac{G(s)}{s}\right]\right\} \tag{3.3.3}$$

where the $z^{-1}$ represents the delay by one sample period as given by Theorem 2.2. Extending this idea to the second gate function, we get the response to the first two gate functions to be

$$\begin{aligned} Y(z) &= u_0\,(1 - z^{-1})\mathcal{Z}\mathcal{L}^{-1}\left[\frac{G(s)}{s}\right] \\ &\quad + u_1(z^{-1} - z^{-2})\mathcal{Z}\mathcal{L}^{-1}\left[\frac{G(s)}{s}\right] \end{aligned} \tag{3.3.4}$$

which can be simplified to be

$$Y(z) = (u_0 + u_1z^{-1})(1 - z^{-1})\mathcal{Z}\mathcal{L}^{-1}\left[\frac{G(s)}{s}\right] \tag{3.3.5}$$

Now for the whole series of gate functions representing $u(t)$, the transform of the sampled output is

$$Y(z) = \left(\sum_{k=0}^{\infty} u_kz^{-k}\right)(1 - z^{-1})\mathcal{Z}\mathcal{L}^{-1}\left[\frac{G(s)}{s}\right] \tag{3.3.6}$$

We may now recognize the first term as the $z$-transform of the $u_k$ sequence that is driving the zero-order hold. We will call this $z$-transform $U(z)$, so we see that the discrete-time transfer function between the input and output sequences is

$$\begin{aligned} G(z) &= \frac{Y(z)}{U(z)} = (1 - z^{-1})\mathcal{Z}\mathcal{L}^{-1}\left[\frac{G(s)}{s}\right] \\ &= \frac{z-1}{z}\mathcal{Z}\mathcal{L}^{-1}\left[\frac{G(s)}{s}\right] \end{aligned} \tag{3.3.7}$$

This is a very important relation in the digital control of continuous-time plants, and it will also be a useful expression in the digital approximation of continuous-time filters, which will be a topic of Chapter 5. The transfer function $G(z)$ given by (3.3.7) is sometimes referred to as the *pulse transfer function* for the system of Fig. 3.3. Several examples should now serve to illustrate the ideas just presented.

**Example 3.1.** A large number of systems, such as thermal and fluid reservoir control problems as encountered in chemical engineering applications, may be described by a first-order differential equation or a transfer function of the form

$$G(s) = \frac{Y(s)}{U(s)} = \frac{K}{s + a}$$

This system will be driven by a zero-order hold and followed by a sampler as illustrated in Fig. 3.6. Relation (3.3.7) may be used to find the $z$-domain transfer function of

$$G(z) = (1 - z^{-1})\mathfrak{Z}\mathcal{L}^{-1}\left[\frac{K}{s(s + a)}\right]$$

and upon making a partial fraction expansion, we get

$$G(z) = (1 - z^{-1})\mathfrak{Z}\mathcal{L}^{-1}\left[\frac{K}{a}\left(\frac{1}{s} - \frac{1}{s + a}\right)\right]$$

Application of a table of $z$-transforms yields

$$G(z) = \frac{z - 1}{z}\left(\frac{z}{z - 1} - \frac{z}{z - e^{-aT}}\right)\frac{K}{a}$$

and upon simplification,

$$G(z) = \frac{K}{a}\,\frac{(1 - e^{-aT})}{z - e^{-aT}}$$

Now let us simplify this expression by choosing $K = a = 1$ and choose a

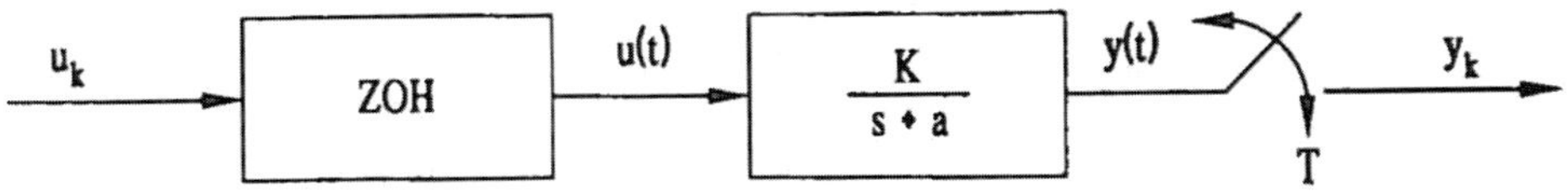


**Figure 3.6.** First-order plant to be digitally controlled.

sampling interval to be $T = 0.2$ to yield

$$G(z) = \frac{1 - e^{-0.2}}{z - e^{-0.2}}$$

and upon evaluating the exponentials,

$$G(z) = \frac{0.1813}{z - 0.8187}$$

**Example 3.2.** Consider another simple example, which is the pure inertial plant driven by zero-order hold as shown in Fig. 3.7. This plant could represent the rotational dynamics of a simple satellite. Relation (3.3.7) gives us the ability to calculate the overall transfer function by

$$G(z) = \frac{z - 1}{z} \mathfrak{Z}\mathcal{L}^{-1}\left[\frac{G(s)}{s}\right]$$

or

$$G(z) = \frac{z - 1}{z} \mathfrak{Z}\mathcal{L}^{-1}\left[\frac{1}{s^3}\right]$$

Consulting a table of $z$-transforms yields

$$G(z) = \frac{z - 1}{z} \frac{T^2}{2} \frac{z(z + 1)}{(z - 1)^3}$$

which after the simplification yields

$$G(z) = \frac{T^2}{2} \frac{(z + 1)}{(z - 1)^2}$$

Note that this transfer function has two poles at $z = 1$ which under the mapping $z = e^{sT}$ correspond to the original double-integrator pole at the origin of the $s$-plane. This sort of correspondence for plant poles (not necessarily zeros) will always hold. In this simple example the $z$-domain pole locations were independent of the sampling period $T$; this is seldom the case in other systems

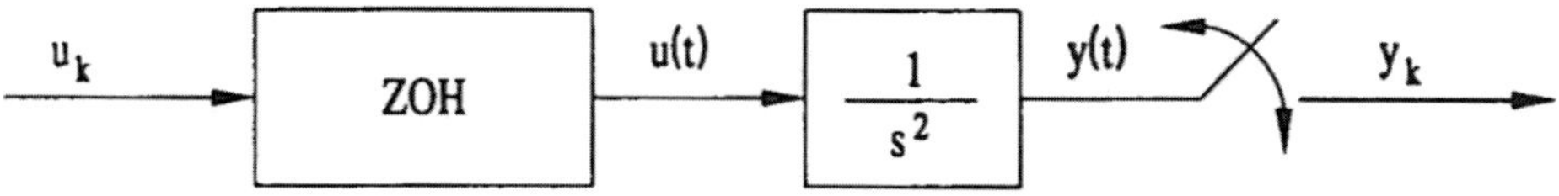


**Figure 3.7.** Inertial plant.

**Example 3.3.** As an example of the technique outlined previously, let us consider the thermal system shown in Fig. 3.8. A heater coil is embedded in medium 2 while it is desired to control the temperature of medium 1 in the presence of a variable environmental temperature $T_0(t)$. If we apply the principle of conservation of energy to each of the two media, the resulting differential equations are

$$m_1c_1 \frac{dT_1}{dt} = k_{12}(T_2 - T_1)$$

and

$$m_2c_2 \frac{dT_2}{dt} = -k_{12}(T_2 - T_1) - k_{20}(T_2 - T_0) + u(t)$$

where the $m_i$ and $c_i$ are the mass and specific heat of the $i$th medium, respectively, and the $k_{ij}$ are the respective interface thermal conductances. Let us now assign the parameters to be $m_1c_1 = 0.5$, $m_1c_2 = 2$, $k_{12} = 1$, and $k_{20} = 0.5$. Substitution of these values into the differential equations gives

$$\frac{dT_1}{dt} = 2(T_2 - T_1)$$

$$\frac{dT_2}{dt} = -0.75T_2 + 0.5T_1 + 0.25T_0 + 0.5u(t)$$

To find a transfer function of this system, take the Laplace transform of

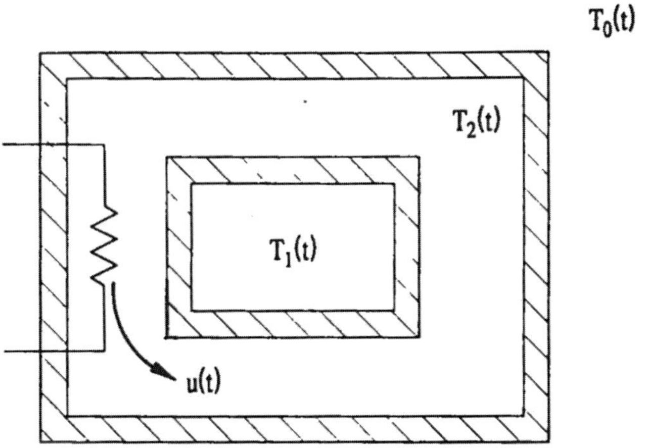


**Figure 3.8.** Thermal plant.

these equations to yield a set of $s$-domain linear equations, which may be written in matrix form as

$$\begin{bmatrix} s + 2 & -2 \\ -0.5 & s + 0.75 \end{bmatrix} \begin{bmatrix} T_1(s) \\ T_2(s) \end{bmatrix} = \begin{bmatrix} 0 \\ 0.25T_0(s) + 0.5U(s) \end{bmatrix}$$

Solving for $T_1(s)$ using Cramer's rule gives

$$T_1(s) = \frac{\begin{vmatrix} 0 & -2 \\ 0.25T_0(s) + 0.5U(s) & s + 0.75 \end{vmatrix}}{\begin{vmatrix} s + 2 & -2 \\ -0.5 & 0.75 \end{vmatrix}}$$

or

$$T_1(s) = \frac{0.5T_0(s)}{s^2 + 2.75s + 0.5} + \frac{U(s)}{s^2 + 2.75s + 0.5}$$

which in block diagram form is shown in Fig. 3.9. If we want to control this system digitally, we need the transfer function between the control effort and the response variable, which after factoring the denominator is

$$G(s)\frac{T_1(s)}{U(s)} = \frac{1}{(s + 0.1957)(s + 2.554)}$$

Now consider this system driven by a zero-order hold with an output sampler as shown in Fig. 3.10. Now let us evaluate the $z$-domain transfer function between the input sequence and the output sequence. We must start with $G(s)/s$, which is

$$\frac{G(s)}{s} = \frac{1}{s(s + 0.1957)(s + 2.554)}$$

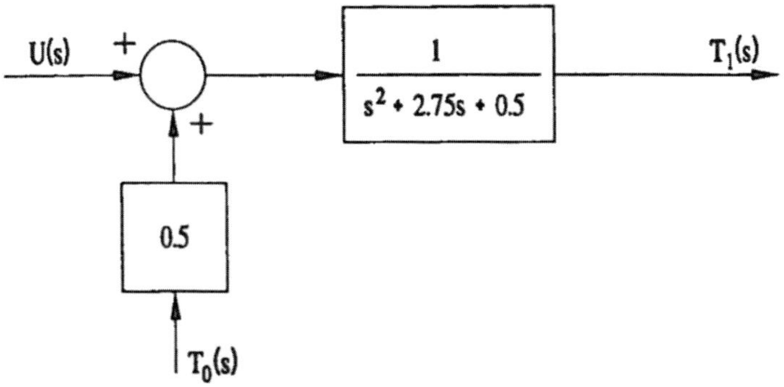


**Figure 3.9.** Block diagram of thermal system.

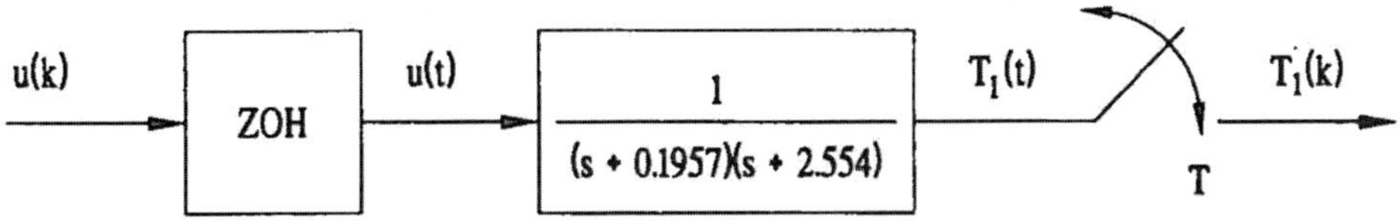


**Figure 3.10.** Thermal plant to be digitally controlled.

Making a partial expansion,

$$\frac{G(s)}{s} = \frac{A}{s} + \frac{B}{s + 0.1957} + \frac{C}{s + 2.554}$$

where the coefficients are given in the usual manner to be

$$A = 2 \qquad B = -2.1667 \qquad C = 0.1667$$

Then the quantity is

$$\frac{G(s)}{s} = \frac{2}{s} - \frac{2.1667}{s + 0.1957} + \frac{0.1667}{s + 2.554}$$

Then from the $z$-transform tables,

$$\mathfrak{Z}\mathcal{L}^{-1}\left[\frac{G(s)}{s}\right] = \frac{2z}{z-1} - \frac{2.1667z}{z - e^{-0.1957T}} + \frac{0.1667z}{z - e^{-2.554T}}$$

Now employing relation (3.3.7) we see that the overall $z$-domain transfer function is

$$G(z) = \frac{z-1}{z}\left[\frac{\alpha z^2 + \beta z}{(z-1)(z - e^{-0.1957T})(z - e^{-2.554T})}\right]$$

where

$$\alpha = -2(e^{-0.1957T} + e^{-2.554T}) + 2.1667(1 + e^{-2.554T}) - 0.1667(1 + e^{-0.1957T})$$

and

$$\beta = 2e^{-0.1957T}e^{-2.554T} - 2.1667e^{-2.554T} + 0.1667e^{-0.1957T}$$

Since the time constants associated with this system are 5.11 and 0.392 s, respectively, the sampling interval should be chosen somewhat smaller than the smallest; for simplicity let us use $T = 0.25$ s. The transfer function now becomes

$$G(z) = \frac{0.025(z + 0.816)}{(z - 0.952)(z - 0.528)} = \frac{T_1(z)}{U(z)}$$

This is the $z$-domain description of the zero-order hold, the continuous-time plant, and the output sampler If we examine carefully the $s$-domain transfer functions of the previous examples, we find that the $z$-domain poles $z_i$ are related to the original poles of the $s$-domain transfer function $s_i$ by

$$z_i = e^{s_i T} \tag{3.3.8}$$

## Systems with Transport Lag

Thus far we have discussed the incorporation of dynamic plants into digital control loops where the plants are described by a transfer function which is the ratio of two polynomials in the Laplace transform variable $s$. Here we shall consider transfer functions that incorporate a transport lag such as that often encountered in chemical process control systems of the form

$$G(s) = G_1(s)e^{-T_d s} \tag{3.3.9}$$

where $G_1(s)$ is a ratio of $s$-domain polynomials and $T_d$ is the transport lag or delay, sometimes referred to as the dead time. The transport delay can be written as the sum of an integer number of sample periods minus some fraction of a sampling period,

$$T_d = mT - fT \tag{3.3.10}$$

where $m$ is an integer and $f$ is between zero and unity. We may still employ relation (3.3.7) to get the $z$-domain transfer function as

$$G(z) = \frac{z-1}{z} \mathscr{Z}\mathscr{L}^{-1}\left[\frac{G(s)}{s} e^{-T_d s}\right] \tag{3.3.11}$$

Employing relation (3.3.10), the transfer function (3.3.9) can be written as

$$G(s) = G_1(s)e^{-(mT-fT)s} \tag{3.3.12}$$

Let us expand the inverse transform of $G(s)/s$,

$$\mathscr{L}^{-1}\left[\frac{G_1(s)}{s}\right] = A_0 + A_1 e^{-p_1 t} + A_2 e^{-p_2 t} + \cdots + A_n e^{-p_n t} \tag{3.3.13}$$

where we have assumed distinct poles of $G(s)$ and no free integrators in $G(s)$. Inverting this transform gives

$$\begin{aligned} \mathscr{L}^{-1}\left[\frac{G_1(s)}{s}\right] e^{-(mT-fT)s} &= A_0 u(t - mT + fT) \\ &\quad + A_1 e^{-p_1(t-mT+fT)} u(t - mT + fT) + \cdots \\ &\quad + A_n e^{-p_n(t-mT+fT)} u(t - mT + fT) \end{aligned} \tag{3.3.14}$$

The delays involved here amount to a shift of the respective functions to the right by $m$ sampling periods and a shift back to the left a fraction of a period. Now taking the $z$-transform of the sampled version of the functions, we get

$$\mathscr{Z}\mathscr{L}^{-1}\left[\frac{G_1(s)}{s}e^{-T_d s}\right]$$
$$= z^{-m}\left(\frac{A_0 z}{z-1} + \frac{A_1 e^{-p_1 fT}z}{z-e^{-p_1 T}} + \cdots + \frac{A_n e^{-p_n fT}z}{z-e^{-p_n T}}\right) \qquad (3.3.15)$$

After finding a common denominator in this expression it is then necessary to multiply by $(z - 1)/z$ in order to get the $z$-domain transfer function.

**Example 3.4.** Consider a transfer function frequently encountered in the chemical process control industry consisting of a first-order system with a transport lag:

$$G(s) = \frac{e^{-0.5s}}{s+1}$$

Let us employ a sampling interval of $T = 0.2$ s, which is a sampling rate of five samples per time constant. For this problem $m = 3$ and $f = 0.5$ as the transport lag is 2.5 sampling intervals. Making a partial fraction expansion of $G(s)/s$, we get

$$\frac{G(s)}{s} = e^{-0.5s}\left(\frac{1}{s} - \frac{1}{s+1}\right)$$

Inverting the transforms, the result is

$$\mathscr{L}^{-1}\left[\frac{G(s)}{s}\right] = u(t-0.5) - e^{-(t-0.5)}u(t-0.5)$$

The sampled version of this signal is

$$\mathscr{L}^{-1}\left[\frac{G(s)}{s}\right]\Bigg|_{t=k0.2} = u(k-3) - e^{-(0.2k-0.5)}u(k-3)$$
$$= u(k-3) - e^{-0.1}e^{-(k-3)0.2}u(k-3)$$

Now take the $z$-transform of this sequence to give the result

$$\mathscr{Z}\mathscr{L}^{-1}\left[\frac{G(s)}{s}\right] = z^{-3}\frac{z}{z-1} - e^{-0.1}z^{-3}\frac{z}{z-e^{-0.2}}$$

Finding a common denominator

$$\mathscr{Z}\mathscr{L}^{-1}\left[\frac{G(s)}{s}\right] = \frac{z^{-3}[(z - e^{-0.2})z - e^{-0.1}(z - 1)z]}{(z - 1)(z - e^{-0.2})}$$

Multiplication by $(z - 1)/z$ gives the $z$-domain transfer function

$$G(z) = \frac{z^{-3}0.095(z + 0.90526)}{z - 0.8187}$$

Where the extra factor of $z^{-3}$ is given by the transport lag. For systems with complex poles this process will be algebraically complicated but can be accommodated in a similar fashion.

## 3.4 IMPLEMENTATION OF DIGITAL CONTROL STRATEGIES

In this section we discuss the implementation of a typical digital control strategy in a microcomputer environment. The typical microcomputer architecture for a control computer is illustrated in Fig. 3.11. In addition to a CPU and associated memory, we see parallel and serial input–output (I/O) and a timer. This timer is used to control the timing between when samples of the input to the A/D converter are taken and is usually programmable as to how high to count from the CPU. The general linear control law to be implemented by the control computer is

$$u(k) = a_n e(k) + a_{n-1}e(k-1) + \cdots + a_0 e(k - n) + b_{n-1}u(k - 1) + \cdots + b_0 u(k - n) \quad (3.4.1)$$

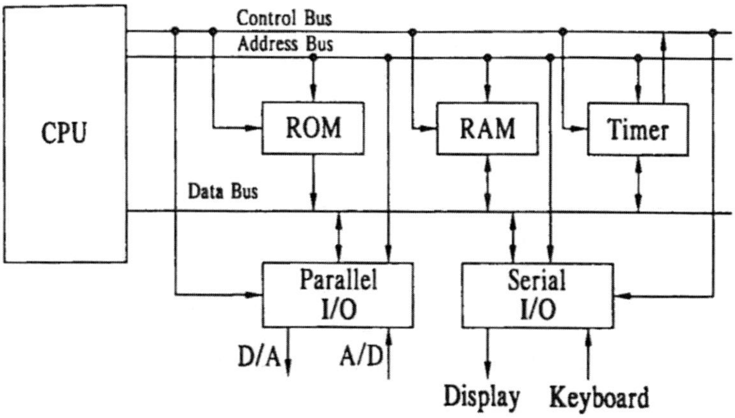


**Figure 3.11.** Architecture of a typical control computer.

This difference equation has an equivalent controller transfer function

$$D(z) = \frac{a_n z^n + a_{n-1} z^{n-1} + \cdots + a_0}{z^n - b_{n-1} z^{n-1} - \cdots - b_1 z - b_0} \tag{3.4.2}$$

In the implementation of (3.4.1) all the past values of $e(k)$ and $u(k)$ are initially set to zero and the parameters $a_i$ and $b_i$ and the sampling interval desired are input. The sampling interval is converted to an integer count desired based on the clock rate to be counted. The timer is started counting down and a start-conversion signal is issued to the A/D converter and the system checks the A/D until conversion is complete. When conversion is complete, the value of $e(k)$ is read from the A/D, the first term of (3.4.1) is evaluated, and the resulting $u(k)$ is output to the D/A converter. Then the data $e(k)$, $e(k - 1)$ . . . and $u(k)$, $u(k - 1)$ . . . is all shifted down one address in the memory and all terms of (3.4.1) but the first are precalculated for the next cycle. At this point the timer is interrogated continuously until it has counted down, at which time control is transferred to the point in the software where the timer was started and the process is started again. The flowchart for such a program is illustrated in Fig. 3.12, which most certainly will clarify the explanation above.

The sequence of events given in the flowchart can be clarified by investigating a single sampling interval of activities, as illustrated in Fig. 3.13.

It is hoped that the discussion above will clarify any confusion about the microcomputer implementation of control laws of the form of (3.4.1). For the remainder of this chapter we shall be concerned with implementation of proportional control strategies, which are simply the first term on the right side of relation (3.4.1) or

$$u(k) = Ke(k) \tag{3.4.3}$$

or, from (3.5.2),

$$D(z) = K \tag{3.4.4}$$

where $a_n$ has been replaced by the controller gain $K$. Of course, this strategy simplifies the programming implied by Fig. 3.12 but results in a control system that often does not have as high a performance as the designer might desire.

## 3.5 CLOSED-LOOP CHARACTERISTIC EQUATION

The common technique for linear closed-loop control of a discrete-time plant represented by transfer function $G(z)$ is to calculate the stepwise

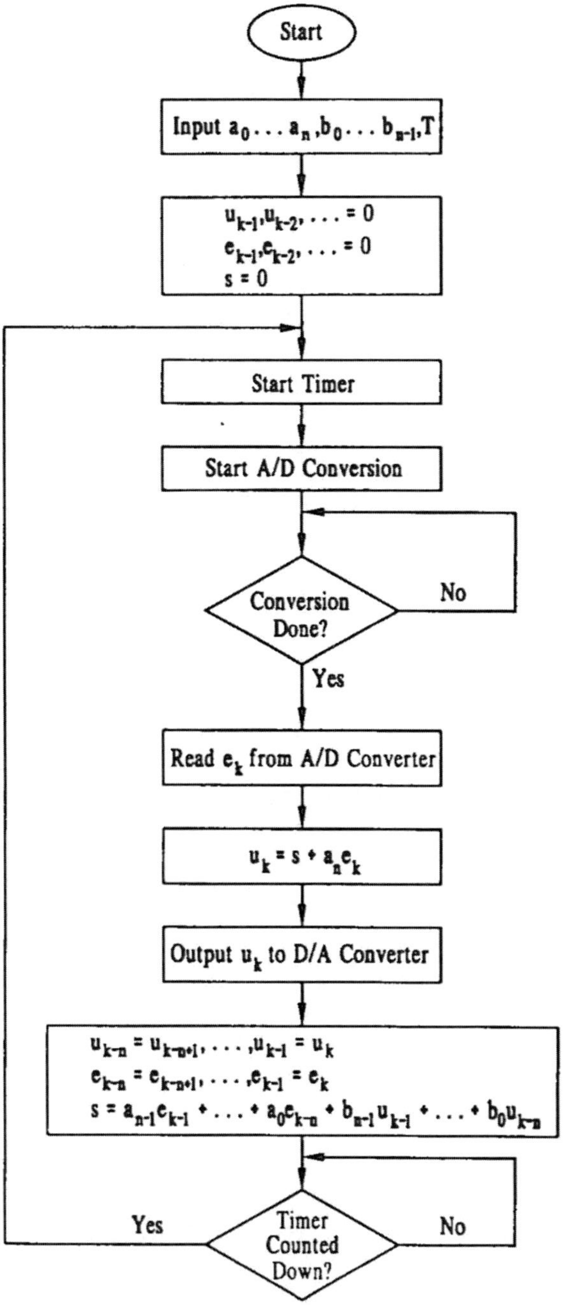


**Figure 3.12.** Flowchart for digital control algorithm implementation.

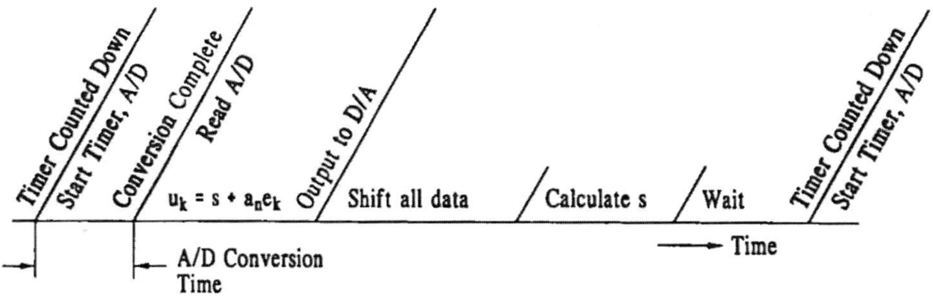


**Figure 3.13.** Timing diagram for a single sample interval.

effort $u_k$ based on a finite number of error measurements $e_k$, $e_{k-1}$, . . . , $e_{k-m}$ and perhaps past values of the control sequence $u_{k-1}$ · · · as discussed in Section 3.4. This situation is illustrated in Fig. 3.14a. If a control algorithm in the form of a linear difference equation with constant coefficients is employed, the entire system can be represented by $z$-domain transfer functions as shown in Fig. 3.14b.

The relation for the summing point is

$$E(z) = R(z) - Y(z) \tag{3.5.1}$$

and for the controller

$$U(z) = D(z)E(z) \tag{3.5.2}$$

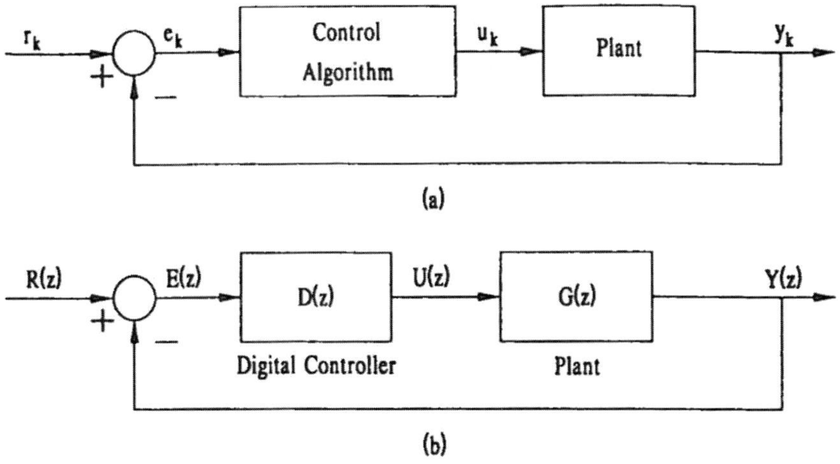


**Figure 3.14.** Closed-loop control system: (a) time domain; (b) $z$-domain.

while that for the plant is

$$Y(z) = G(z)U(z) \tag{3.5.3}$$

Substitution of (3.5.1) into (3.5.2) yields

$$U(z) = D(z)[R(z) - Y(z)] \tag{3.5.4}$$

Multiplication of (3.5.4) by $G(z$ will give $Y(z)$ according to expression (3.5.3) or

$$Y(z) = G(z)D(z)[R(z) - Y(z)] \tag{3.5.5}$$

Consolidating like terms gives

$$Y(z)[1 + G(z)D(z)] = G(z)D(z)R(z) \tag{3.5.6}$$

The closed-loop transfer function $M(z)$ is now the ratio of system output to reference input, or

$$M(z) = \frac{Y(z)}{R(z)} = \frac{G(z)D(z)}{1 + G(z)D(z)} \tag{3.5.7}$$

The quantity $1 + G(z)D(z)$, after rationalization, is the ratio of two polynomials, the numerator of which is the denominator polynomial of $M(z)$, which is the closed-loop system characteristic equation. The location of the roots of this characteristic equation in the $z$-plane determines the dynamic character of the closed-loop system, as discussed in Chapter 2. If one or more of these roots lie outside the unit circle, the closed-loop system is unstable, which is undesirable. A simple illustration is the best way to demonstrate these ideas.

**Example 3.5.** Consider the closed-loop feedback control system, consisting of the first-order plant given in Example 3.1, which is to be controlled with a proportional control algorithm with a gain denoted as $K_p$. The complete feedback system is shown in Fig. 3.15.

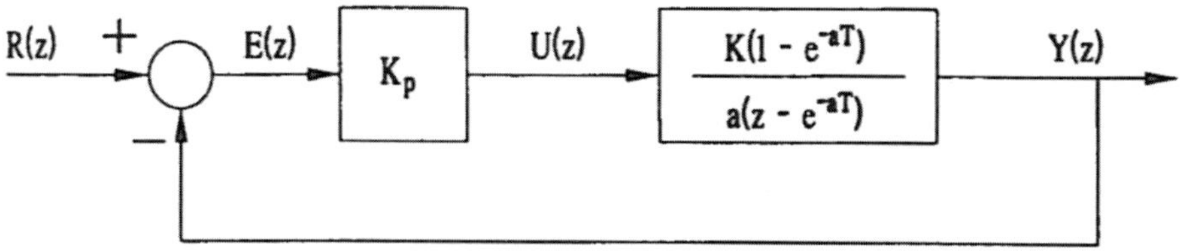


**Figure 3.15.** Proportional control of a first-order plant.

The closed-loop characteristic equation is given from relation (3.5.7) to be

$$1 + G(z)D(z) = 1 + \frac{K_pK}{a}\left(\frac{1 - e^{-aT}}{z - e^{-aT}}\right) = 0$$

or

$$z - e^{-aT} + \frac{K_pK}{a}(1 - e^{-aT}) = 0 \qquad \text{(a)}$$

The complete root locus is shown in Fig. 3.16 as a function of the parameter $K_pK/a$. It is clear that the system becomes marginally stable when the closed-loop characteristic root is at $z = -1$. If we let $z = -1$ in (a) we may evaluate the value of the gain parameter for which marginal stability occurs.

$$\left(\frac{K_pK}{a}\right)_{\text{crit}} = \frac{1 + e^{-aT}}{1 - e^{-aT}} \qquad \text{(b)}$$

If we multiply both numerator and denominator of relation (b) by $e^{aT/2}$, we get

$$\left(\frac{K_pK}{a}\right)_{\text{crit}} = \frac{e^{aT/2} + e^{-aT/2}}{e^{aT/2} - e^{-aT/2}} = \coth\left(\frac{aT}{2}\right)$$

The stability boundary for this system as a function of the parameter $aT/2$ is illustrated in Fig. 3.17.

If we wish to locate the closed-loop pole at an arbitrary location $z = b$ we may do so by letting $z = b$ in relation (a) and solve for the gain parameter to get

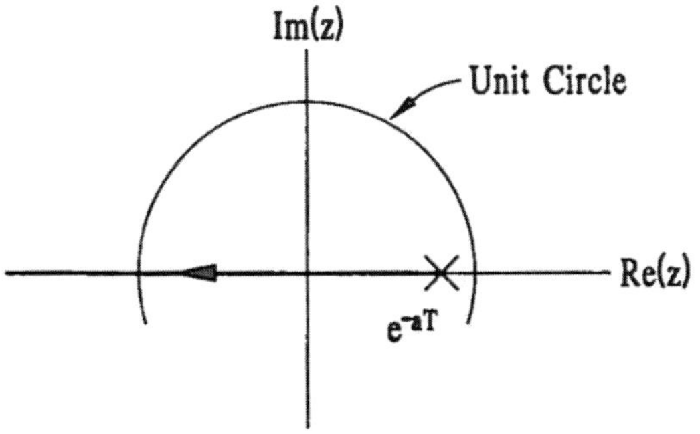


**Figure 3.16.** Root locus for proportionally controlled first-order plant.

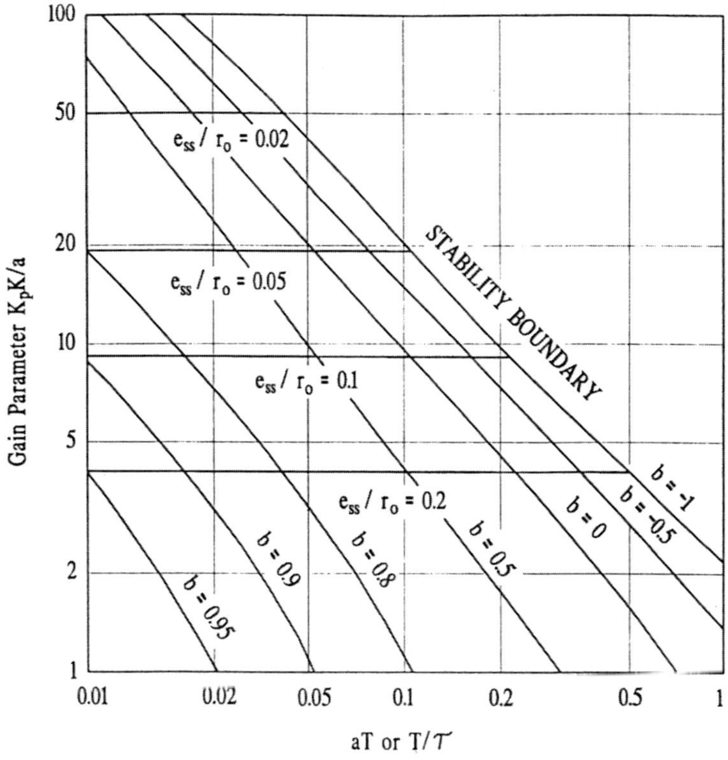


**Figure 3.17.** Stability boundary, closed-loop pole loci, and steady-state error loci for the proportionally controlled first-order system.

$$\frac{K_pK}{a} = \frac{-b + e^{-aT}}{1 - e^{-aT}} \tag{c}$$

The loci for various values of $b$ are also illustrated in Fig. 3.17. If we wish to evaluate the effect of the gain parameter on the steady-state error to a reference input, we can use the transfer function between the reference input and the error, which is

$$\frac{E(z)}{R(z)} = \frac{1}{1 + G(z)D(z)} = \frac{z - e^{-aT}}{z - e^{-aT} + \dfrac{K_pK}{a}(1 - e^{-aT})}$$

If the reference input $r(k)$ is a step of height $r_0$, then $R(z) = r_0z/(z - 1)$. The steady-state error then can be given by the final value theorem to be

$$e_{ss} = \lim_{z\to 1} \frac{z - 1}{z} \frac{r_0z}{z - 1} \frac{z - e^{-aT}}{z - e^{-aT} + \dfrac{K_pK}{a}(1 - e^{-aT})}$$

If we carry out the indicated limiting process, then

$$e_{ss} = r_0 \frac{1}{1 + \frac{K_p K}{a}}$$

Note that this steady-state error is independent of the sampling interval $T$ and thus that loci of constant steady-state error are presented as horizontal lines in Fig. 3.17.

It is interesting to note that in the example just given the system can be unstable, but the same plant under the action of continuous-time control is always stable. The reason for this is that digital control is achieved with only samples of the output variable, so the control policy is formulated based on a finite amount of information, while in the continuous-time case an infinite amount of information is available in the continuous-time error signal. In other words, toward the end of a sampling interval we are still using the control strategy that was being employed at the start of the interval. Another way of looking at this problem is that the system tends to get "out of hand" between the sampling instants, and when the controller does take action, it overcompensates by virtue of its high gain. We also note from Fig. 3.17 that as the sampling interval is decreased the gain for marginal stability increases, thus approaching the continuous-time case.

## 3.6 CONVENTIONAL DIGITAL CONTROL DESIGN TECHNIQUES

In Example 3.5 we saw how the root-locus technique could be employed in the design of a digital control system. Since we are considering a closed-loop system that has a closed-loop characteristic equation given by

$$1 + G(z)D(z) = 0 \tag{3.6.1}$$

the root-locus technique as developed in one of many conventional continuous-time control systems texts is directly applicable to this problem in the $z$-plane. See the excellent texts of Dorf (1992) or Kuo (1987) for the details of this method.

One can also design in the frequency response domain by using the open-loop frequency response functions and the concepts of gain margin and phase margin as discussed in the two previously cited continuous-time control systems texts. An alternative is to design to meet specifications on the closed-loop frequency response function, such as the closed-loop resonant frequency or the closed-loop bandwidth.

## 3.7 CONVENTIONAL CONTROL SYSTEM DESIGN SPECIFICATIONS

Past experience of control system designers has indicated that there are certain gross system properties that are indicative of acceptable system performance. These performance specifications are generally categorized as either time-domain or frequency-domain specifications. The advantage of these simple specifications is that they are easy to apply and have historically given good designs. The disadvantage is that an attempt to meet more than one of these specifications may lead to conflicting design strategies, in which case a compromise design or specification must be sought. This conflict may be alleviated by the use of modern optimal control theory as presented in Chapter 9, although this theory is accompanied by its own inherent difficulties.

### Time-Domain Specifications

Often, the measure of performance of a control system is made in terms of its response to a step reference input—in other words, how well the system follows the step. The step input is used for a number of reasons: (1) it is an easily generated input; (2) it is sufficiently severe to excite all natural responses of the system; and (3) it gives the system some steady nonzero input to track. The response of a typical high-performance control system to a unit-step input is illustrated in Fig. 3.18, and several critical specifications are noted in the figure.

Commonly specified performance measures in terms of the step response are:

1. Percent overshoot $P$
2. Peak time $T_p$
3. Settling time $T_s$
4. Rise time $T_r$
5. Steady-state error $e_{ss}$

### Open-Loop Frequency Response Specifications

Some of the earliest specifications were based on the margins from instability based on the Nyquist stability criterion (Nyquist, 1932; James et al., 1947; Dorf, 1992). The Nyquist stability criterion states that for a minimal-phase system to be stable, the following must be true: the loop gain at the frequency where the phase shift is 180° must be less than unity (0 dB). A typical open-loop plot of magnitude and phase for a stable system is illustrated in Fig. 3.19 with the gain margin $G_m$ in dB and the phase margin $\phi_m$ clearly labeled.

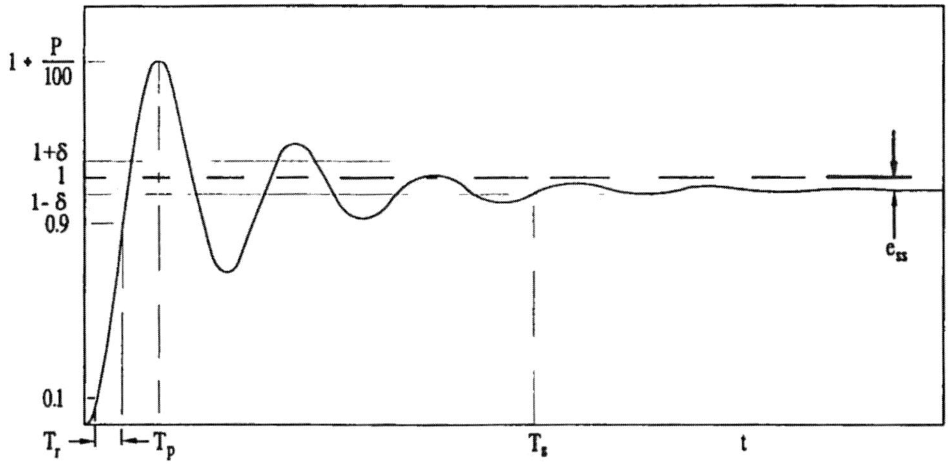


**Figure 3.18.** Step response of a high-performance control system.

For purposes of system classification a number that characterizes the system type is assigned based on the number of poles of the open-loop transfer function, which are at $z = 1$. Type zero systems have no poles at $z = 1$, type one systems have one pole at $z = 1$, and so on.

With many years of experience, system designers, have found that gain margins and phase margins which yield good designs differ for systems of different type and the results of that experience are tabulated in Table 3.1.

## Closed-Loop Frequency Response Specifications

In Chapter 2 we discussed the relationship between the discrete-time transfer function and the response of the system to a sampled sinusoid of radian frequency $\omega$. If the closed-loop transfer function as derived in Section 3.4 is denoted as $M(z)$, the frequency response function, which gives both attenuation and phase shift information about the closed-loop system, is

$$M(e^{j\omega T}) = M(z)|_{z=e^{j\omega T}} \tag{3.7.1}$$

A plot of $|M(e^{j\omega T})|$ for a typical high-performance control system is given in Fig. 3.20. It is common to specify quantitatively the desired properties of this magnitude function.

Commonly specified quantities are indicated in Fig. 3.20 and are:

1. System bandwidth BW
2. Resonant peak gain $M_m$
3. Peak frequency $\omega_m$

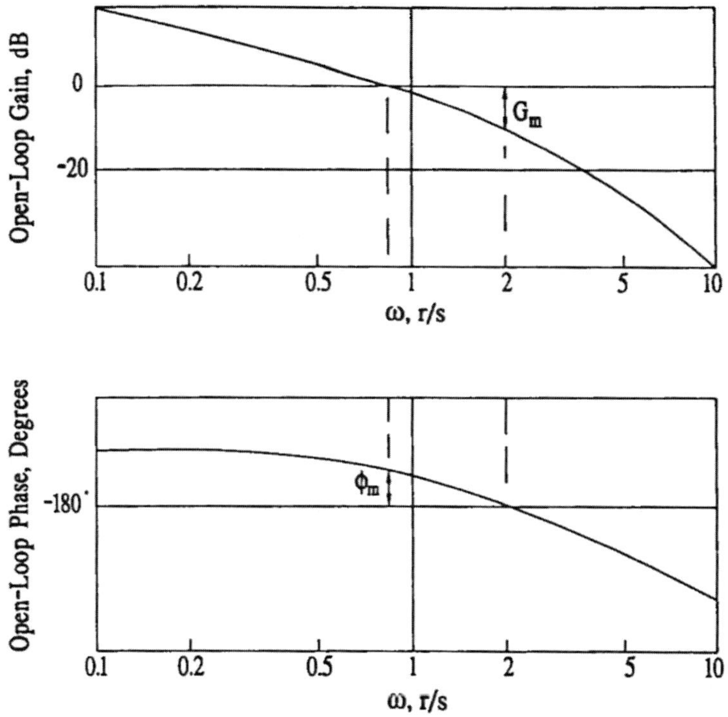


**Figure 3.19.** Open-loop magnitude and phase plots for a typical system with gain margin and phase margin shown.

**Table 3.1.** Gain and Phase Margins for Different Types of Systems

| System type | Gain margin (dB) | Phase margin (deg.) |
|---|---|---|
| 0 | 5–10 | 30 |
| 1 | 10–15 | 45–50 |
| 2 | 10–15 | 45–50 |

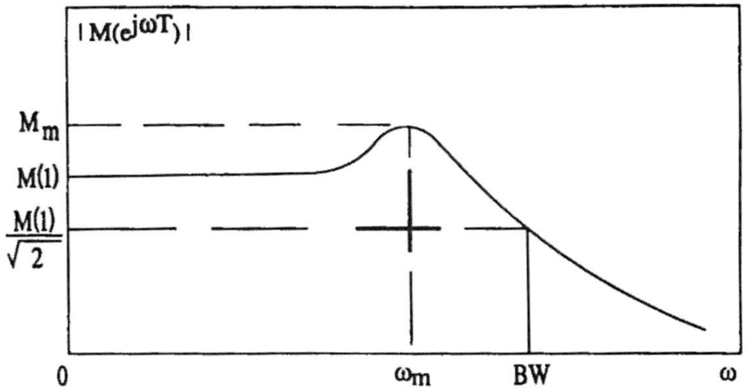


**Figure 3.20.** Typical high-performance control system frequency response.

It should be noted that speed of response can be correlated to bandwidth for simple systems; hence the bandwidth is a common specification for control systems. It should be noted, however, that the greater the bandwidth, the more susceptible the system will be to noise injected into the loop. It is also known that for simple underdamped second-order systems, the peak $M_m$ is related to the complex-conjugate pole locations. Also, it is almost always clear that there exists a direct relation between the peak frequency $\omega_m$ and the bandwidth BW.

We shall investigate in detail the step response and frequency response of a simple second-order system because in a large number of design cases for high-order systems there are often a pair of dominant poles that dominate either the frequency or the step response. In continuous-time systems these are those poles closest to the $j\omega$-axis of the $s$-plane, and in discrete-time systems these are the poles that lie closest to the unit circle (which is the image of the $j\omega$-axis under the operation of sampling). This is the justification of the thorough analysis of second-order system response that follows.

### Second-Order System Step Response

Let us consider a second-order system with transfer function

$$M(s) = \frac{\omega_n^2}{s^2 + 2\zeta\omega_n s + \omega_n^2} \tag{3.7.2}$$

If the input to this system is a unit-step function, the response function

$y(t)$ for this input and zero initial conditions is

$$y(t) = 1 + \frac{e^{-\zeta\omega_n t}}{\sqrt{1 - \zeta^2}} \sin\left(\omega_n t\sqrt{1 - \zeta^2} - \alpha\right) \tag{3.7.3}$$

where

$$\alpha = \sin^{-1}\left(\sqrt{1 - \zeta^2}\right) \tag{3.7.4}$$

By differentiating this function, we can solve for the peak time $T_p$ to be

$$T_p = \frac{\pi}{\omega_n\sqrt{1 - \zeta^2}} \tag{3.7.5}$$

Also, evaluation of the function $y(t)$ at time $T_p$ yields the percent overshoot, which is

$$P = 100 \exp\left(\frac{-\zeta\pi}{\sqrt{1 - \zeta^2}}\right) \tag{3.7.6}$$

If the value $\delta$ is chosen as 0.02 or if settling is defined as being within 2% of the final value, the settling time may be evaluated as

$$T_s = \frac{4}{\zeta\omega_n} \tag{3.7.7}$$

Plots of percent overshoot and dimensionless settling and peak times as functions of the damping ratio $\zeta$ are given in Fig. 3.21. In this case, there is no steady-state error as occurs in the case of a transfer function of higher order with more complicated numerator dynamics.

### Second-Order System Frequency Response

A simple second-order system transfer function is given in expression (3.7.2). If the variable $s$ is evaluated at $j\omega$, this becomes the frequency response function $M(j\omega)$, which is

$$M(j\omega) = \frac{\omega_n^2}{-\omega^2 + \omega_n^2 + j2\zeta\omega_n\omega} \tag{3.7.8}$$

The magnitude character of this function is similar to that shown in Fig. 3.20. Differentiation of the magnitude function with respect to $\omega$ and equating to zero will show that the peak frequency is

$$\omega_m = \omega_n\sqrt{1 - 2\zeta^2} \tag{3.7.9}$$

and if the frequency response function of (3.7.8) is evaluated at this fre-

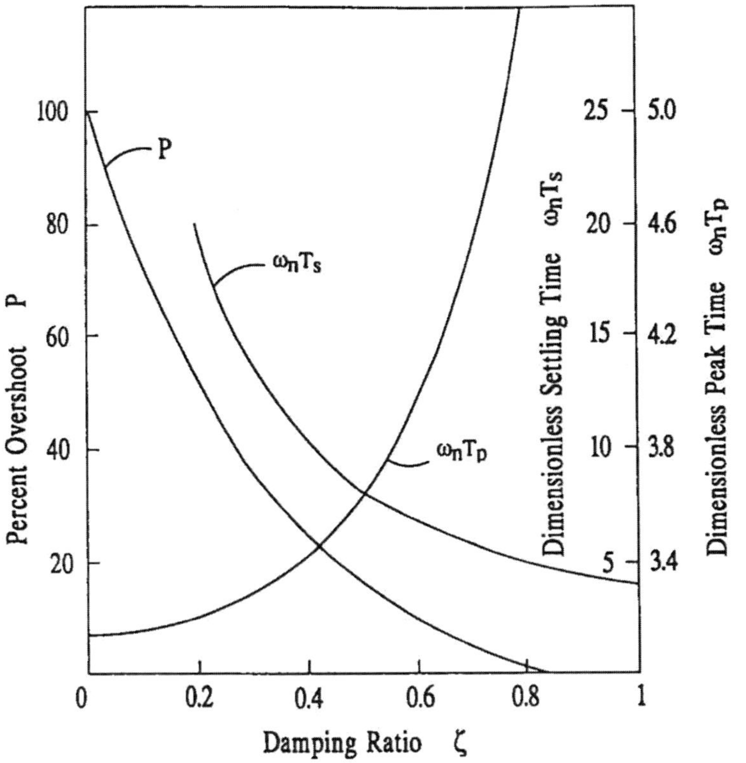


**Figure 3.21.** Common step response performance specifications for a second-order system.

quency, the value for $M_m$ is

$$M_m = |M(j\omega_m)| = \frac{1}{2\zeta\sqrt{1 - \zeta^2}} \tag{3.7.10}$$

If we now equate the magnitude of (3.7.8) to $\sqrt{2}/2$ and solve for the frequency at which this occurs, this will fit the definition of the bandwidth as illustrated in Fig. 3.20, and after considerable algebra the resulting bandwidth BW is

$$\text{BW} = \omega_n\sqrt{(1 - 2\zeta^2) + \sqrt{4\zeta^4 - 4\zeta^2 + 2}} \tag{3.7.11}$$

In the analysis above, we have found how the parameters $\omega_n$ and $\zeta$ influence the commonly used performance specifications in a simple second-order system. These relationships are presented graphically in Fig. 3.22.

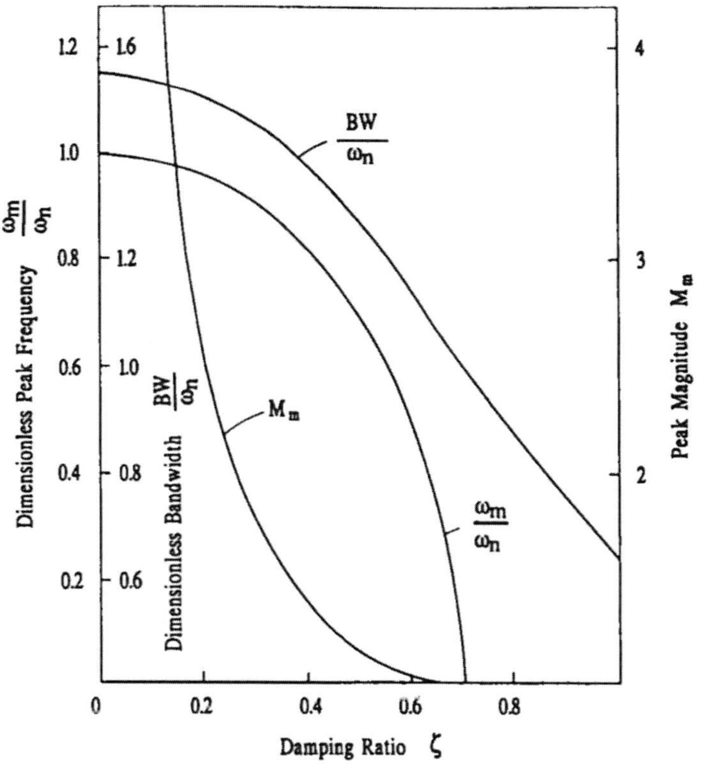


**Figure 3.22.** Common frequency response specifications for a simple second-order system.

## Summary of Performance Specifications

In the previous sections we have shown that for a simple second-order system the performance specifications in both the time and frequency domains are specified completely by two parameters: the undamped natural frequency $\omega_n$ and the damping ratio $\zeta$. These two parameters are sufficient to locate the poles.

A real control system is seldom of second order, but often by specifying the poles closest to the imaginary axis (dominant poles) of the $s$-plane, the performance can be controlled, and hence the time spent in this vein is justified. By understanding the relation of $s$-plane pole locations to $z$-plane pole locations for discrete-data systems, these design techniques can be carried over to the digital control systems.

## 3.8 ELEMENTARY z-DOMAIN DESIGN CONSIDERATIONS

In simple continuous-time systems, the performance specifications discussed in Section 3.7 can be correlated to $s$-domain pole locations for the closed-loop transfer function. There are similar correlations to $z$-domain pole locations, but since we are not as familiar with the $z$-plane, we should first correlate $s$-plane locations to $z$-plane locations to allow the use of concepts that are well understood for design purposes.

We have shown that a continuous-time plant driven by a zero-order hold and followed by an output sampler can be represented by a discrete transfer function $G(z)$. The process of conventional digital control system design amounts to the synthesis of the control algorithm, reflected in the compensator transfer function $D(z)$, to yield acceptable closed-loop dynamics or error character.

In Section 3.3 we have shown that the plant pole locations in the $z$-plane are related to those of the continuous-time plant in the $s$-plane by

$$z_i = e^{s_i T} \tag{3.8.1}$$

We also showed in Chapter 2 that under this mapping, vertical lines in the $s$-plane became concentric circles in the $z$-plane, while horizontal lines became radial lines in the $z$-plane.

In the design process we are often interested in systems with complex-conjugate poles whose $s$-plane locations are given by the roots of the quadratic equation

$$s^2 + 2\zeta\omega_n s + \omega_n^2 = 0 \tag{3.8.2}$$

where $\omega_n$ is the undamped natural frequency and $\zeta$ is the damping ratio. For $\zeta < 1$ these roots are given by the quadratic formula to be

$$s_{1,2} = -\zeta\omega_n \pm j\omega_n\sqrt{1 - \zeta^2} \tag{3.8.3}$$

which are illustrated in Fig. 3.23. The $z$-plane images of such $s$-plane poles are given by the mapping (3.8.1), or

$$z_{1,2} = e^{-\zeta\omega_n T \pm j\omega_n T\sqrt{1-\zeta^2}} \tag{3.8.4}$$

In the $s$-plane we are commonly interested in fixing the parameter $\zeta$ so as to control overshoot and settling time of the closed-loop control system. Lines of constant damping ratio $\zeta$ in the $s$-plane are radial lines in the left half of the $s$-plane, as illustrated in Fig. 3.23. The angle $\beta$ is related to the damping ratio by

$$\zeta = \cos\beta \tag{3.8.5}$$

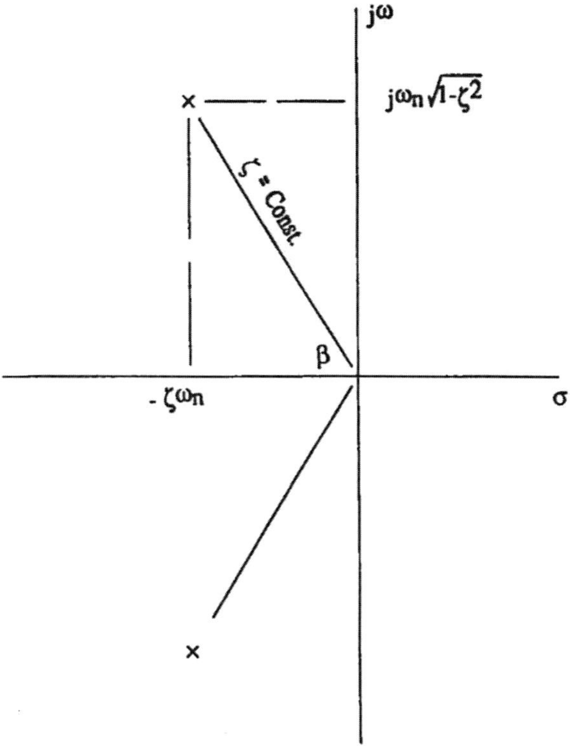


**Figure 3.23.** Complex-conjugate poles in the $s$-plane with a line of constant damping ratio shown.

The radial location of the poles in the $z$-plane is given by (3.8.4) to be

$$R = e^{-\zeta\omega_n T} \tag{3.8.6}$$

and the angular location is given to be

$$\theta = \omega_n T\sqrt{1 - \zeta^2} \tag{3.8.7}$$

Now if we solve (3.8.7) for $\omega_n T$ and substitute into (3.8.6), we get the radial pole location as a function of the angular location and the parameter $\zeta$, so

$$R = \exp\left(\frac{-\zeta\theta}{\sqrt{1 - \zeta^2}}\right) \tag{3.8.8}$$

This is the equation of a logarithmic spiral, the "tightness" of which is

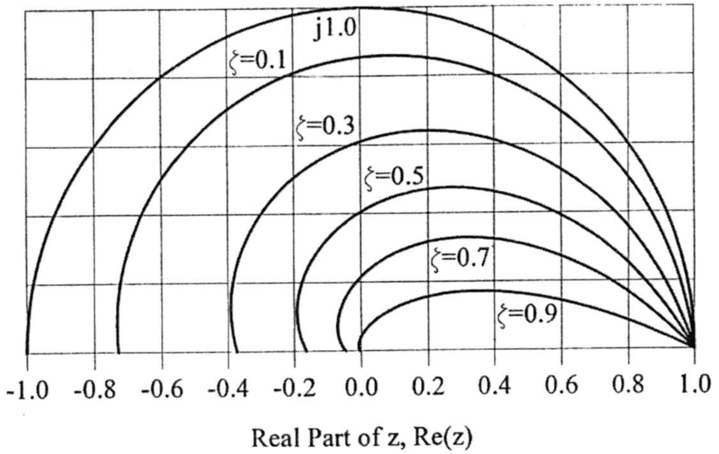


**Figure 3.24.** Constant damping ratio (ζ) loci in the $z$-plane.

controlled by the parameter ζ. These loci for various values of ζ are shown in Fig. 3.24.

The process of control algorithm design is often that of selecting the parameters of the controller to give the closed-loop system poles desirable $z$-plane locations. With a known plant and specified compensator, the effect of varying a gain parameter on the closed-loop roots may be examined by the conventional root-locus plot. If some other parameter is to be varied, an extension of the method of root locus is also possible, as shown by Kuo (1987).

**Example 3.6.** Consider the thermal control system considered in Example 3.3, in which the transfer function from the input of the zero-order hold to the sampler output was found to be

$$G(z) = \frac{0.025(z + 0.816)}{(z - 0.952)(z - 0.528)} = \frac{T_1(z)}{U(z)}$$

For a proportional controller $D(z) = K$, the closed-loop characteristic equation is

$$(z - 0.952)(z - 0.528) + K(0.025)(z + 0.816) = 0$$

On expanding and combining like terms in $z$, the result is

$$z^2 - z(1.48 - 0.025K) + (0.5026 + 0.0204K) = 0$$

The closed-loop pole locations are then

$$z_{1,2} = 0.74 - 0.0125K \pm \sqrt{(0.74 - 0.0125K)^2 - (0.5026 + 0.0204K)}$$

and if we consider only the complex roots,

$$z_{1,2} = 0.74 - 0.0125K \pm j\sqrt{0.5026 + 0.204K - (0.74 - 0.0125K)^2}$$

For a pole location on the unit circle $|z_i|^2 = 1$, or

$$1 = (0.74 - 0.0125K)^2 + (0.5026 + 0.0204K) - (0.74 - 0.0125K)^2$$

or

$$1 = 0.5026 + 0.020K$$

Then the critical value of the gain parameter is

$$K_{\text{crit}} = 24.38$$

The root locus for this system is shown in Fig. 3.25, with several values of the parameter $K$ being given. For a damping ratio of about 0.7, it appears that the value of $K$ is 2.5. The proportional control algorithm corresponding to this set of closed-loop locations is then given by

$$u(k) = 2.5[r(k) - T_1(k)] = 2.5e(k)$$

We have, in fact, adjusted the damping ratio such that there would not be excessive overshoot according to Fig. 3.21. We have not, however, guaranteed that the speed of response is adequate. If a faster response is desired, a more complicated control law in the form of a discrete-time

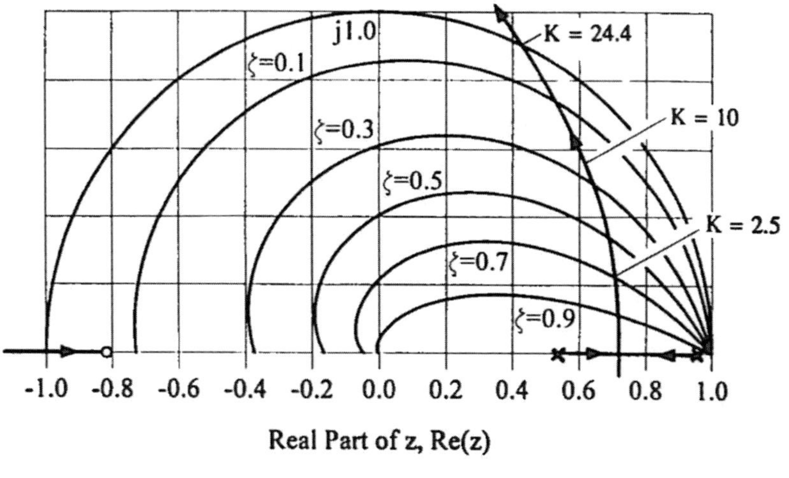


**Figure 3.25.** Root locus for proportionally controlled thermal plant.

compensator will be required. This problem is considered in Section 3.10 and Chapters 4 and 5. The step responses and control efforts for three values of the proportional controller gain are illustrated in Fig. 3.26.

## 3.9 EFFECT OF DISTURBANCES ON THE CLOSED-LOOP SYSTEM

In Section 3.8 we considered a control system with only discrete-time inputs; however, we know that continuous-time systems under the action

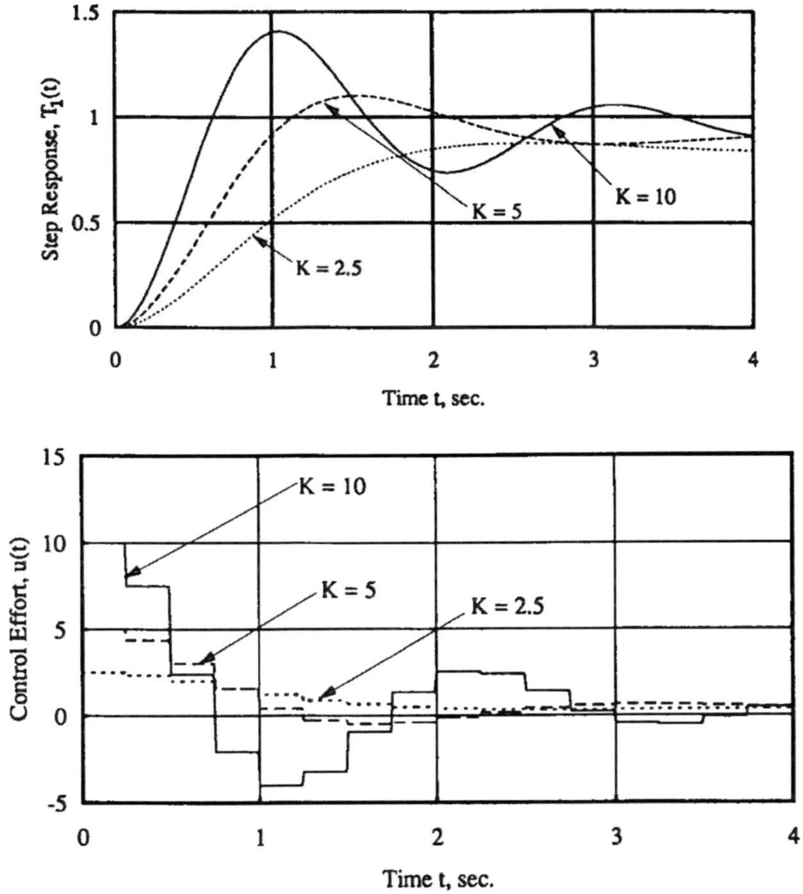


**Figure 3.26.** Control effort and step response of proportionally controlled thermal system.

of digital controllers are often subject to disturbances that are continuous-time functions. Examples of such disturbances are random ambient temperature changes in a thermal control problem and turbulent gust disturbances to an aircraft controlled to fly a straight and level path.

Let us consider now the system of Fig. 3.27, in which the reference input is zero, which makes this system a regulator. The system is, however, subject to a disturbance $W(s)$ which may enter the control loop directly or through some dynamics represented by transfer function $G_2(s)$. A discrete-time control algorithm is represented by transfer function $D(z)$, while the continuous-time plant dynamics are represented by $G_p(s)$. The discrete-time transfer function between the discrete controller and the sampler output will be represented by $G(z)$ and can be found by the methods of Section 3.2. The output of this system $Y(z)$ can be thought of as being composed of two parts, one due to $U(z)$ and the other due to $W(s)$. Since the system is linear, the principle of superposition is applicable and the output $Y(z)$ is

$$Y(z) = G(z)U(z) + \mathfrak{Z}\mathcal{L}^{-1}\,[G_2(s)G_p(s)W(s)] \tag{3.9.1}$$

But the control effort $U(z)$ is given for zero reference input as

$$U(z) = -D(z)Y(z) \tag{3.9.2}$$

and substitution of this into (3.9.1) gives

$$Y(z) = -G(z)D(z)Y(z) + \mathfrak{Z}\mathcal{L}^{-1}\,[G_1(s)G_p(s)W(s)] \tag{3.9.3}$$

and solution for the $z$-domain output yields

$$Y(z) = \frac{\mathfrak{Z}\mathcal{L}^{-1}[G_2(s)G_p(s)W(s)]}{1 + G(z)D(z)} \tag{3.9.4}$$

If the time-domain response sequence is desired, this must be inverted to yield the $y(k)$ sequence. If, however, only the final value of the output

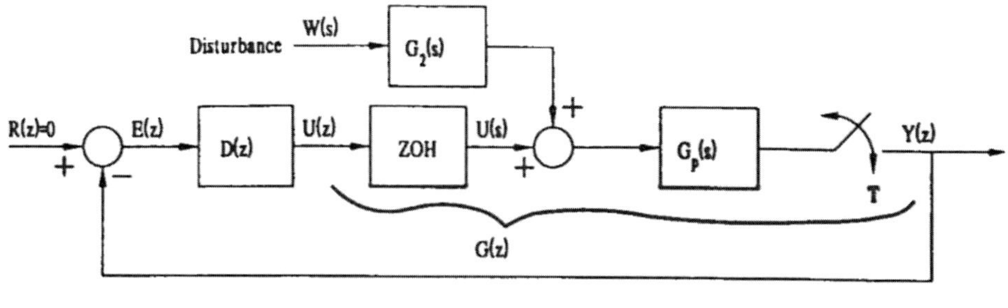


**Figure 3.27.** Closed-loop digital control system with continuous disturbance.

sequence is desired to a step disturbance, the final value theorem is applicable. These concepts are best illustrated by means of an example.

**Example 3.7.** Consider now the step disturbance of the first-order plant with transfer function

$$G_p(s) = \frac{1}{s + 1}$$

with a sampling interval of $T = 0.2$ s. The transfer function $G(z)$ was given in Examples 3.1 and 3.5 to be

$$G(z) = \frac{0.1813}{z - 0.8187}$$

The step disturbance of magnitude $A$ is

$$W(s) = \frac{A}{s}$$

The disturbance input transfer function is unity or

$$G_2(s) = 1$$

and for a proportional controller the transfer function is

$$D(z) = K$$

The $z$-domain response is given by expression (3.9.4) to be

$$Y(z) = \frac{\mathcal{Z}\mathcal{L}^{-1}\,[A/s(s + 1)]}{1 + K(0.1813)/(z - 0.8187)}$$

and carrying out the operations indicated in the numerator gives

$$Y(z) = \frac{A[z/(z - 1)] - z/(z - 0.8187)}{1 + K(0.1813)/(z - 0.8187)}$$

Rationalizing this fraction yields

$$Y(z) = \frac{A(0.1813)z}{(z - 1)[(z - 0.8187) + K(0.1813)]}$$

We know from Example 3.5 that the closed-loop pole position is controlled by selection of the gain parameter $K$. The steady-state error due to the step disturbance can be calculated by the application of the final value theorem, or

$$y(\infty) = \lim_{z \to 1} \frac{A(0.1813)}{(z - 0.8187) + K(0.1813)}$$

and carrying out the indicated operations we get

$$y(\infty) = \frac{A}{1 + K}$$

and hence the larger the value of controller gain $K$, the smaller the steady-state error. There is, however, an upper limit on gain $K$ due to the instability problem encountered in Example 3.5.

## 3.10 CONCEPT OF THE DYNAMIC CONTROLLER OR COMPENSATOR

In previous sections of this chapter we have explored only the possibility of the controller being one of the proportional nature where the control law is of the form

$$u(k) = Ke(k) \tag{3.10.1}$$

It was found that such a controller often yields a system that has a steady-state error to a step reference input and that there can be little done to improve the speed of response without an accompanying increase in oscillatory response.

As one contemplates the control problem, the question arises whether the system performance could be improved by implementing a control law closer to that of equation (3.4.1). Let us consider the case where we allow the controller to use the current and previous measurement $(ek)$ and $e(k - 1)$ and the previous control effort $u(k - 1)$ to calculate the current control effort. In other words, the controller is of the form

$$u(k) = a_1 e(k) + a_0 e(k - 1) + b_0 u(k - 1) \tag{3.10.2}$$

If we take the $z$-transform of this difference equation, the resulting controller transfer function is

$$D(z) = \frac{U(z)}{E(z)} = \frac{a_1 + a_0 z^{-1}}{1 - b_0 z^{-1}} \tag{3.10.3}$$

or in positive powers of $z$

$$D(z) = \frac{a_1(z + a_0/a_1)}{z - b_0} \tag{3.10.4}$$

Clearly, this transfer function has a pole at $z = b_0$ and a zero at $z = -a_0/a_1$. The question arises as to the selection of the three coefficients in the control law $a_0$, $a_1$ and $b_0$. We could rewrite the transfer function in a more standard form as

$$D(z) = \frac{K(z - \alpha)}{z - \beta} \tag{3.10.5}$$

where the pole location is $z = \beta$, the zero location is $z = \alpha$, and $K$ is a gain factor, so there are still three parameters for the designer to select. The process we shall use is to choose $\alpha$ and $\beta$ and then adjust the gain factor $K$ to give acceptable dynamics. As in most design processes we may have to revise our choice of $\alpha$ and $\beta$ after we examine the dynamic behavior of the closed-loop system. The design process is a trial-and-error process and it is common to have to redesign, hence computational assistance from a digital computer will greatly ease the pain. The closed-loop characteristic equation of the system with unity feedback was shown in Section 3.5 to be

$$1 + G(z)D(z) = 0 \tag{3.10.6}$$

The pole and zero of $D(z)$ can thus be selected to improve the dynamic character. For example, the zero of $D(z)$ could be selected to cancel an undesirable pole of $G(z)$, and similarly, the pole of $D(z)$ could be selected to be very fast and hence improve the speed of the closed-loop system.

The cancellation strategy works well as long as the undesirable pole of $G(z)$ is not an unstable pole exterior to the unit circle. Since cancellation is never exact, an unstable pole would remain even if there is a nearby zero. To illustrate the process, let us return to the design of a controller for the thermal plant of Example 3.6.

**Example 3.8.** Design a controller for the thermal plant of Example 3.6 that will dramatically speed up the step response while maintaining a damping ratio of the closed loop poles at $\zeta = 0.7$.

The plant pulse transfer function is

$$G(z) = \frac{0.025(z + 0.816)}{(z - 0.952)(z - 0.528)}$$

The plant response is controlled by the locations of the poles. The pole at $z = 0.952$ is referred to as the "slow pole" while that at $z = 0.528$ is the "fast pole." The slow pole at $z = 0.952$ acts somewhat as an integrator in the closed loop and thus helps to minimize the steady-state error, so it might be unwise to cancel it to improve performance. On the other hand, it might be good to cancel the fast pole and make it even faster by adding a pole of $D(z)$ at $z = 0.2$, which is a somewhat arbitrary choice. Select $\alpha = 0.528$ and $\beta = 0.2$. Now all that remains is selection of the gain factor $K$ to give a closed-loop damping ratio of $\zeta = 0.7$.

The closed loop characteristic equation is, from equation (3.10.6)

$$1 + G(z)D(z) = 1 + \frac{K(0.025)(z + 0.816)(z - 0.528)}{(z - 0.952)(z - 0.528)(z - 0.2)}$$

where we see the pole–zero cancellation taking place. The root locus for the resulting closed-loop as a function of the gain factor $K$ is shown in Fig. 3.28.

Examination of the root locus indicates that the pole at $z = 0.2$ has had the effect of pulling the complex branches of the locus to the left, hence resulting in a faster system than for proportional control. The value of $K$ that yields closed-loop roots with a damping ratio of 0.7 is $K = 7$. The resulting compensator is

$$D(z) = \frac{U(z)}{E(z)} = \frac{7(z - 0.528)}{z - 0.2}$$

and the resulting control law is

$$u(k) = 7e(k) - 3.696e(k - 1) + 0.2u(k)$$

The step response is illustrated in Fig. 3.29 along with that for the proportionally controlled system. The step response is illustrated in Fig. 3.29 along with that for the proportionally controlled system with the same damping ratio in the closed-loop poles.

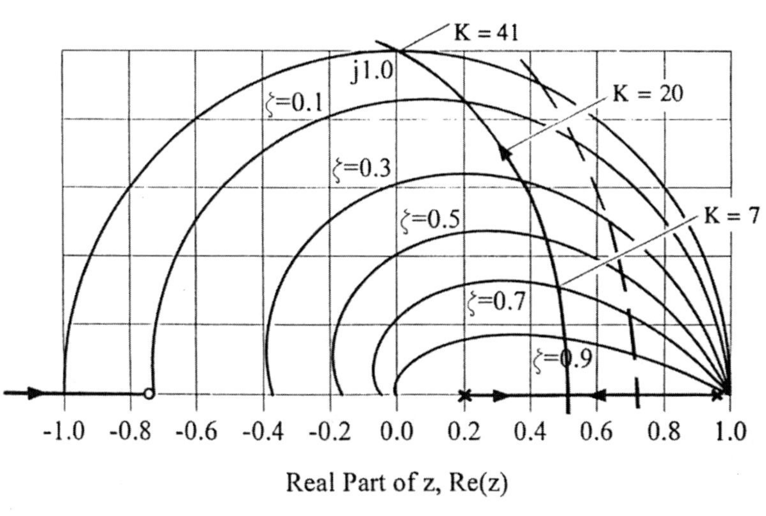


**Figure 3.28.** Root locus for system with dynamic controller. The dashed line is for the proportional controller.

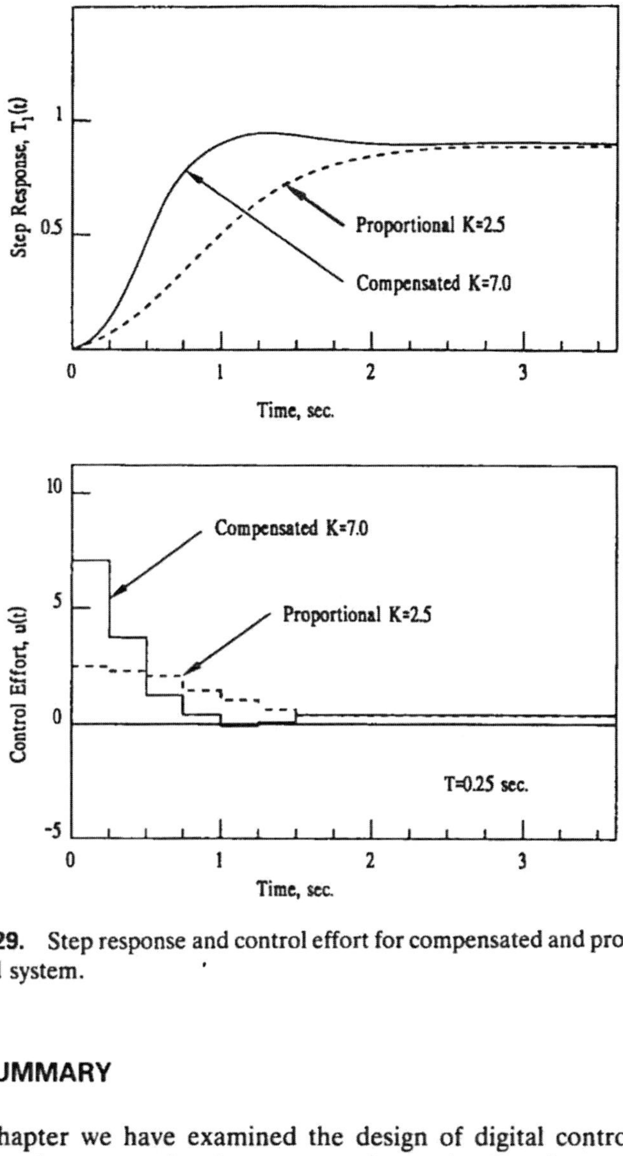


**Figure 3.29.** Step response and control effort for compensated and proportionally controlled system.

## 3.11 SUMMARY

In this chapter we have examined the design of digital control systems employing the conventional concepts of transfer function, root locus, and other frequency-domain techniques. We first examined how the continuous-time plant would be represented in discrete-time form if the analog-to-digital and digital-to-analog converters were considered with the plant, thus yielding a discrete-time input and output for the plant. Control system design specifications are discussed for both time response and frequency response, and these are given in detail for second-order systems. The closed-loop characteristic equation is developed and root-locus design is discussed. The effect of continuous-time disturbances is considered and a technique developed to include these in the discrete-time mathematics.

## PROBLEMS

3.1. Find the $z$-domain pulse transfer function for the system shown for $T = 0.2$ s.

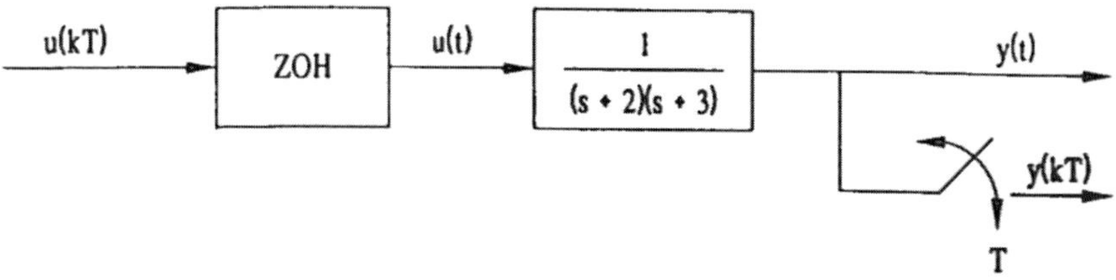


3.2. Use the final value theorem to find the steady-state output sequence $y(\infty)$ when the system of Problem 3.1 is subjected to a constant input sequence $u_k = R(k \geq 0)$.

3.3. A continuous-time plant with transfer function

$$G(s) = \frac{Y(s)}{U(s)} = \frac{1}{s(s + 1)}$$

is driven by a zero-order hold and followed by a sampler.

(a) Find the overall pulse transfer function $G(z)$ for $T = 0.1$ s.

(b) Evaluate the gain of a proportional controller $[D(z) = K]$ which will yield marginal stability.

(c) By a trial-and-error procedure, determine the value of proportional controller gain that will yield a closed-loop damping ratio of $\zeta = 0.7$.

3.4. A plant has the following pulse transfer function:

$$G(z) = \frac{z + 1}{z^2 - 1.6z + 0.89}$$

Proportional control in the form of a simple gain $K$ will be used to control this device.

(a) Sketch the root locus for this system.

(b) Find the value of $K$ that will make the system marginally stable. Note that in quadratic equations of the form $z^2 + bz + c =$

0, $\sqrt{c}$ is the distance of the complex roots from the origin of the $z$-plane.

3.5. A positioning servo has a typical power amplifier/motor transfer function of

$$G(s) = \frac{5}{s(s + 10)}$$

Since this transfer function has a free integrator, there will be a zero steady-state error to step input with proportional control, but a steady-state error will develop to a ramp input.

(a) Find the pulse transfer function $G(z)$ for this plant driven by a zero-order hold with an output sampler for $T = 0.02$ s.

(b) Since this plant has a free integrator, it will have zero steady-state error to a step input. Design a compensator to improve step response and give a peak time of 0.2 s ($k = 10$) and a closed-loop damping ratio of $\zeta = 0.7$.

3.6. A fluid reservoir system typical of that found in the chemical-processing industry is illustrated in the figure. Assume that the flow between the tanks is proportional to the head difference in the tanks.

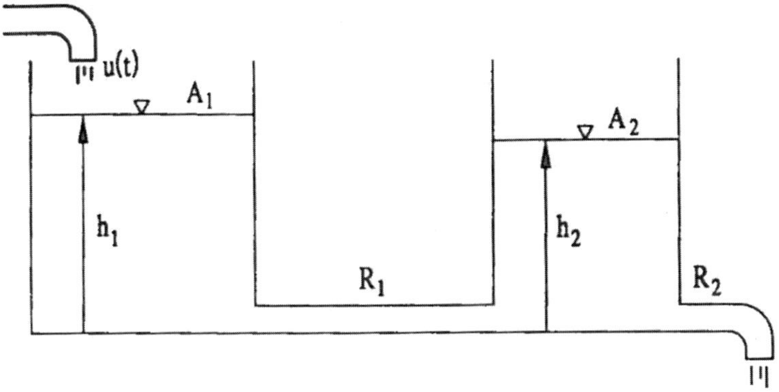


(a) Show that by application of the principle of conservation of mass (volume if the fluid is incompressible), the governing equations are

$$A_1 \frac{dh_1}{dt} = \frac{1}{R_1}(h_2 - h_1) + u(t)$$

$$A_2 \frac{dh_2}{dt} = -\frac{1}{R_1}(h_2 - h_1) - \frac{1}{R_2} h_2$$

where $u(t)$ is the volume flow rate into tank 1.

(b) For parameters $A_1 = A_2 = 1$ and $R_1 = \frac{1}{2}$, $R_2 = \frac{1}{3}$, show that these equations reduce to

$$\frac{dh_1}{dt} = 2h_1 + 2h_2 + u(t)$$

and

$$\frac{dh_2}{dt} = 2h_1 - 5h_2$$

(c) Find the transfer function between $H_2(s)$ and flow rate $U(s)$:

$$G(s) = \frac{H_2(s)}{U(s)}$$

(d) If the valve controlling the flow rate into the tank is driven by a D/A converter and the response $h_2(t)$ is sampled at a rate of 10 Hz ($T = 0.1$ s), show from part (c) that the discrete data pulse transfer function is

$$G(z) = \frac{0.008(z + 0.790)}{z^2 - 1.4536z + 0.4966}$$

3.7. Consider the plant of Problem 3.6 with the discrete-time transfer function

$$G(z) = \frac{0.008(z + 0.790)}{z^2 - 1.4536z + 0.4966}$$

(a) If proportional digital control is employed, sketch the root locus as a function of the controller gain $K$.

(b) Find the value of $K$ that will yield poles on the unit circle. Note that for a quadratic equation of the form $z^2 + bz + c = 0$, the distance to any complex roots is $\sqrt{c}$.

(c) Find a value of $K$ that will yield a closed-loop damping ratio of 0.7.

3.8. Consider the thermal control system of Example 3.3, which is subject to environmental disturbances $T_0(t)$ as shown below. Example

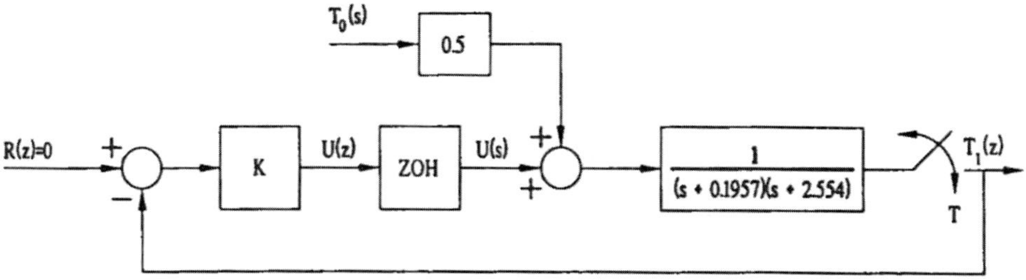

3.6 resulted in a good value of controller gain $K$ to be $K = 5$. Find the steady-state ouput temperature for a unity temperature disturbance. Assume as previously that the sampling interval is $T = 0.25$ s.

3.9 A problem that arises in computer line printers is that of micropositioning a sprung mass as shown below. The catilever springs may be represented by equivalent coil springs with modulus $k$, and the energy dissipation is represented by an equivalent damper $b$. Newton's law for the mass yields the differential equation $m\ddot{y} + b\dot{y} + ky = u(t)$.

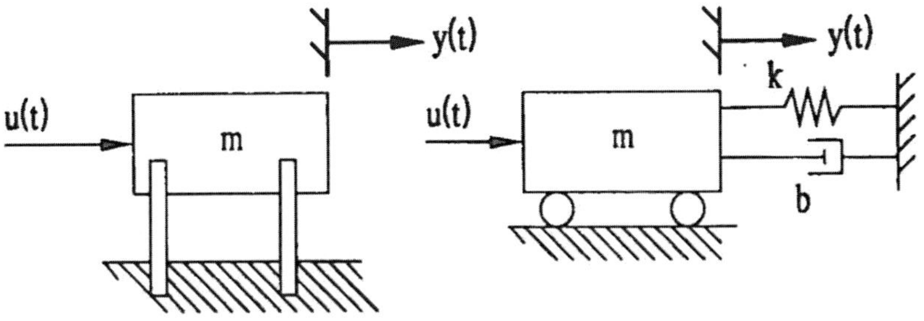


(a) Find the transfer function $Y(s)/U(s)$ and evaluate it numerically for $m = 0.5$, $k = 200$, $b = 2$, all with appropriate units.
(b) Since the undamped natural frequency $\omega_n = (k/m)^{1/2} = 20$ rad/s, it would seem reasonable to sample this system at a frequency of 20 Hz or $T = 0.05$ s. Find the pulse transfer function for this system driven by a zero-order hold and followed by an ideal sampler.
(c) Sketch the $z$-domain root locus for a proportional control algorithm with gain $K$.
(d) Find the gain $K$ ($K > 0$) at which the system becomes unstable as the gain is increased from zero. Note that for a quadratic equation of the form $z^2 + bz + c$ with complex roots, the distance from the origin to the roots is $\sqrt{c}$.
(e) From the results of parts (c) and (d), make a comment on the feasibility of proportional control and propose a remedy for the situation.

## REFERENCES


Astrom, K. J., and B. Whittenmark, 1984. *Computer Controlled Systems*, Prentice Hall, Englewood Cliffs, NJ.

Bode, H. W., 1945. *Network Analysis and Feedback Amplifier Design*, Van Nostrand, Princeton, NJ.

Dorf, R. C., 1992. *Modern Control Systems*, 6th Ed., Addison-Wesley, Reading, MA.

Evans, W. R., 1948. Graphical analysis of control systems, *Trans. AIEE*, *67*: 547–551.

Franklin, G. F., J. D. Powell, and M. L. Workman, 1990. *Digital Control of Dynamic Systems*, 2nd Ed., Addison-Wesley, Reading, MA.

Houpis, C. H., and G. B. Lamont, 1992. *Digital Control Systems: Theory, Hardware, Software*, 2nd Ed., McGraw-Hill, New York.

James, H. M., N. B. Nichols, and R. S. Phillips, 1947. *Theory of Servomechanisms*, McGraw-Hill, New York.

Katz, P., 1981. *Digital Control Using Microprocessors*, Prentice Hall, Englewood Cliffs, NJ.

Kuo, B. C., 1987. *Automatic Control Systems*, 5th Ed., Prentice Hall, Englewood Cliffs, NJ.

Kuo, B. C., 1992. *Digital Control Systems*, 2nd Ed., Saunders, Chicago.

Nyquist, H., 1932. Regeneration theory, *Bell System Technical Journal*, *11*: 126–147.

Phillips, C. L., and H. T. Nagle, 1990. *Digital Control Systems: Analysis and Design*, Prentice Hall, Englewood Cliffs, NJ.

Routh, E. J., 1877. *A Treatise on the Stability of a Given State of Motion*, Macmillan, London.

VanLandingham, H. F., 1985. *Introduction to Digital Control Systems*, Macmillan, New York.

Yeung, K. S., and H. M. Lai, 1988. A reformulation of Nyquist criterion for discrete systems, *IEEE Trans. on Education*, *31*(1): 32–34.

# 4

# Advanced Digital Control System Design Techniques Employing the *z*-Transform

## 4.1 INTRODUCTION

In Chapter 3 we explored the proportional control strategy and a simple compensator as the means of system control. We saw that response speed and quality were limited in the case of proportional control. High proportional controller gain was shown to decrease steady-state errors to step reference inputs while affecting closed-loop stability in a negative manner. This is the usual trade-off with the proportional control strategy. In this chapter we pursue techniques for designing higher-order controllers/compensators which will handle steady-state error problems while retaining reasonable system speed of response and tracking fidelity.

In this chapter we pursue the techniques for the design of controllers which are governed by a difference equation of the form

$$u(k) = a_n e(k) + a_{n-1}e(k-1) + \cdots + a_0 e(k-n) + b_{n-1}u(k-1) + \cdots + b_0 u(k-n) \tag{4.1.1}$$

The design portion of this problem is in the selection of the coefficients $a_i$ and $b_i$ once the sampling interval $T$ is specified.

One other point of interest is that the designer of control system hardware needs to know whether the actuating elements of the control system will be able to respond to the magnitude of the control effort generated by the algorithm (4.1.1) in the digital processor. This question is perhaps best answered by digital simulation, wherein the designer can readily change the initial conditions and sizes of reference inputs and examine the control efforts required.

## 4.2 GENERAL PID DIRECT DIGITAL CONTROL ALGORITHM

In the study of continuous-time control systems it was found that if proportional control is employed, a steady-state error was necessary in order to have a steady-state output. It was also found that if an integrator replaces the proportional controller, the steady-state error can be made zero for a steady output. Often, the introduction of the integrator into the loop will create instability or, at best, poor dynamic character, which manifests itself as overshoot and excessive ringing in the output. Several compromises are possible, one of which uses an actuating signal that has one component proportional to the error signal and the other proportional to the integral of the error. This combination reduces the steady-state error to zero and often yields acceptable dynamics. If further improvement in dynamics is required, a differentiator that is sensitive to error rate can be included in parallel with the other two devices. This continuous-time control scheme is shown in Fig. 4.1 with a continuous-time plant. The time-domain relation for the controller is

$$u(t) = K_p e(t) + K_i \int_0^t e \, dt + K_d \frac{de}{dt} \tag{4.2.1}$$

The associated $s$-domain transfer function for the controller is given by Laplace transformation of (4.2.1) to yield

$$D(s) = \frac{U(s)}{E(s)} = \frac{K_d s^2 + K_p s + K_i}{s} \tag{4.2.2}$$

where the choice of the constants will determine the system dynamics.

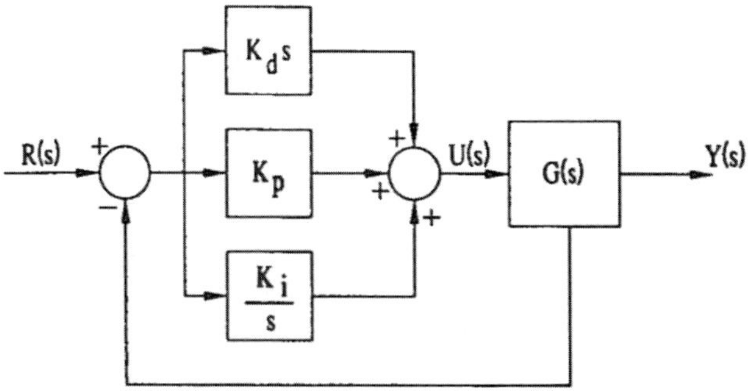


**Figure 4.1.** Proportional plus integral plus derivative (PID) control.

Since this technique has proven so useful for continuous-time control systems, it is desirable to develop a digital control algorithm that will be of similar character to the continuous-time scheme given above. We shall approximate the integral with trapezoidal integration and the derivative with a backward difference equation, or

$$u_k = K_p e_k + K_i T\left[\frac{1}{2}(e_0 + e_1) + \frac{1}{2}(e_1 + e_2) + \cdots + \frac{1}{2}(e_{k-1} + e_k)\right] + \frac{K_d}{T}(e_k - e_{k-1}) \tag{4.2.3}$$

and the algorithm for the previous step in time is written with the appropriate shift in subscripts, or

$$u_{k-1} = K_p e_{k-1} + K_i T\left[\frac{1}{2}(e_0 + e_1) + \cdots + \frac{1}{2}(e_{k-2} + e_{k-1})\right] + \frac{K_d}{T}(e_{k-1} - e_{k-2}) \tag{4.2.4}$$

Subtraction of (4.2.4) from (4.2.3) yields

$$u_k - u_{k-1} = K_p(e_k - e_{k-1}) + \frac{K_i T}{2}(e_{k-1} + e_k) + \frac{K_d}{T}(e_k - 2e_{k-1} + e_{k-2}) \tag{4.2.5}$$

or combining like terms yields

$$u_k = u_{k-1} + \left(K_p + \frac{K_i T}{2} + \frac{K_d}{T}\right)e_k + \left(\frac{K_i T}{2} - K_p - \frac{2K_d}{T}\right)e_{k-1} + \frac{K_d}{T}e_{k-2} \tag{4.2.6}$$

which is the direct digital control algorithm. By taking the $z$-transform of the difference equation [Eq. (4.2.6)] we can determine the compensator transfer function which will perform the proportional plus integral plus derivative (PID) control function:

$$D(z) = \frac{U(z)}{E(z)} = \frac{\alpha + \beta z^{-1} + \gamma z^{-2}}{1 - z^{-1}} = \frac{\alpha z^2 + \beta z + \gamma}{z(z - 1)} \tag{4.2.7}$$

where

$$\alpha = K_p + \frac{K_i T}{2} + \frac{K_d}{T} \tag{4.2.8}$$

$$\beta = \frac{K_i T}{2} - K_p - \frac{2K_d}{T} \tag{4.2.9}$$

and

$$\gamma = \frac{K_d}{T} \tag{4.2.10}$$

It is interesting to note that the transfer function of (4.2.7) has a quadratic numerator that may be chosen such as to cancel two slow, troublesome poles of a plant. As long as those poles are interior to the unit circle, the cancellation need not be exact and the resultant root-locus branches will contribute little to the closed-loop response because of the zeros of the closed-loop transfer function being coincident with the zeros of the controller.

If only proportional plus integral action is required, it is a simple matter to let $K_d$ be zero, which yields a simplified transfer function for the compensator:

$$D(z) = \frac{U(z)}{E(z)} = \frac{\alpha z + \beta}{z - 1} \tag{4.2.11}$$

where

$$\alpha = K_p + \frac{K_i T}{2} \tag{4.2.12}$$

and

$$\beta = \frac{K_i T}{2} - K_p \tag{4.2.13}$$

In the derivation above we used the trapezoidal approximation of integration and the backward difference for differentiation. Other approximations will, of course, yield different algorithms for the digital controller. In fact, if one is not careful in choosing these approximations, the algorithm could be noncausal, with future values of error being used to calculate the current control effort. It should be noted, however, that as long as proportional and integral action are present, the compensator $D(z)$ will have a pole at $z = 1$ that is the image of the continuous-time integrator pole in relation (4.2.2).

**Example 4.1.** Consider the thermal system of Examples 3.3 and 3.6, for which we would like to design a proportional plus integral (PI) controller so we will have zero steady-state error to a step input. The plant transfer function is

$$G(z) = \frac{0.025(z + 0.816)}{(z - 0.952)(z - 0.528)}$$

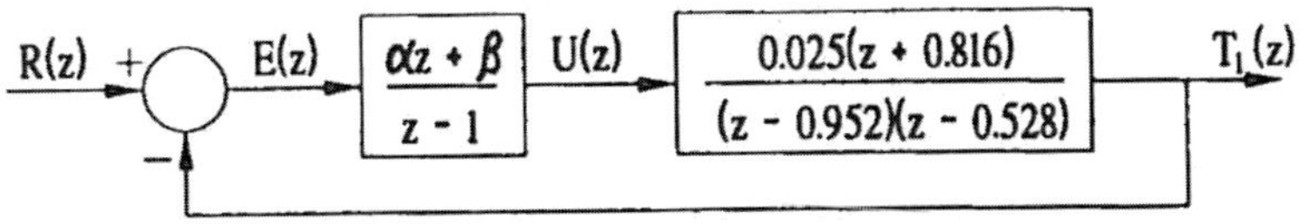


**Figure 4.2.** Proportional plus integral (PI) control of thermal plant.

We shall choose to locate the zero of the PI controller so as to cancel the slow pole of the plant, so the compensator transfer function is

$$D(z) = \frac{\alpha z + \beta}{z - 1} = \frac{\alpha(z + \beta/\alpha)}{z - 1}$$

where $\beta/\alpha = -0.952$. The whole feedback system structure is shown in Fig. 4.2, and the root-locus diagram for this example for variable controller gain $\alpha$ is shown in Fig. 4.3.

A trial-and-error design gives a closed-loop damping ratio of $\zeta = 0.7$ for a controller gain of $\alpha = 2.5$. The resulting control algorithm is

$$u_k = u_{k-1} + 2.5(e_k - 0.952e_{k-1})$$

The step response and associated control effort for this system are given in Fig. 4.4(a) and (b), respectively. It should be noted that due to the

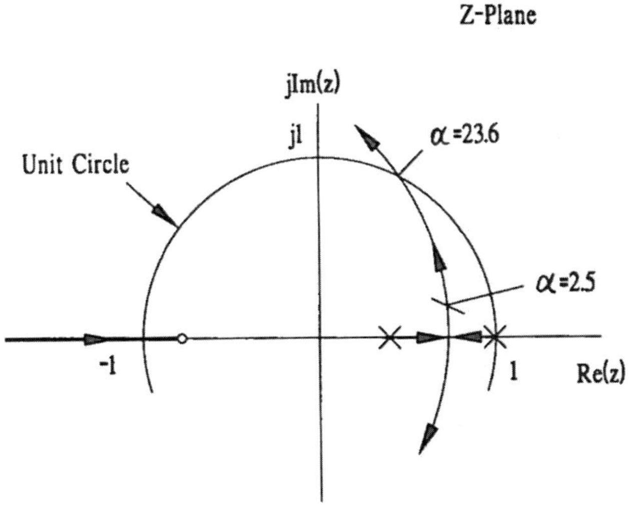


**Figure 4.3.** Root locus for PI control of the thermal system.

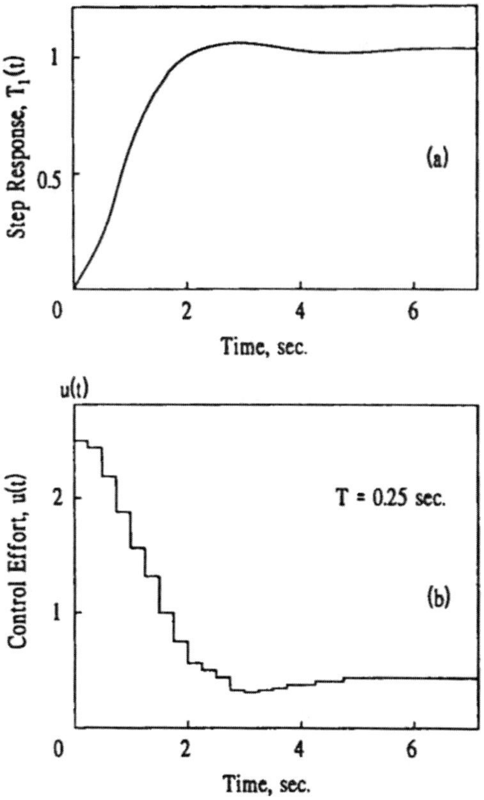


**Figure 4.4.** Step response (a) and control effort (b) for the PI control of the thermal plant.

integral action the steady-state error is zero and that the step response is reasonable with an overshoot of 4%; however, because the root locus was shifted to the right, the response is slightly slower than in proportionally controlled cases. To have both fast response and zero steady-state error a higher-order compensator will be necessary.

## 4.3 ZIEGLER–NICHOLS TUNING PROCEDURE FOR PID CONTROL

In their early papers, Ziegler and Nichols (1942, 1943) explored a process by which the system designer could decide what the values of $K_p$, $K_i$, and $K_d$ should be. Once these parameters are evaluated, relations (4.2.8),

(4.2.9), and (4.2.10) could be employed to get the parameters α, β, and γ in the PID control algorithm.

The technique is based on experimental evaluation of several parameters associated with the step response of the plant to be controlled. The first task is to evaluate the plant response to a sudden constant change in the variable to be used as the control effort. Such an experimental response is illustrated in Fig. 4.5.

From this step response it is necessary to find two quantities. The first is $R$, the slope of the response curve at the inflection point, which is an indication of the speed of response. The second is the time $L$, which is a measure of the lag of the plant. For a PI controller the tuning strategy relates $K_p$ and $K_i$ to the values of $R$ and $L$ as

$$K_p = \frac{0.9}{RL} \tag{4.3.1}$$

and

$$K_i = \frac{1}{3.3L} K_p = \frac{0.272}{RL^2} \tag{4.3.2}$$

If the PID control strategy is chosen, the tuning equations are

$$K_p = \frac{1.2}{RL} \tag{4.3.3}$$

$$K_i = \frac{1}{2L} K_p = \frac{0.6}{RL^2} \tag{4.3.4}$$

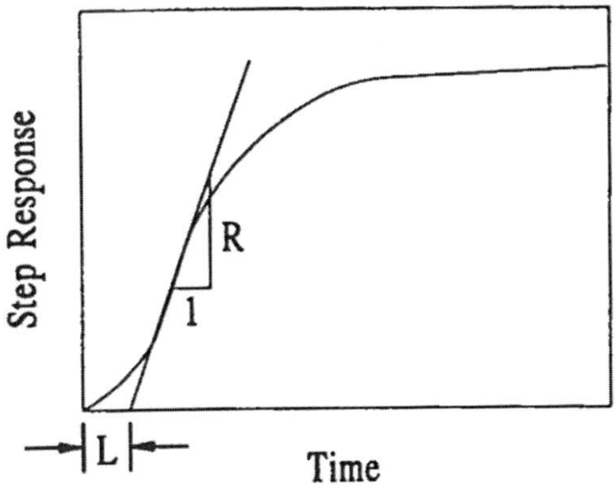


**Figure 4.5.** Step response of a typical plant.

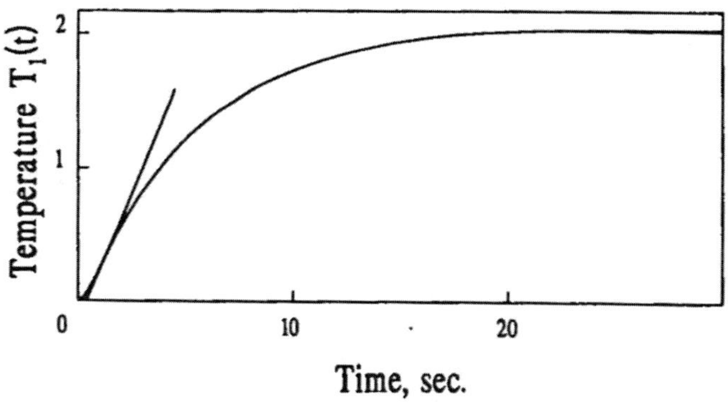


**Figure 4.6.** Step response of thermal plant.

and

$$K_d = 0.5LK_p = \frac{0.6}{R} \tag{4.3.5}$$

From these expressions it is possible to evaluate the parameters α, β, and γ of the controller.

**Example 4.2.** Use the Ziegler–Nichols tuning strategy to design a PI controller for the thermal plant of Example 4.1 for which the continuous-time transfer function is

$$G(s) = \frac{1}{s^2 + 2.75s + 0.5} = \frac{T_1(s)}{U(s)}$$

The step response of this system is illustrated in Fig. 4.6.

Evaluation of the slope at the inflection point yields a slope of $R = 0.37$ and the lag parameter indicated is $L = 0.47$ s. Calculation of $K_p$ and $K_i$ yield $K_p = 5.175$ and $K_i = 3.336$. From these values and a sampling interval of $T = 0.25$ s the two controller coefficients are $\alpha = 5.592$ and $\beta = -4.758$. And thus the resultant controller is

$$D(z) = \frac{5.592(z - 0.85)}{z - 1}$$

The associated control algorithm is

$$u(k) = u(k - 1) + 5.592e(k) - 4.758e(k - 1)$$

The sampled unit step closed-loop response for the system is shown in Fig. 4.7. It should be noted that the procedure at this point may require some

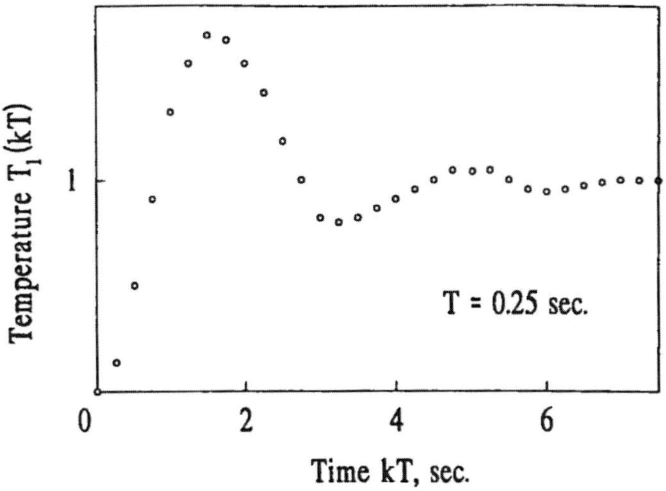


**Figure 4.7.** Closed-loop step response of PI controlled thermal system with Ziegler–Nichols tuning.

further tuning since we have simply implemented a discrete version of a continuous-time design. In fact, if we evaluate the damping ratio for the closed-loop poles, we find that it is $\zeta = 0.32$, which may be too small for many applications where we want to limit the overshoot. Decreasing the gain of the controller will decrease the overshoot with an accompanying slowing of the step response.

## 4.4 DIRECT DESIGN METHOD OF RAGAZZINI

We have shown that for a single-input/single-output plant with transfer function $G(z)$ in a unity feedback loop with a cascade controller $D(z)$, the closed-loop transfer function $M(z)$ is

$$M(z) = \frac{Y(z)}{R(z)} = \frac{G(z)D(z)}{1 + G(z)D(z)} \tag{4.4.1}$$

If we know the desired closed-loop transfer function $M(z)$ and $G(z)$, we may solve (4.4.1) for the needed controller transfer function $D(z)$, or

$$D(z) = \frac{1}{G(z)} \frac{M(z)}{1 - M(z)} \tag{4.4.2}$$

Note that $D(z)$ cancels the plant dynamics $G(z)$ and adds whatever is necessary to give the specified $M(z)$. We must be careful that we do not ask the impossible in $D(z)$, and hence we address the following fundamental issues.

### Causality

A causal controller is one for which the current control effort depends only on the current and past inputs and past control efforts. If the controller transfer function $D(z)$ is causal, it does not have a pole at infinity. If we examine (4.4.2) we see that if $G(z)$ has a zero at infinity, then $D(z)$ would have a pole at infinity unless we specify $M(z)$ to cancel it. For $M(z)$ to be causal it must have a zero at infinity of the same order as the zero of $G(z)$ at infinity. In the time domain a zero of $G(z)$ at infinity implies that the pulse response of $G(z)$ has a delay of at least one sample interval. If there is any transport lag in the plant $G(z)$ it will have a numerator factor in $z^{-m}$ as discussed in Section 3.3. The causality requirement on $M(z)$ is that the closed-loop system must have at least as many pure delays as the plant. No causal controller can improve this situation. A causal controller is one for which the *current* control effort $u(k)$ is not dependent on future values of the error.

### Stability

Let us consider what is necessary to have a stable system. The closed-loop characteristic equation is

$$1 + G(z)D(z) = 0 \tag{4.4.3}$$

Let $D(z) = c(z)/d(z)$ and $G(z) = b(z)/a(z)$, the closed-loop characteristic polynomial is

$$a(z)d(z) + b(z)c(z) = 0 \tag{4.4.4}$$

Let us suppose that there is a common factor $(z - \alpha)$ in the numerator of $D(z)$ and the denominator of $G(z)$ as would happen if we attempted to cancel a pole of $G(z)$. Suppose that the factor is a pole of $G(z)$ or $a(z) = (z - \alpha)\bar{a}(z)$ and to cancel it we write $c(z) = (z - \alpha)\bar{c}(z)$, then (4.4.4) becomes

$$(z - \alpha)\bar{a}(z)d(z) + (z - \alpha)\bar{c}(z)b(z) = 0$$

or

$$(z - \alpha)[\bar{a}(z)d(z) + \bar{c}(z)b(z)] = 0 \tag{4.4.5}$$

A common factor remains a factor of the closed-loop characteristic polynomial. If the $\alpha$ is outside the unit circle, the system is unstable. To avoid this unfortunate situation, we must consider relation (4.4.2), wherein we see that if $D(z)$ is not to cancel a pole of $G(z)$, then the factor of $a(z)$ must also be a factor of $1 - M(z)$. Similarly, if $D(z)$ is not to cancel a zero of $G(z)$, such zeros must be factors of $M(z)$.

The constraints may be summarized as follows:

(a) $1 - M(z)$ must contain as zeros all poles of $G(z)$ that lie on or outside the unit circle.
(b) $M(z)$ must contain as zeros all the zeros of $G(z)$ lying on or outside the unit circle.

Another problem in the cancellation of unstable poles is that exact cancellation can never be accomplished since the poles of the plant are known only approximately. An inexact cancellation of an unstable pole will lead to a short root-locus branch between the pole and the "canceling" zero which will yield a closed-loop characteristic root that is unstable. In general, cancellation of unstable poles is not a feasible control strategy.

### Steady-State Accuracy

Since $M(z)$ is the overall transfer function and the error is

$$E(z) = R(z)[1 - M(z)] \tag{4.4.6}$$

For a unit-step input we may employ the final value theorem to get the steady-state error of

$$e_{ss} = \lim_{z \to 1} \frac{(z - 1)}{z} \frac{z}{z - 1} [1 - M(z)] \tag{4.4.7}$$

Then the steady-state error to a unit-step reference input is

$$e_{ss} = 1 - M(1) \tag{4.4.8}$$

For zero steady-state error to a step input, the dc gain of the closed-loop system must be unity, or

$$M(1) = 1 \tag{4.4.9}$$

If the steady-state error to a ramp input is to be $1/K_v$ ($K_v$ is the system velocity constant), the steady-state error is

$$e_{ss} = \lim_{z \to 1} \frac{z - 1}{z} \frac{Tz}{(z - 1)^2} [1 - M(z)] = \frac{1}{K_v} \tag{4.4.10}$$

For the ramp error to be a constant, instead of increasing, we know that first the step error must be zero and hence $M(1) = 1$. This leaves us with an indeterminate system with the evaluation of relation (4.4.10) and we must employ L'Hospital's rule to evaluate the limit $z \to 1$, which gives us

$$-T\frac{dM}{dz}\bigg|_{z=1} = \frac{1}{K_v} \tag{4.4.11}$$

or

$$\frac{dM}{dz}\bigg|_{z=1} = \frac{-1}{K_vT} \tag{4.4.12}$$

It is sometimes convenient to note that $(dM/dz)|_{z=1} = -(dM/dz^{-1})|_{z=1}$.

**Example 4.3.** Consider the thermal plant of Example 3.3 for $T = 0.25$ s with transfer function

$$G(z) = \frac{0.025(z + 0.816)}{z^2 - 1.48z + 0.5026}$$

The design specifications are to have pole locations of $0.4 \pm j0.4$ ($\omega_d = \pi$ rad/s $\zeta = 0.58$), zero steady-state error to a step input, and a velocity constant of $K_v = 10/T$. From these specifications the closed-loop transfer function is

$$M(z) = \frac{b_0 + b_1z^{-1} + b_2z^{-2} + b_3z^{-3} + \cdots}{1 - 0.8z^{-1} + 0.32z^{-2}}$$

The causality requirement dictates that $b_0 = 0$ since the plant has one step of pure delay. From the steady-state step response error the requirement is that

$$M(1) = \frac{b_1 + b_2 + b_3 + \cdots}{0.52} = 1$$

The closed-loop transfer function is

$$M(z) = \frac{b_1z + b_2 + b_3z^{-1} + \cdots}{z^2 - 0.8z + 0.32}$$

Since we have only two specifications to meet other than the pole locations, we may truncate the numerator after $b_2$ and can still meet the specifications. The velocity constant requirement is

$$\frac{dM}{dz}\bigg|_{z=1} = \frac{(z^2 - 0.8z + 0.32)b_1 - (b_1z + b_2)(2z - 0.8)}{(z^2 - 0.8z + 0.32)^2}$$

$$= -\frac{1}{K_vT} = -0.1$$

or

$$\frac{0.52b_1 - (b_1 + b_2)(1.2)}{(0.52)^2} = -0.1$$

But $b_1 + b_2 = 0.52$, so $b_1 = 1.148$; then $b_2 = 0.52 - b_1 = -0.628$. Thus the closed-loop transfer function is

$$M(z) = \frac{1.148z - 0.628}{z^2 - 0.8z + 0.32}$$

From relation (4.4.2) the controller is of the form

$$D(z) = \frac{z^2 - 1.48z + 0.5026}{0.025(z + 0.816)} \frac{1.148z - 0.628}{(z^2 - 0.8z + 0.32) - 1.148z + 0.628}$$

or

$$D(z) = \frac{1.148}{0.025}\left[\frac{(z - 0.528)(z - 0.952)}{z + 0.816} \frac{z - 0.547}{(z - 1)(z - 0.948)}\right]$$

This part cancels the plant dynamics (first factor)

This part gives a type one system and the correct closed-loop poles and error constant $K_v$ (second factor)

Multiplying out the compensator factor yields the following third-order controller transfer function:

$$D(z) = \frac{45.92(z^3 - 2.027z^2 + 1.3122z - 0.2749)}{z^3 - 1.132z^2 - 0.6416z + 0.7736}$$

The closed-loop step response for this compensated system is illustrated in Fig. 4.8. Note that although the samples of the step response seem to be well behaved, the continuous-time response between the samples has sustained oscillations of relatively high amplitude, which may not be desirable. The control effort $u(t)$ also exhibits this oscillatory tendency. Of course, this transfer function could be converted to a time-domain difference equation to be implemented in the digital processor.

### Finite-Settling-Time Systems

We now discuss what we must do to make the sampled output $y(kT)$ settle to zero steady-state error in a finite number of steps. For this to happen

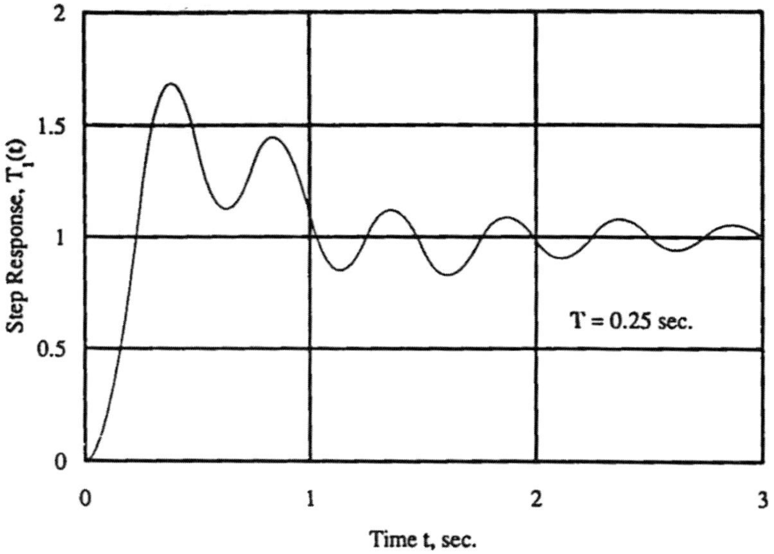


**Figure 4.8.** Step response of the thermal system with a third-order controller.

the closed-loop transfer function must be a finite polynomial in $z^{-1}$. This implies that the closed-loop impulse response sequence $m(k)$ must be of finite duration or be of the FIR (finite impulse response) type. In other words,

$$M(z) = a_0 + a_1z^{-1} + a_2z^{-2} + \cdots + a_Nz^{-N} \qquad (4.4.13)$$

For zero steady-state error to a step reference input $M(1) = 1$, so

$$1 = a_0 + a_1 + \cdots + a_N \qquad (4.4.14)$$

Without further specifications there are many choices of $a_0, \ldots, a_N$ which will satisfy (4.4.14), and any arbitrary choice should be substantiated with a closed-loop simulation. Generally, the shorter the duration of the impulse response sequence (smaller $N$), the larger the control efforts required to establish this performance.

## Minimum Prototype Systems

If the plant has one step of pure delay, then for a causal controller to result, $a_0$ must be zero. With this restriction the simplest possible case for finite settling to a step input is to synthesize $D(z)$ such that the closed-loop transfer function $M(z)$ will be a pure delay of one time step, or

$$M(z) = \frac{1}{z} \tag{4.4.15}$$

The resulting samples of the plant output will be a step sequence delayed by one step from the reference input; however, it is not unusual for the continuous-time step response to oscillate excessively between the samples.

For the system to settle to zero error in a finite number of steps to a polynomial type input $[r(k) = (kT)^n]$ in a minimum number of steps, the required closed-loop transfer function is

$$M(z) = 1 - (1 - z^{-1})^{n+1} \tag{4.4.16}$$

For a polynomial input of degree $n$, the system will generally require $n + 1$ sampling intervals to settle to zero error. Such a system is referred to as a *minimal prototype system*.

The difficulty with requiring the system to settle to zero error in a minimum number of steps is that the control effort $u(k)$ has a tendency to oscillate violently. Although the output tracks at the sample times, the continuous-time response of the plant output often oscillates unacceptably between the samples. The shorter the sampling interval relative to the plant characteristic times (time constants and natural periods), the more violent these continuous-time oscillations will be.

## Ripple-Free Design

One way to handle this excessive oscillation of the control effort and the continuous-time response is to design what is commonly referred to as a ripple-free controller. For a ripple-free design the plant must be stable [i.e., the poles of $G(z)$ cannot be on or outside the unit circle]. If we want a finite settling time to a step response as before, $M(z)$ must be a finite polynomial in $z^{-1}$. Also, the transfer function from the reference input $R(z)$ to the control effort must be of the finite settling type, or $U(z)/R(z)$ is

$$\frac{U(z)}{R(z)} = \frac{M(z)}{G(z)} \tag{4.4.17}$$

is a finite polynomial in $z^{-1}$. So that there are no poles of $M(z)/G(z)$ except at $z = 0$, $M(z)$ must be selected such that it has a factor to cancel the finite zeros of $G(z)$.

**Example 4.4.** Consider the attitude control of the satellite outlined in Example 3.2 for a sampling interval of $T = 0.1$ s. The plant transfer function is then

$$G(z) = \frac{1}{200} \frac{z + 1}{(z - 1)^2}$$

Design a ripple-free finite-settling controller for this plant. Since there is a zero at $z = -1$ on the unit circle and we want $M(z)$ to be of the finite settling type, we should pick $M(z)$ to be

$$M(z) = (1 + z^{-1})(c_0 + c_1 z^{-1} + c_2 z^{-2}) \qquad \text{(a)}$$

Since the term $(1 + z^{-1})$ has been included to avoid the cancellation problem, the result will be a ripple-free system. Also, the $1 - M(z)$ must be of polynomial form in $z^{-1}$ and have a factor of $(1 - z^{-1})^2$ because of the poles on the unit circle

$$1 - M(z) = (1 - z^{-1})^2(b_0 + b_1 z^{-1} + b_2 z^{-2}) \qquad \text{(b)}$$

If we expand $M(z)$, we note for causality that $c_0 = 0$ and thus

$$M(z) = c_1 z^{-1} + (c_1 + c_2)z^{-2} + c_2 z^{-3} \qquad \text{(c)}$$

Also, because $1 - M(z)$ in (a) has no term in $z^{-4}$, then $b_2$ in (b) must be zero

$$1 - M(z) = b_0 + (b_1 - 2b_0)z^{-1} + (b_0 - 2b_1)z^{-2} + b_1 z^{-3} \qquad \text{(d)}$$

and from (c),

$$1 - M(z) = 1 - c_1 z^{-1} - (c_1 + c_2)z^{-2} - c_2 z^{-3} \qquad \text{(e)}$$

Equating (d) and (e) results in a set of linear equations for the coefficients:

$$\begin{aligned} z^0&: \quad 1 = b_0 \\ z^{-1}&: \quad -c_1 = b_1 - 2b_0 \\ z^{-2}&: \quad -c_1 - c_2 = b_0 - 2b_1 \\ z^{-3}&: \quad -c_2 = b_1 \end{aligned}$$

Since $b_0 = 1$, we may solve the remaining three equations to yield

$$\begin{aligned} b_1 &= 0.75 \\ c_1 &= 1.25 \\ c_2 &= -0.75 \end{aligned}$$

Thus

$$M(z) = (1 + z^{-1})(1.25z^{-1} - 0.75z^{-2}) = 1.25z^{-1} + 0.5z^{-2} - 0.75z^{-3}$$

Now let us solve for the compensator transfer function using relation (3.8.2):

$$D(z) = \frac{200(z - 1)^2}{z + 1} \frac{(1 + z^{-1})(1.25z^{-1} - 0.75z^{-2})}{(1 - z^{-1})^2(1 + 0.75z^{-1})}$$

Canceling the appropriate terms, the result is

$$D(z) = \frac{250(z - 0.6)}{z + 0.75}$$

and the corresponding control algorithm is

$$u(k) = -0.75u(k - 1) + 250e(k) - 150e(k - 1)$$

A unit-step response for the system is illustrated in Fig. 4.9.

Note that the response settles to zero error in three steps; however, the overshoot may be more than tolerable. This could be relaxed by allowing the duration of the closed-loop impulse response to be longer and hence more terms in polynomial in $z^{-1}$ for $M(z)$. This will make it necessary to arbitrarily select some of the constants, which will probably have to be done by a trial-and-error process. Another example should be illustrative of ripple-free controller design process.

**Example 4.5.** Design a zero-ripple, finite-settling-time controller for the thermal plant with a pulse transfer function of

$$G(z) = \frac{0.025(z + 0.816)}{z^2 - 1.48z + 0.5026} = \frac{0.025z^{-1}(1 + 0.816z^{-1})}{1 - 1.48z^{-1} + 0.5026z^{-2}}$$

which will yield zero steady-state error to a step reference input.

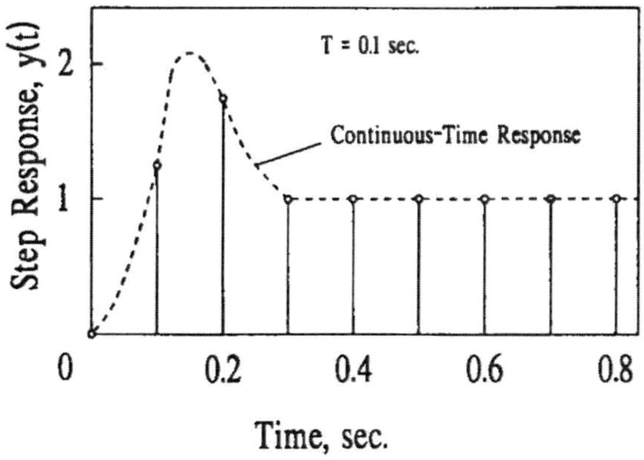


**Figure 4.9.** Finite settling-time response of satellite dynamics.

Since we want zero steady-state error to a step reference input and there are no free integrators in $G(z)$ [i.e., no factors of the form $(1 - z^{-1})$], then $1 - M(z)$ must have such a factor or

$$1 - M(z) = (1 - z^{-1})(a_0 + a_1 z^{-1}) \qquad \text{(a)}$$

The plant is stable and hence it is possible to design a zero ripple controller so that $M(z)$ must have as a factor the numerator term from $G(z)$, so

$$M(z) = (1 + 0.816z^{-1})d_1 z^{-1}$$

Then

$$1 - M(z) = 1 - d_1 z^{-1} - 0.816 d_1 z^{-2} \qquad \text{(b)}$$

Equation relations (a) and (b) term by term results in the linear equations

$$a_0 = 1$$

$$a_1 - a_0 = -d_1$$

$$-a_1 = -0.816 d_1$$

Solving these equations yields the following constants:

$$a_0 = 1 \qquad a_1 = 0.4493 \qquad d_1 = 0.5507$$

The closed-loop transfer function is

$$M(z) = 0.5507z^{-1} + 0.4493z^{-2}$$

The controller/compensator then is given by relation (4.4.2) to be

$$D(z) = \frac{1 - 1.48z^{-1} + 0.5026z^{-2}}{0.025z^{-1}} \frac{0.5507z^{-1}}{(1 - z^{-1})(1 + 0.4493z^{-1})}$$

Thus the compensator is

$$D(z) = \frac{0.5507(z^2 - 1.48z + 0.5026)}{0.025(z - 1)(z + 0.4493)} = \frac{22.028(z^2 - 1.48z + 0.5026)}{z^2 - 0.5507z - 0.4493}$$

It is also interesting to calculate the transfer function from $R(z)$ to $U(z)$, which is

$$\frac{U(z)}{R(z)} = \frac{M(z)}{G(z)} = \frac{0.5507(z + 0.816)}{z^2} \frac{z^2 - 1.48z + 0.5026}{0.025(z + 0.816)} = \frac{22.028(z^2 - 1.48z + 0.5026)}{z^2}$$

which is a finite-settling-time transfer function which settles in two steps also to a control effort of 0.5.

## 4.5 SUMMARY

In this chapter we have explored the design of compensator/controllers in the $z$-domain. PID controllers and their tuning were explored in detail. Also, the direct method of Ragazzini was developed and the synthesis of finite-settling controllers and ripple-free designs was discussed and examples were presented.

## PROBLEMS

4.1. Shown below is a simple first-order plant with a PI controller.

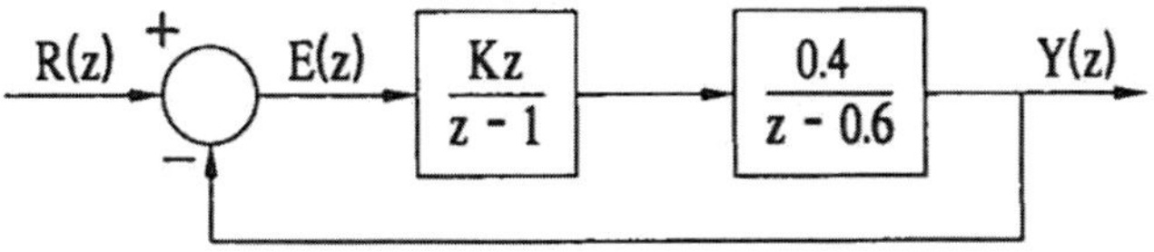


(a) Find the steady-state error to a unit step input $R(z)$.
(b) If the input is $R(z) = T(z)/(z - 1)^2$ (a sampled ramp). Find the steady-state error to this input by employing the final value theorem.

4.2. Derive a proportional plus integral (PI) control algorithm using the forward rectangular approximation for the integrator, or $i_k = i_{k-1} + e_{k-1}T$, which is illustrated below. One check of your work will be that the pole of $D(z)$ lies at $z = 1$.

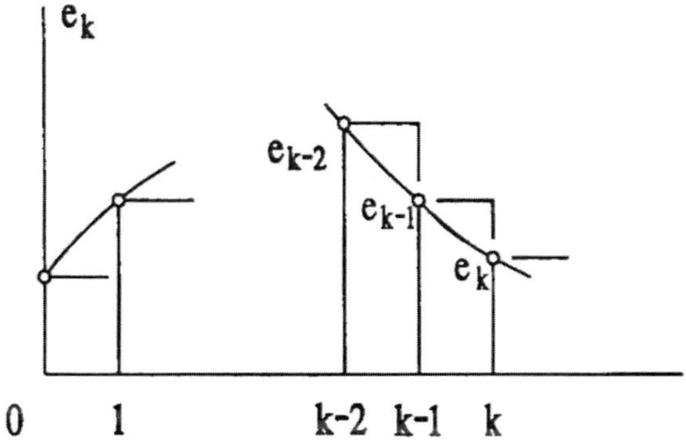

4.3. The two-tank problem of Problem 3.6 has a pulse transfer function of

$$G(z) = \frac{0.008(z + 0.790)}{z^2 - 1.4536z + 0.4966} = \frac{H_2(z)}{U(z)}$$

Design a PI controller yielding a closed-loop damping ratio of $\zeta = 0.8$ for this system and evaluate the peak time for a closed-loop step response.

4.4. (a) For the control system shown below, sketch the root locus and find the value of $K$ for marginal stability.
(b) Find the value of $K$ that will yield a closed-loop damping ratio of $\zeta = 0.5$.
(c) For $R(z) = 0$, find the transfer function $Y(z)/W(z)$.
(d) What steady-state deviation in the output will be caused by a step disturbance $W(z)$?

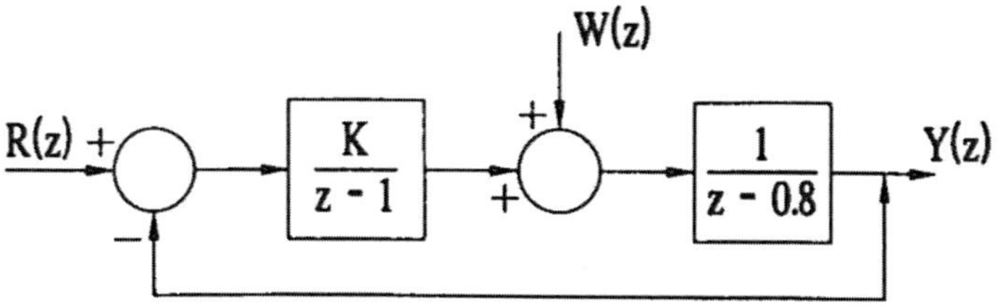


4.5. For the printhead positioning problem posed in Problem 3.9, the pulse tranfer function is

$$G(z) = \frac{0.002155(z + 0.9334)}{z^2 - 0.9854z + 0.8187}$$

for a sampling interval of $T = 0.05$ s. Design a PID controller for the system that will result in a damping ratio of $\zeta = 0.7$. Evaluate the step response to find the peak time. You may want to consider using the numerator of the PID controller transfer function to cancel the complex poles.

4.6. The continuous-time transfer function for a plant to be controlled is

$$G(s) = \frac{1}{s^2 + 5s + 6} = \frac{Y(s)}{U(s)}$$

The step response of this system is shown below. Design a PI controller via the Ziegler–Nichols procedure for a sampling interval of $T = 0.2$ s.

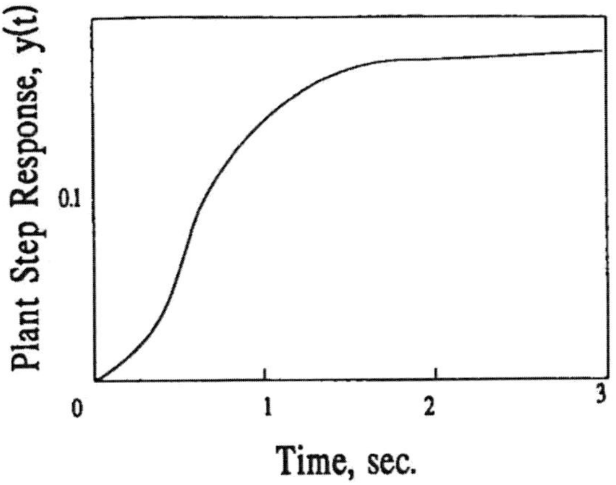


4.7. A discrete-time process is governed by the difference equation $y_{k+1} = 0.8y_k + 0.1u_k$.

(a) Find the cascade controller $D(z)$ that will make the system settle to a step reference input in one sampling interval. In other words, make $M(z) = 1/z$.

(b) Write the difference equation, which will be the control algorithm.

(c) Could there be any practical problems associated with implementation of the difference equation of part (b)?

4.8. For the two-tank problem given in Problem 3.6, the resulting pulse transfer function is

$$G(z) = \frac{0.008(z + 0.790)}{z^2 - 1.4536z + 0.4966}$$

Design a controller $D(z)$ that will result in a closed-loop transfer function of

$$M(z) = 0.25z^{-1} + 0.5z^{-2} + 0.25z^{-3}$$

How many time steps will it take for the system output to settle to zero error for a step response input?

4.9. Consider the two-tank problem as posed in Problem 3.6, which for $T = 0.1$ has a pulse transfer function

$$G(z) = \frac{0.008(z + 0.79)}{z^2 - 1.4536z + 0.4966} = \frac{H_2(z)}{U(z)}$$

Design, by means of the Ragazzini direct method, a controller that will have zero steady-state error to a ramp input which will achieve that zero error situation in a minimum number of time steps, so $1 - M(z)$ should be of the finite impulse response type and have two zeros at $z = 1$, so $D(z)$ will have double integrator character, so

$$1 - M(z) = (1 - z^{-1})^2(b_0 + b_1z^{-1})$$

The controller should also be of the zero-ripple type, so the transfer function $U(z)/R(z)$ should be of the finite-settling type. The requirement for this to occur is that $M(z)$ contain as a factor the zeros of $G(z)$, or it should be of the form

$$M(z) = (1 + 0.79z^{-1})(a_1z^{-1} + a_2z^{-2})$$

In the course of the design you should accomplish the following steps:

(a) Find constants $a_1$, $a_2$, $b_0$, and $b_1$.
(b) Find $M(z)$ and plot the closed-loop impulse $m(k)$. A helpful check is that the sum of weights of $m(k)$ be unity.
(c) By graphical convolution evaluate the unit-step response sequence to verify zero steady-state error and finite settling time.
(d) Find the compensator transfer function $D(z)$, being sure to cancel common numerator and denominator terms so that the transfer function will be of minimal order.
(e) Find the difference equation for the control algorithm.

## REFERENCES

Franklin, G. F., J. D. Powell, and M. L. Workman, 1990. *Digital Control of Dynamic Systems*, 2nd Ed., Addison-Wesley, Reading, MA.

Houpis, C. H., and G. B. Lamont, 1992. *Digital Control Systems: Theory, Hardware, Software*, 2nd Ed., McGraw-Hill, New York.

Katz, P., 1981. *Digital Control Using Microprocessors*. Prentice Hall, Englewood Cliffs, NJ.

Kuo, B. C., 1992. *Digital Control Systems*, 2nd Ed., Holt, Rinehart and Winston, New York.

Phillips, C. L., and H. T. Nagle, Jr., 1990. *Digital Control Systems: Analysis and Design*, 2nd Ed., Prentice Hall, Englewood Cliffs, NJ.

Ragazzini, J. R., and G. F. Franklin, 1958. *Sampled Data Control Systems*, McGraw-Hill, New York.

Ziegler, J. G., and N. B. Nichols, 1942. Optimum settings for automatic controllers, *Trans. ASME*, *64*(8): 759–768.

Ziegler, J. G., and N. B. Nichols, 1943. Process lags in automatic control circuits, *Trans. ASME*, *65*(5): 433–444.

# 5
# Digital Filtering and Digital Compensator Design

## 5.1 INTRODUCTION

The student of digital control theory might ask what a chapter on digital filtering is doing in the initial portion of a book on control theory. The reason for this is that digital control algorithms take the form of difference equations in which the coefficients are fixed by design techniques. An inhomogeneous difference equation or set of difference equations could be $z$-transformed to yield a $z$-domain transfer function or matrix of transfer functions, as in the case of a multiple-input and/or multiple-output system. Any $z$-domain transfer function could then be thought of as a digital filter.

## 5.2 CONVENTIONAL DESIGN TECHNIQUES FOR DIGITAL COMPENSATORS

The design of digital control systems is the process of choosing the difference equations or equivalent $z$-domain transfer functions for either cascade or feedback compensators, which will, when combined with the dynamics of the continuous-time plant, yield acceptable performance from the completed closed-loop system. As in the continuous-time problem, the specification of performance could take many forms, such as rise time, settling time, percent overshoot, closed-loop frequency response magnitude, bandwidth, or damping ratio. These specifications were discussed in detail in Chapter 3.

Design techniques for continuous-time control systems employ the frequency response methods of Nyquist, Bode, and Nichols, in which the gain and phase margin concepts are very important. Alternatively, we know that often the system can be designed based simply on the location or the dominant closed-loop poles in the $s$-plane. The tool for this pole placement technique is the root-locus method of Evans. We know that by insertion of additional dynamic systems (compensators) with specific properties into the control loop, the closed-loop system dynamics may be greatly improved. This amounts to reshaping the open-loop frequency response plot or the root-locus plot.

In the design of digital compensators, two paths are generally taken. The first is to ignore any zero-order holds and samplers in the control loop and do a preliminary design in the $s$-domain as if one were building a continuous-time control system. This design must then be converted to a discrete-time design by some approximate technique to yield a discrete-time compensator. Once these steps have been accomplished, a $z$-domain analysis can be employed to verify whether the original design goals have been met. The second method is to design the compensator directly in the $z$-domain using $z$-domain frequency response methods or the $z$-domain root-locus method; this technique was the topic of Chapters 3 and 4.

The first method has an advantage since engineers are more accustomed to thinking clearly in the $s$-plane than in the $z$-plane. It has a disadvantage since in the process of conversion to a discrete-time design, by whatever method, the $z$-plane poles are distorted from where they are needed, and hence a trial-and-error design procedure may be in order.

The second method has the advantage that the poles and zeros of the compensator are located directly, and the designer can pick these locations *a priori*. A disadvantage is that it is difficult for the designer to visualize exactly where he might want the $z$-domain poles and zeros to improve system performance.

The author's experience indicates that the first method is usually best for all but experienced $z$-domain designers. We shall assume for the remainder of this chapter that the design has been done in the $s$-domain and then is to be implemented in discrete form on the digital computer. In this chapter we discuss conversion of an $s$-domain transfer function $H(s)$ into a discrete-time transfer function $H(z)$ or an equivalent difference equation.

Let us review some of the properties of continuous-time filters that will yield methods of approximation. The $s$-domain transfer function is defined as the ratio of the $s$-domain output $Y(s)$ to the $s$-domain input $U(s)$ for zero initial conditions. The transfer function is the ratio of two polynomials in $s$, or

$$H(s) = \frac{Y(s)}{U(s)} = \frac{b_n s^n + b_{n-1} s^{n-1} + \cdots + b_1 s + b_0}{s^n + a_{n-1} s^{n-1} + \cdots + a_1 s + a_0} \tag{5.2.1}$$

The coefficients $a_i$ and $b_i$ determine both the frequency- and time-domain character of the filter.

Expression (5.2.1) could be cross-multiplied and inverse Laplace-transformed to yield an equivalent differential equation

$$\frac{d^n y}{dt^n} + a_{n-1} \frac{d^{n-1} y}{dt^{n-1}} + \cdots + a_0 y = b_n \frac{d^n u}{dt^n} + \cdots + b_0 u \tag{5.2.2}$$

which is the time-domain description of the filter.

An alternative way of describing the filter is by the impulse response function $h(t)$, which is defined as the inverse Laplace transform of transfer function $H(s)$. The filter response $y(t)$ to an arbitrary input function $u(t)$ can be given by convolution of the impulse response function into the input function, or

$$y(t) = \int_0^t h(t - \tau) u(\tau) \, d\tau \tag{5.2.3}$$

When considering sinusoidal and periodic signals, the concept of the frequency response function is useful. The frequency response function is defined as the transfer function evaluated along the imaginary axis of the $s$-plane and is denoted as $H(j\omega)$.

These four alternative means of describing the continuous filter will be the means by which discrete-time approximations are obtained.

**Example 5.1.** Consider a second-order bandpass filter with a bandpass frequency of 10 rad/s and a $Q$ of 5 with the continuous-time transfer function

$$H(s) = \frac{Y(s)}{U(s)} = \frac{2s}{s^2 + 2s + 100} \tag{a}$$

If we cross-multiply the transfer function, we get

$$(s^2 + 2s + 10)Y(s) = 2sU(s)$$

Inversion of the indicated transform yields a differential equation description

$$\frac{d^2 y}{dt^2} + 2\frac{dy}{dt} + 100y = 2\frac{du}{dt} \tag{b}$$

An alternative description is that of the impulse response function, which is the inverse Laplace transform of the transfer function of (a). This inversion can be accomplished by adding a constant term to the numerator of (a) and then subtracting a term of the same size upon completing the square in the denominator

$$H(s) = \frac{2(s + 1)}{(s + 1)^2 + 99} - \frac{2}{\sqrt{99}} \frac{\sqrt{99}}{(s + 1)^2 + 99}$$

Using the appropriate table entries in Appendix A, the impulse response function is

$$h(t) = \begin{cases} 2e^{-t} \cos \sqrt{99}t - \dfrac{2}{\sqrt{99}} e^{-t} \sin \sqrt{99}\, t & t \geq 0 \\ 0 & t < 0 \end{cases} \tag{c}$$

The frequency response of this filter is

$$H(j\omega) = H(s)|_{s=j\omega} = \frac{2j\omega}{-\omega^2 + 2j\omega + 100} \tag{d}$$

All of these descriptions are equivalent, and any one of them may be obtained from any other with the aid of a table of Laplace transforms.

If the filter described by relation (5.2.1) has numerator and denominator polynomials of equal order, long division must be carried out to give a constant plus a remainder proper polynomial ratio. The result is of the form

$$H(s) = b_n + \frac{c_{n-1}s^{n-1} + \cdots + c_1 s + c_0}{s^n + a_{n-1}s^{n-1} + \cdots + a_1 s + a_0} = \frac{Y(s)}{U(s)} \tag{5.2.4}$$

If we now define a new response variable $W(s)$ by the transfer function relation

$$\frac{W(s)}{U(s)} = \frac{1}{s^n + a_{n-1}s^{n-1} + \cdots + a_1 s + a_0} \tag{5.2.5}$$

and thus the output relation is

$$Y(s) = b_n U(s) + (c_{n-1}s^{n-1} + \cdots + c_1 s + c_0)W(s) \tag{5.2.6}$$

Relation (5.2.5) could be realized as a cascade of integrators with feedback to create the denominator dynamics as illustrated in the lower half of Fig. 5.1. Relation (5.2.6) can be thought of as a feedforward of the input $U(s)$ and the outputs of the integrators as illustrated in the upper portion of Fig. 5.1. One common way to convert the continuous-time filtering of Fig. 5.1 to a digital filter would be to change the signals in the block diagram to $z$-domain signals and to replace the integrator ($1/s$) blocks by approximate discrete-time integrators. In the following section we explore several approximate integration algorithms.

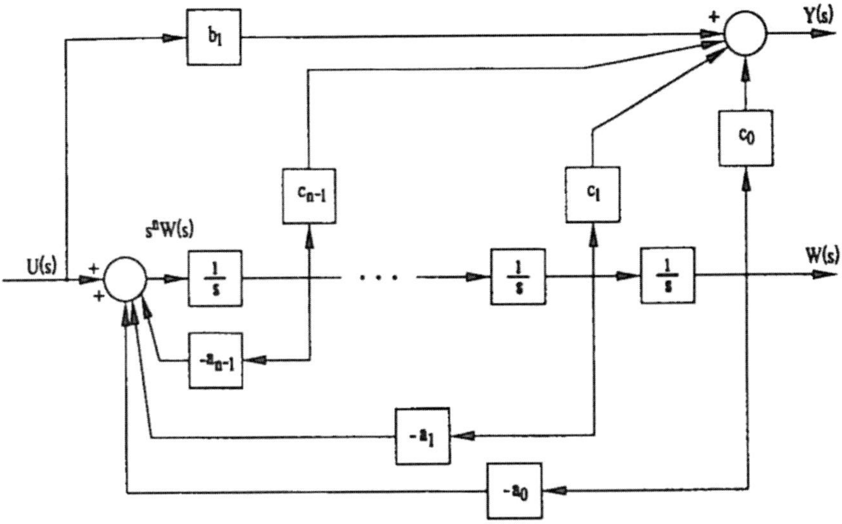


**Figure 5.1.** Phase-variable filter realization.

**Example 5.2.** The filter examined in Example 5.1 had a transfer function

$$H(s) = \frac{2s}{s^2 + 2s + 100} = \frac{Y(s)}{U(s)}$$

From relation (5.2.5) the forward dynamics are described by

$$\frac{W(s)}{U(s)} = \frac{1}{s^2 + 2s + 100}$$

and the output equation from (5.2.6) is $Y(s) = 2sW(s)$. These dynamics are represented in Fig. 5.2.

## 5.3 APPROXIMATE NUMERICAL INTEGRATION TECHNIQUES

In this section we explore several techniques for numerical integration with the purpose in mind of approximation of filter transfer functions. The function we wish to approximate is represented in Fig. 5.3. The input–output relation is

$$y(t) = \int_0^t x(\tau)\, d\tau \qquad (5.3.1)$$

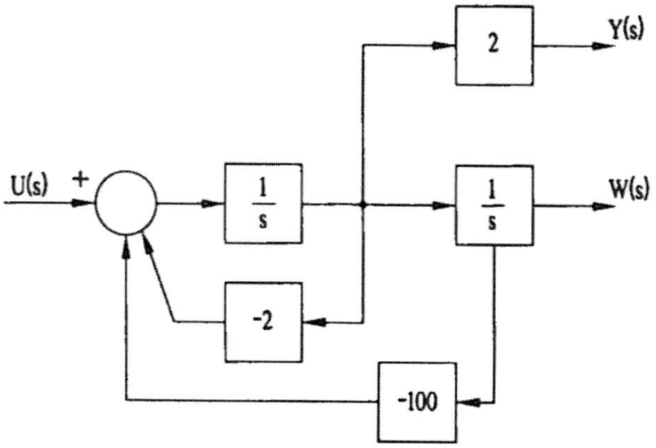


**Figure 5.2.** Bandpass filter realization.

which can be thought of the area under the $x(\tau)$ curve between $\tau = 0$ and $\tau = t$.

### Forward Rectangular Integration

In Fig. 5.4 we note that the integral at time $kT$ will be denoted as $y(kT)$ and that at time $(k - 1)T$ will be denoted as $y((k - 1)T)$.

The area accumulated on the interval $((k - 1)T, kT)$ is then $x((k - 1)T) \cdot T$. Then the area at time $kT$ can be thought of as the area at $(k - 1)T$ plus the rectangular area shown in Fig. 5.4. The integration algorithm is then

$$y(kT) = y((k - 1)T) + x((k - 1)T) \cdot T \tag{5.3.2}$$

This difference equation algorithm for numerical integration can be transformed to give a $z$-domain transfer function

$$H(z) = \frac{Y(z)}{X(z)} = \frac{T}{z - 1} \tag{5.3.3}$$

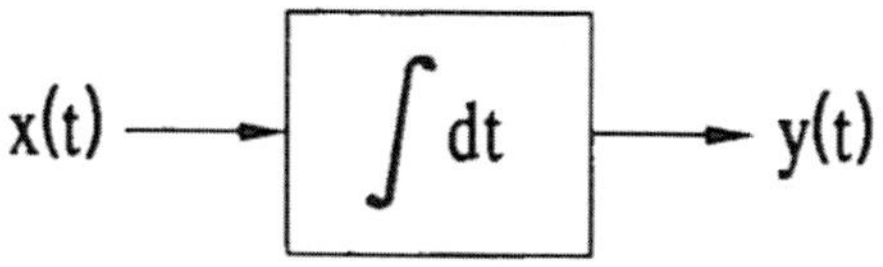


**Figure 5.3.** Continuous time-domain integrator.

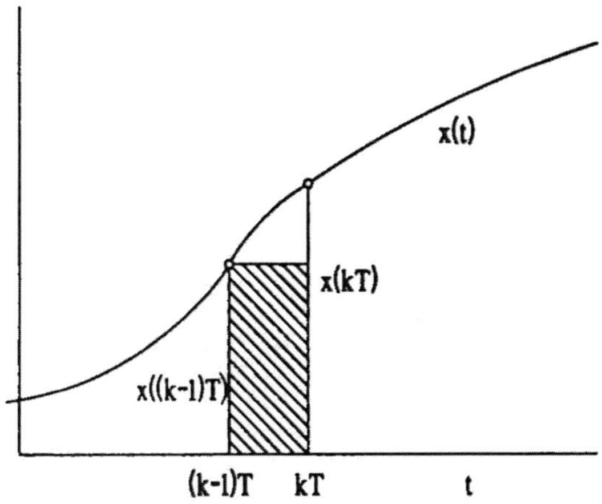


**Figure 5.4.** Forward rectangular integration.

### Backward Rectangular Integration

Consider now an alternative definition for the integral approximation illustrated in Fig. 5.5. With this definition the integration algorithm becomes

$$y(kT) = y((k-1)T) + x(kT) \cdot T \tag{5.3.4}$$

Upon $z$-transformation the transfer function for this discrete-time integra-

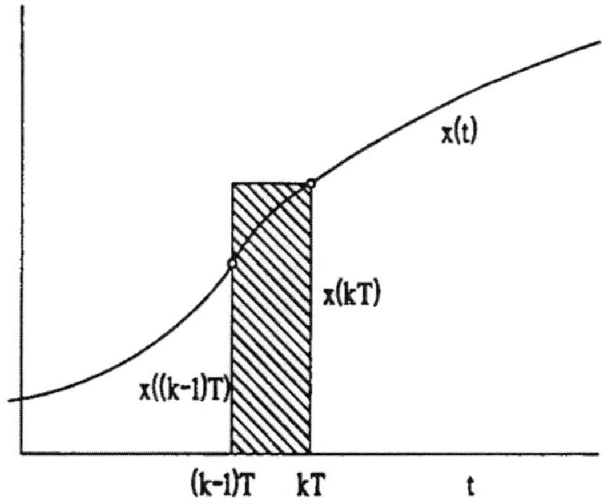


**Figure 5.5.** Backward rectangular integration.

tor is

$$H(z) = \frac{Y(z)}{X(z)} = \frac{Tz}{z - 1} \tag{5.3.5}$$

which differs in the numerator from that given by forward rectangular integration.

### Trapezoidal Integration

Perhaps a better approximation to the integral can be obtained by using both samples of the $x(t)$ function in computation of the additional accumulated area by using the trapezoidal area illustrated in Fig. 5.6.

For this approximation the additional accumulated area is the area of the trapezoid or the algorithm becomes

$$y(kT) = y((k - 1)T) + \frac{T}{2}[x((k - 1)T) + x(kT)] \tag{5.3.6}$$

and the associated $z$-domain transfer function is

$$H(z) = \frac{T}{2}\left(\frac{z + 1}{z - 1}\right) \tag{5.3.7}$$

This is commonly referred to as the bilinear transformation, and the method of filter synthesis using this method is sometimes called Tustin's method

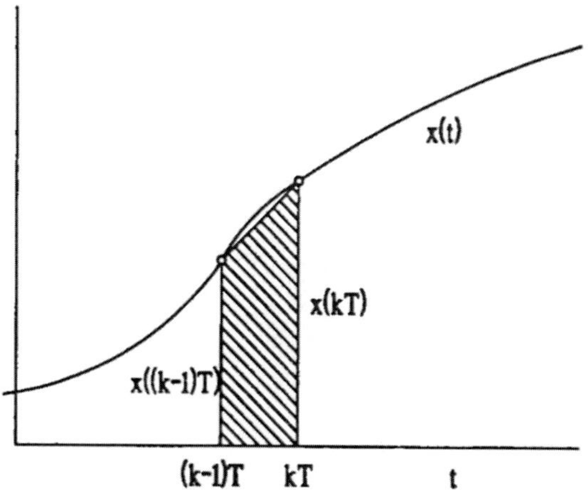


**Figure 5.6.** Trapezoidal integration.

(Tustin, 1947). It is interesting to note that each of these transfer functions has a pole at $z = 1$ which is the exponential mapping of a pole at $s = 0$, which is the case for a continuous-time integrator. These three integrator transfer functions differ in that they each have different zeros.

A technique for digital filter synthesis is now to substitute for the integrators the appropriate $z$-domain transfer function. This is equivalent to replacing the $s$ in the original continuous-time transfer function by the reciprocal of the appropriate $z$-domain transfer function for the particular integration method chosen. These functions are given in Table 5.1. The theory just developed is best illustrated by several examples that follow.

**Example 5.3.** Consider the second-order bandpass filter with the passband centered at 10 rad/s and a quality factor of $Q = 5$, which was illustrated in Example 5.1. The $s$-domain transfer function is

$$H(s) = \frac{Y(s)}{U(s)} = \frac{2s}{s^2 + 2s + 100} \qquad (^*)$$

Synthesize an "equivalent" discrete-time filter using the backward rectangular integration method. Since the critical frequency is at 10 rad/s (1.59 Hz), a reasonable sampling frequency would be 10 Hz or the sampling period $T$ will be 0.1 s. The appropriate substitution from Table 5.1 would be

$$s = \frac{10(z - 1)}{z}$$

Substitution of this relation into relation (*) yields

$$H(z) = \frac{2\left[\dfrac{10(z - 1)}{z}\right]}{\left[\dfrac{10(z - 1)}{z}\right]^2 + 2\left[\dfrac{10(z - 1)}{z}\right] + 100}$$

**Table 5.1.** Substitutions for Various Integration Methods

| Method | |
|---|---|
| Forward rectangular | $s = (z - 1)/T$ |
| Backward rectangular | $s = (z - 1)/Tz$ |
| Trapezoidal (bilinear transformation) | $s = 2(z - 1)/T(z + 1)$ |

Cleaning up the algebra yields the transfer function

$$H(z) = \frac{0.0909z(z - 1)}{z^2 - 1.0z + 0.4545}$$

This filter has zeros at $z = 0$ and $z = 1$, while there is a complex pair of poles at $z = 0.5 \pm j0.4522$.

The difference equation, which is equivalent to the transfer function, is

$$y(k + 2) = y(k + 1) - 0.4545y(h) + 0.0909u(k + 2) - 0.0909u(k + 1)$$

We examine the frequency response of this filter later.

**Example 5.4.** Consider the bandpass filter considered in Example 5.1 with transfer function

$$H(s) = \frac{2s}{s^2 + 2s + 100}$$

Synthesize an "equivalent" digital filter employing the bilinear transformation method and a sampling integral of $T = 0.1$ s.

The appropriate bilinear transformation is

$$s = \frac{20(z - 1)}{z + 1}$$

Substitution of this relation into the original $s$-domain transfer function yields

$$H(z) = \frac{\dfrac{40(z - 1)}{z + 1}}{400\left(\dfrac{z - 1}{z + 1}\right)^2 + (2)(20)\left(\dfrac{z - 1}{z + 1}\right) + 100}$$

Upon cleaning up the algebra yields a $z$-domain transfer function is

$$H(z) = \frac{0.0740(z - 1)(z + 1)}{z^2 - 1.111z + 0.8519}$$

This filter has zeros at $z = 1$ and $z = -1$ and poles at $z = 0.5556 \pm j0.737$. The difference equation algorithm for real-time filtering is

$$y(k + 2) = 1.111y(k + 1) - 0.8519y(k) + 0.0740u(k + 2) - 0.0740u(k)$$

We explore the frequency response of this filter later.

## 5.4 ANOTHER LOOK AT THE BILINEAR TRANSFORMATION

In Section 5.3 we derived the bilinear transformation or Tustin's method from the point of view of numerical integration; however, there is another path that we should discuss. In our discussion of the $z$-transform we noted the exponential relationship between the poles of the $z$-transform of a sequence and the poles of the Laplace transform of the function from which the sequence was taken or

$$z = e^{Ts} \tag{5.4.1}$$

or the inverse relation

$$s = \frac{1}{T} \ln z \tag{5.4.2}$$

The logarithm function can be expanded in a series (see any math handbook) of the form

$$\ln z = 2\left[\frac{z-1}{z+1} + \frac{1}{3}\left(\frac{z-1}{z+1}\right)^3 + \cdots\right] \tag{5.4.3}$$

If we choose as an approximation the first item, the result is

$$s \cong \frac{2}{T}\frac{z-1}{z+1} \tag{5.4.4}$$

If we examine this mapping, we see that this maps the unit circle in the $z$-plane to the left half of the $s$-plane just as does (5.4.2); however, the mapping within and outside the unit circle is not the same as that of (5.4.1) or (5.4.2). The technique is simply to substitute (4.6.4) into a given $H(s)$ to yield $H(z)$, which will require some simplification to get the form of a simple ratio of two polynomials. It should be noted also that when $\omega = \omega_s/2$, the resulting frequency response function is forced to zero. The technique is best illustrated by an example, which follows. If we evaluate the frequency response, we will find that it will have been warped by this approximation, and this is the topic of the next section.

## 5.5 BILINEAR TRANSFORMATION WITH PREWARPING

We made reference in Section 5.4 to the fact that the bilinear transformation warped the frequency response at the critical frequencies of the filter. To overcome this problem, a technique has been developed by which the critical frequencies of the original $s$-domain filter are prewarped such that the critical frequencies of the $z$-domain filter formed by bilinear transformation end up where they belong. The technique is the topic of this section. In the $s$-domain a typical term of the numerator or denominator of the transfer function has the form

$$T(s) = \frac{s}{\omega_0} + 1 \tag{5.5.1}$$

When we are interested in frequency response, we simply let $s$ approach $j\omega$, so the term of interest is

$$T(j\omega) = \frac{j\omega}{\omega_0} + 1 \tag{5.5.2}$$

which at the critical frequency $\omega = \omega_0$ is

$$T(j\omega_0) = j + 1 \tag{5.5.3}$$

If we prewarp $\omega_0$ to some yet unknown $\omega^*$ and employ the bilinear transformation as before by letting $s$ be

$$\frac{2}{T}\frac{z-1}{z+1}$$

the result is

$$T(z) = \frac{2}{\omega^* T}\frac{z-1}{z+1} + 1 \tag{5.5.4}$$

and for frequency response we let $z = e^{j\omega T}$:

$$T(e^{j\omega T}) = \frac{2}{\omega^* T}\frac{e^{j\omega T}-1}{e^{j\omega T}+1} + 1 \tag{5.5.5}$$

If we are interested in the frequency response at the critical frequency $\omega_0$, we let $\omega$ approach $\omega_0$ and equate (5.5.5) to (5.5.3) to yield

$$T(e^{j\omega_0 T}) = \frac{2}{\omega^* T}\frac{e^{j\omega_0 T}-1}{e^{j\omega_0 T}+1} + 1 = j + 1 \tag{5.5.6}$$

Recalling that $e^{j\omega_0 T} = \cos \omega_0 T + j \sin \omega_0 T$ and equating both sides of (5.5.6), we get

$$\frac{2}{\omega^* T}(\cos \omega_0 T + j \sin \omega_0 T - 1) = j(\cos \omega_0 T + j \sin \omega_0 T + 1) \tag{5.5.7}$$

Equating the real parts, the result is

$$\frac{2}{\omega^* T}(\cos \omega_0 T - 1) = -\sin \omega_0 T \tag{5.5.8}$$

and similarly for the imaginary parts,

$$\frac{2}{\omega^* T} \sin \omega_0 T = \cos \omega_0 T + 1 \tag{5.5.9}$$

Substitution of (5.5.8) into (5.5.9) yields

$$\left(\frac{2}{\omega^* T}\right)^2 (1 - \cos \omega_0 T) = (1 + \cos \omega_0 T) \tag{5.5.10}$$

or

$$\left(\frac{\omega^* T}{2}\right)^2 = \frac{1 - \cos \omega_0 T}{1 + \cos \omega_0 T} \tag{5.5.11}$$

and with the appropriate double-angle trigonometric identities applied to the right side, we get

$$\left(\frac{\omega^* T}{2}\right)^2 = \frac{\sin^2(\omega_0 T/2)}{\cos^2(\omega_0 T/2)} \tag{5.5.12}$$

Taking the square root on both sides, we get the prewarped critical frequency to be

$$\omega^* = \frac{2}{T} \tan \frac{\omega_0 T}{2} \tag{5.5.13}$$

With this equation we can "prewarp" the critical $s$-domain frequencies prior to the bilinear transformation to the $z$-domain. With considerable algebraic effort second-order factors with complex roots may be shown to require the same prewarping. In general, an $s$-domain pole at complex location $s_i$ should be relocated radially to a new location $s_{i^*}$ given by

$$|s_{i^*}| = \frac{2}{T} \tan \left(\frac{|s| T}{2}\right) \tag{5.5.14}$$

**Example 5.5.** Consider the bandpass filter discussed in all previous examples of this chapter with a transfer function

$$H(s) = \frac{2s}{s^2 + 2s + 100}$$

It is clear that the radius to the complex poles is 10 rad/s thus for a sampling interval of $T = 0.1$ s. The new pole location is

$$|s_{i^*}| = \frac{2}{0.1} \tan \left[\frac{(10)(0.1)}{2}\right] = 10.93$$

So the prewarped $s$-domain transfer function is

$$H(s) = \frac{2.185s}{s^2 + 2.185s + 119.46}$$

Now let $s$ be replaced by

$$s = 20\,\frac{z-1}{z+1}$$

to yield

$$H(z) = \frac{43.72\left(\dfrac{z-1}{z+1}\right)}{400\left(\dfrac{z-1}{z+1}\right)^2 + 43.72\left(\dfrac{z-1}{z+1}\right) + 119.46}$$

If we proceed carefully with the simplification of the algebra, the result is

$$H(z) = \frac{0.07759(z-1)(z+1)}{z^2 - 0.09963z + 0.8448} = \frac{Y(z)}{U(z)}$$

The poles of this filter are $z_{1,2} = 0.49815 \pm j0.7750$. The difference equation for the filtering algorithm is

$$\begin{aligned} y(k+2) &= 0.9963y(k+1) - 0.8448y(k) \\ &\quad + 0.07759u(k+2) - 0.07759u(k) \end{aligned}$$

## 5.6 MATCHED POLE–ZERO TECHNIQUE

We have seen that the operation of sampling of time domain has the effect of directly mapping the poles of the $s$-domain functions to the $z$-domain according to the relation

$$z = e^{Ts} \tag{5.6.1}$$

where $T$ is the sampling interval. Since there are usually more poles than zeros in $H(s)$, the $H(s)$ is said to have zeros at infinity. Under the action of the mapping (5.6.1), the zeros at infinity are mapped to the point $z = -1$. This technique maps all the poles and zeros directly and then chooses an arbitrary gain constant such that the $z$-domain frequency response exactly matches the $s$-domain filter. This should match the dc response, while in a bandpass filter the midband gain should be matched.

Given a continuous time filter for which we know all the poles and zeros, we could write its transfer function as

$$H(s) = \frac{K \prod_{i=1}^{m} (s + \alpha_i) \prod_{i=1}^{n} [(s + a_i)^2 + b_i^2]}{\prod_{j=1}^{r} (s + \beta_j) \prod_{j=1}^{s} [(s + c_j)^2 + d_j^2]} \qquad (5.6.2)$$

Performing the mapping of (5.6.1), we get a transfer function

$$H(z) = \frac{K'(z + 1)^k \prod_{i=1}^{m} (z - e^{-\alpha_i T}) \prod_{i=1}^{n} (z^2 - 2e^{-a_i T} \cos b_i Tz + e^{-2a_i T})}{\prod_{j=1}^{r} (z - e^{-\beta_j T}) \prod_{j=1}^{s} (z^2 - 2e^{-c_j T} \cos d_j Tz + e^{-2c_j T})} \qquad (5.6.3)$$

where

$$k = 2s + r - 2n - m \qquad (5.6.4)$$

The design technique could be summarized as follows:

1. Map all poles and zeros of $H(s)$ to the $z$-plane by (5.6.1).
2. Add enough zeros at $z = -1$ to make the number of poles equal to the number of zeros.
3. Match the gain of the two filters at some critical frequency.

Let us now apply this technique to the bandpass filter of the previous examples.

**Example 5.6.** Synthesize the bandpass filter of Example 5.1 using the matched pole–zero technique. The original $s$-domain transfer function is

$$H(s) = \frac{2s}{s^2 + 2s + 100}$$

The poles of this filter are at $s_{1,2} = -1 \pm j9.95$. For a sampling interval of $T = 0.1$ s these $s$-domain poles become after the exponential mapping

$$z_{1,2} = 0.4927 \pm j0.7588$$

Thus the discrete-time transfer function becomes

$$H(z) = \frac{K'(z + 1)(z - 1)}{z^2 - 0.9853z + 0.8186}$$

The bandpass frequency is at 10 rad/s and we want the gain at $\omega = 10$ rad/s for this filter to be the same as that of the continuous time filter, which is unity:

$$H(s)|_{s=j10} = H(z)|_{z=e^{j1}} = 1$$
$$1 = K'(11.02)$$

or the factor $K'$ is

$$K' = 0.09074$$

so the final form of the $z$-domain filter is

$$H(z) = \frac{0.09074(z^2 - 1)}{z^2 - 0.9853z + 0.8186} = \frac{Y(z)}{U(z)}$$

The poles of this filter are at $z = 0.4927 \pm j0.7588$ and the zeros are located at $z = 1$ and $z = -1$, respectively. The difference equation algorithm is

$$\begin{aligned} y(k + 2) = {} & 0.9853y(k + 1) - 0.8186y(k) \\ & + 0.09074u(k + 2) - 0.09074u(k) \end{aligned}$$

## 5.7 ZERO-ORDER-HOLD APPROXIMATION

The zero-order-hold approximation is one of the most popular techniques used by designers to get from a continuous design to a discrete design. This technique simply assumes that the continuous filter we want to approximate is preceded by a zero-order hold (ZOH) and followed by a sampler such that the input and output are sequences of numbers, as shown in Fig. 5.7.

In Section 3.3 we discussed a situation that involved the derivation of a $z$-domain transfer function for a continuous-time plant driven by a zero-order hold and followed by a sampler. The result of that derivation indicated that the $z$-domain transfer function from the input of the zero-order hold to the output of the sampler was given by

$$H(z) = \frac{Y(z)}{U(z)} = (1 - z^{-1})\mathfrak{Z}\mathcal{L}^{-1}\left[\frac{H(s)}{s}\right] \tag{5.7.1}$$

This is sometimes referred to as the step-invariant method because the step response samples will be identical.

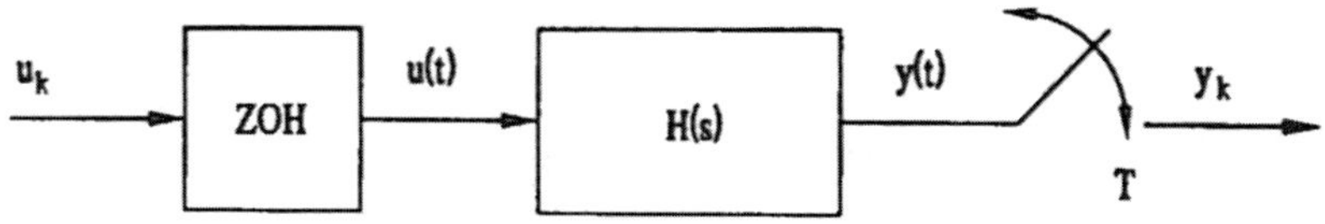


**Figure 5.7.** Zero-order-hold approximation.

Let us now consider the application of (5.7.1) to the design of a digital filter.

**Example 5.7.** Given the bandpass transfer function of Example 5.1, let us generate an approximately equivalent digital filter transfer function by the zero-order-hold method for a sampling interval of $T = 0.1$ s.

$$\frac{H(s)}{s} = \frac{1}{s}\frac{2s}{s^2 + 2s + 100}$$

or

$$\frac{H(s)}{s} = \frac{2}{s^2 + 2s + 100} = \frac{2}{(s + 1)^2 + 9.95^2}$$

the $z$-transform of the samples of the inverse can be established by noting that $b = 9.95$ and thus $bT = 0.995$ and $a = 1$ and thus $aT = 0.1$, so

$$e^{-aT} = 0.9048$$
$$\sin bT = 0.8388$$
$$\cos bT = 0.5445$$

Using entry 12 from Table A.2, the resulting $z$-transform is

$$\mathscr{Z}\mathscr{L}^{-1}\left[\frac{H(s)}{s}\right] = \frac{2(0.9048)(0.8388)z}{9.95(z^2 - 0.9853z + 0.8186)}$$

Employing relation (5.7.1) the $z$-domain transfer function is

$$H(z) = \frac{0.1526(z - 1)}{z^2 - 0.9853z + 0.8186} = \frac{Y(z)}{U(z)}$$

The $z$-domain poles are at $z = 0.4927 \pm j0.7588$ while the zero is at $z = 1$. The filtering algorithm difference equation is

$$y(k + 2) = 0.9853y(k + 1) - 0.8186y(k) + 0.1526u(k + 1) - 0.1526u(k)$$

## 5.8 IMPULSE INVARIANT METHOD

The impulse invariant method revolves around the fact that a continuous-time filter may be specified by its impulse response function $h(t)$. This approximation is accomplished by making the impulse response sequence for the digital filter proportional to samples of the continuous-time filter impulse response function. To preserve the magnitude character of the filter frequency response, the proportionality constant should be the sampling period $T$. In other words, we can simply write that

$$H(z) = T\,\mathcal{Z}\mathcal{L}^{-1}[H(s)] \tag{5.8.1}$$

where the operator $\mathcal{Z}\mathcal{L}^{-1}[\cdot]$ means finding the $z$-transform in the same row of the table as the given $H(s)$. We now apply (5.8.1) to Example 5.7.

**Example 5.8.** Consider the synthesis of the bandpass filter of Example 5.1 by the impulse invariant method. The impulse response was given in Example 5.1 to be

$$h(t) = \begin{cases} 2e^{-t}\cos\sqrt{99}t - \dfrac{2}{\sqrt{99}}e^{-t}\sin\sqrt{99}\,t & t \geq 0 \\ 0 & t < 0 \end{cases}$$

Using transform entries 12 and 13 from Table A.2 for a sampling interval of $T = 0.1$ s yields

$$H(z) = 0.1\left[\frac{2(z^2 - 0.49266z)}{z^2 - 0.9853z + 0.8186} - \frac{2}{\sqrt{99}}\frac{0.7589z}{z^2 - 0.9853z + 0.8186}\right]$$

Combining the two terms gives

$$H(z) = \frac{0.2z(z - 0.5689)}{z^2 - 0.9853z + 0.8186}$$

The poles of this filter are located at $z = 0.4927 \pm j0.7588$, with zeros located at $z = 0$ and $z = 0.5689$. Clearly, the lack of a zero at $z = 1$ allows the frequency response to be nonzero at $\omega = 0$.

To this point we have considered a number of techniques for synthesizing digital filters and have applied most of them to an example second-order bandpass filter. No method will exactly preserve the fidelity of the frequency response of the original continuous-time filter over all frequencies since the digital filter frequency response magnitude must fold about half the sampling frequency (the Nyquist frequency). This can also be shown by considering a factor from either the denominator or numerator of an $s$-domain transfer function or $T_s(s) = (s - a)$ and a term from a discrete-time filter transfer function $T_z(z) = (z - e^{aT})$. To evaluate the frequency response, we let $s = j\omega$ and $z = e^{j\omega T}$, respectively,

$$T_s(j\omega) = j\omega - a$$

and

$$T_z(e^{j\omega T}) = e^{j\omega T} - e^{aT}$$

It is clear that at some arbitrary frequency $\omega$ the term $T_z(e^{j\omega T})$ is not the exponential mapping of $(j\omega - a)$ since $e^{j\omega T} - e^{aT}$ is not the same as $e^{(j\omega - a)T}$.

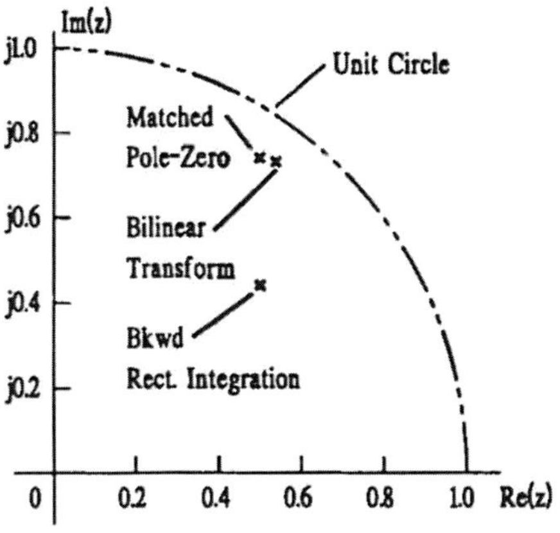


**Figure 5.8.** Poles of several digital bandpass filters.

The poles for the filters of several of the examples are presented in Fig. 5.8. One notes the small discrepancy in poles given by the bilinear transformation from that of the matched pole–zero filter, while that given by the backward rectangular integration method is located at a considerable distance from the others. These mismatched pole locations will result in frequency responses that will not match. These associated frequency response magnitude (gain) functions are presented in Fig. 5.9 along with that of the original continuous-time bandpass filter. The matched pole–zero

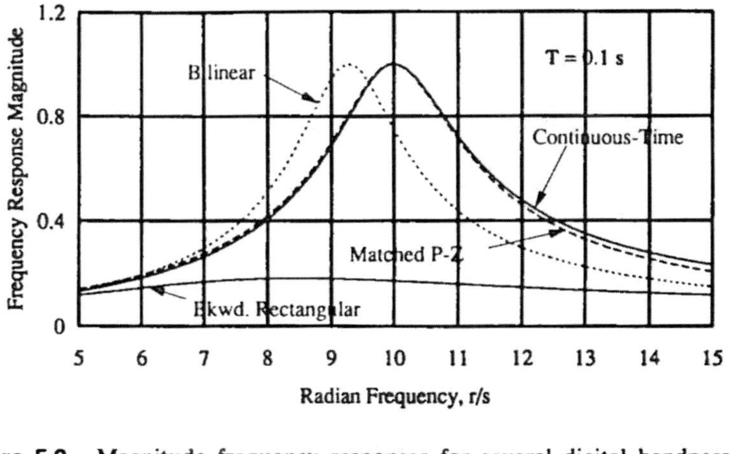


**Figure 5.9.** Magnitude frequency responses for several digital bandpass filter realizations.

filter seems to have good quality, while that of the bilinear transformation has a passband at a frequency lower than specified in the original filter, indicating the need for prewarping.

## 5.9 USING PROTOTYPES TO DESIGN DIGITAL FILTERS

If the filter desired is one of the standard configurations (lowpass, bandpass, or highpass), it is common to begin with a prototype $s$-domain filter transfer function $H(s)$ and apply one of the methods just discussed to generate the "equivalent" digital filter transfer function.

One good prototype filter is the Butterworth filter, which has minimum ripple in the passband and stopband, although it does not make the transition from passband to stopband as crisply as other possible filters. The best way to start the process is to begin with a unity-gain, unity-bandwidth lowpass Butterworth filter of a fixed order. Typical low-order unity bandwidth (1 rad/sec) Butterworth transfer functions are given in Table 5.2. There are standard techniques for choosing the filter order which are presented in the excellent text of Van Valkenberg (1982).

For a lowpass filter with arbitrary cutoff frequency $\omega_0$, we simply employ the lowpass transformation whereby $s$ is replaced by $s/\omega_0$, which essentially frequency scales the lowpass filter:

$$s \rightarrow \frac{s}{\omega_0} \tag{5.9.1}$$

For a bandpass filter with the passband centered at $\omega_0$ and a bandwidth of BW rad/s we employ the bandpass transformation, which is

$$s \rightarrow \frac{1}{\mathrm{BW}}\left(\frac{s^2 + \omega_0^2}{s}\right) \tag{5.9.2}$$

**Table 5.2.** Butterworth Unity-Gain, Unity-Cutoff Frequency Transfer

| Order | $H_B(s)$ |
|---|---|
| 2 | $\dfrac{1}{s^2 + 1.414s + 1}$ |
| 3 | $\dfrac{1}{s^3 + 2s^2 + 2s + 1}$ |
| 4 | $\dfrac{1}{s^4 + 2.6133s^3 + 3.414s^2 + 2.6133s + 1}$ |

If a highpass filter is desired, we use the highpass transformation, which is

$$s \rightarrow \frac{\omega_0}{s} \tag{5.9.3}$$

We shall now illustrate the procedure with an example.

**Example 5.9.** Use the matched pole–zero method to synthesize a fourth-order digital bandpass filter with a passband at 1000 Hz and a bandwidth of 100 Hz. Assume that the sampling frequency is 10,000 Hz, so that $T = 100\ \mu s$.

For a fourth-order bandpass filter we should start with a second-order lowpass filter because the bandpass transformation of (5.9.2) will double the order, or

$$H_B(s) = \frac{1}{s^2 + 1.414s + 1} \tag{a}$$

The correct bandpass transformation with BW = 628.32 rad/s and $\omega_0$ = 6283.2 rad/s is

$$s \rightarrow 1.59 \times 10^{-3}\left(\frac{s^2 + 3.948 \times 10^7}{s}\right) \tag{b}$$

Substitution of (b) into (a) gives

$$H(s) = \frac{1}{2.533 \times 10^{-6}\left(\frac{s^2 + 3.948 \times 10^7}{s}\right)^2 + 2.25 \times 10^{-3}\left(\frac{s^2 + 3.948 \times 10^7}{s}\right) + 1}$$

Simplification of the algebra yields the $s$-domain bandpass prototype

$$H(s) = \frac{3.948 \times 10^5 s^2}{s^4 + 8.883 \times 10^2 s^3 + 8.025 \times 10^7 s^2 + 3.507 \times 10^{10} s + 1.558 \times 10^{15}}$$

If we factor the denominator using a computer program for finding the roots and factors of a polynomial, the result is

$$H(s) = \frac{3.948 \times 10^5 s^2}{(s^2 + 4.8 \times 10^2 s + 4.667 \times 10^7)(s^2 + 4.076 \times 10^2 s + 3.337 \times 10^7)}$$

Upon completion of the square in each of the quadratic factors of the denominator

$$H(s) = \frac{3.948 \times 10^5 s^2}{[(s + 240)^2 + 4.66 \times 10^7][(s + 203.8)^2 + 3.333 \times 10^7]}$$

Now apply relation (5.6.3) to yield

$$H(z) = \frac{K'(z - 1)^2(z + 1)^2}{(z^2 - 1.515z + 0.953)(z^2 - 1.6372z + 0.9549)}$$

Multiplication of the two denominator factors yields

$$H(z) = \frac{K'(z - 1)^2(z + 1)^2}{z^4 - 3.152z^3 + 4.388z^2 - 3.007z + 0.910}$$

Now evaluate this frequency response at the bandpass frequency $\omega$ = 6283.2 rad/s and find that the magnitude is 305.22.

If we want the gain at the bandpass frequency to be that of the continuous-time prototype the value of $K'$ must be 0.001036, so the transfer function is

$$H(z) = \frac{0.001036(z - 1)^2(z + 1)^2}{z^4 - 3.152z^3 + 4.388z^2 - 3.007z + 0.910}$$

The frequency response of this filter is illustrated in Fig. 5.10, and it is clear that the digital filter matches the analog filter to a very high degree.

## 5.10 *z*-PLANE DESIGN OF DIGITAL COMPENSATORS

Thus far in this chapter we have discussed how we may approximate an *s*-domain compensator design as a discrete-time control algorithm or com-

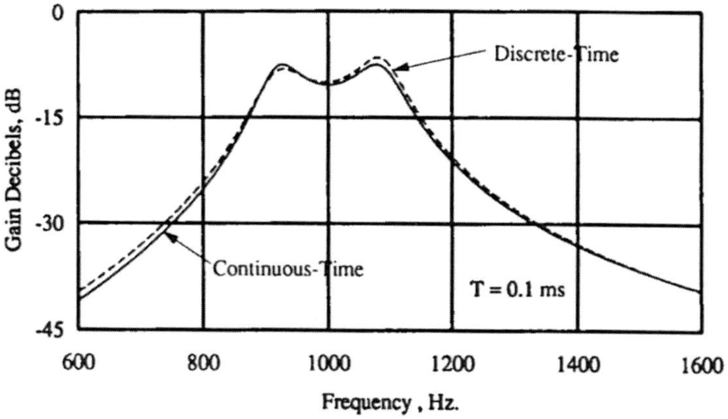


**Figure 5.10.** Frequency response for a fourth-order digital bandpass filter.

pensator. In this section we discuss briefly how the closed-loop design, either in the $s$-domain or directly in the $z$-domain, may be accomplished.

The simplest possible strategy to generate control efforts is to make them proportional to the error in the system output. The disadvantage to this is that in order to reduce steady-state errors to acceptable levels, the proportional controller gain must be so high that unacceptable transient response results. We saw in Chapter 4 that if an integrator (either continuous-time or discrete-time) is inserted into the control loop such that the control effort is proportional to the integral of output error, the steady-state error could be reduced to zero. The insertion of such an integrator often leads to stability problems, which can be somewhat alleviated by using a proportional plus integral (PI) control strategy. Often, the proportionally controlled plant with acceptable dynamics will not have sufficient speed; hence we must seek a more sophisticated control algorithm. If the plant has a slow pole (one near the $s$-plane origin or near $z = 1$), one technique is to design a compensator with a zero that will cancel the slow pole. To make the resulting control scheme realizable, it is necessary to insert a pole, which can, in theory, be as fast as need be. In practice, however, this pole location will be limited by the control effort, which tends to saturate the output hardware in either the digital or analog implementation of the compensator.

If the $z$-domain transfer function of the plant has a pole, say $p_1$, which we wish to cancel, the compensator we would first employ would be of the form

$$D(z) = \frac{K_c(z - p_1)}{z - p_c} \tag{5.10.1}$$

where the numerator cancels the pole of the plant and $p_c$ is selected as far to the left of $z = 1$ as is physically possible.

The problem that commonly occurs with this type of design is that if $K_c$ is selected to give acceptable dynamics, large steady-state error to step reference inputs or disturbances will result. In this case, the cure for the steady-state error is to insert an integrator pole (at $z = 1$) in the compensator, then select the gain to give acceptable loop dynamics. These ideas are best illustrated with detailed examples, which follow.

**Example 5.10.** As an example of designing in the $s$-domain and converting this to a discrete design, consider the thermal control system considered first in Example 3.3, where the plant $s$-domain transfer function was derived to be

$$G(s) = \frac{1}{(s + 0.1957)(s + 2.554)}$$

In this example let us examine the compensation in the $s$-domain, and once a design is fixed, convert this design to a discrete-time compensator employing the method of bilinear transformation.

To improve the dynamics, design a compensator to place a zero at $s = -0.1957$ to cancel the slow plant pole. To make the design realizable, we need to have a pole as far left as possible in the $s$-plane. We shall arbitrarily place the pole at $s = -4$, so the continuous-time compensator transfer function is

$$D(s) = \frac{K(s + 0.1957)}{s + 4}$$

The closed-loop characteristic equation is given by

$$1 + G(s)D(s) = 1 + \frac{K(s + 0.1957)}{(s + 4)(s + 0.1957)(s + 2.544)} = 0$$

The closed-loop characteristic polynomial is then

$$s^2 + 6.544s + 10.216 + K = 0$$

so the undamped natural frequency is

$$\omega_n = \sqrt{10.216 + K}$$

and from the middle term of the characteristic equation,

$$2\zeta\omega_n = 6.544$$

or the damping ratio is

$$\zeta = \frac{6.544}{2\sqrt{10.216 + K}}$$

which can be solved for $K$.

Let us choose $\zeta = 0.707$ in order to limit system oscillations; this will fix $K$ to be 11.259 and yield a characteristic equation of

$$s^2 + 6.544s + 21.475 = 0$$

In this equation the undamped natural frequency is 4.63 rad/s and the damped natural frequency is 3.274 rad/s. For a design with simple proportional control,

$$D(s) = K_p$$

and the characteristic equation is

$$s^2 + 2.7495s + K_p + 0.5 = 0$$

To fix $\zeta = 0.707$, then $K_p = 3.28$, $\omega_n = 1.94$ rad/s, and $\omega_d = 1.37$ rad/s. We see that the compensated system is faster by more than a factor of 2 by virtue of its undamped natural frequency.

For the compensated system the compensator transfer function is

$$D(s) = \frac{11.259(s + 0.1957)}{s + 4}$$

For a sampling interval of $T = 0.25$ s we can convert to a discrete design employing the bilinear transformation

$$s = \frac{2}{T}\frac{z - 1}{z + 1} = 8\frac{z - 1}{z + 1}$$

which yields a discrete-data compensator of

$$D(z) = \frac{11.259[8(z - 1)/(z + 1) + 0.1957]}{8(z - 1)/(z + 1) + 4}$$

Simplifying all the algebra yields a discrete-time compensator with transfer function

$$D(z) = 7.69\left(\frac{z - 0.952}{z - 0.333}\right)$$

Evidently, the sampling interval is sufficiently small to give good cancellation of the discrete-data slow pole in the zero-order-hold plant pulse transfer function

$$G(z) = \frac{0.025(z + 0.816)}{(z - 0.952)(z - 0.528)}$$

The closed-loop discrete-time characteristic equation is now

$$(z - 0.333)(z - 0.528) + 7.69(0.025)(z + 0.816) = 0$$

or

$$z^2 - 0.699z + 0.332 = 0$$

The roots of this equation are at

$$z_{1,2} = 0.3345 \pm j0.469$$

which are shown in Fig. 5.11. It is evident by examination of the damping ratio $\zeta$ associated with these roots that it much too low for acceptable performance. Let us now consider redesign by making the gain of the discrete compensator variable, or

$$D(z) = \frac{K_1(z - 0.952)}{z - 0.333}$$

For this compensator the characteristic equation is

$$z^2 = z(0.861 - 0.025K_1) + (0.176 + 0.0204K_1) = 0$$

Trying several values of $K_1$ gives the roots shown in Fig. 5.11, and this shows that $K_1 = 4$ gives a damping ratio which is the same as originally specified. The associated damped natural frequency is $(0.111)\omega_s$ or $(0.111)2\pi/(0.25) = 2.79$ rad/s, which is still twice as fast as proportional control. The final discrete-time transfer function can be inverted to give a control algorithm of

$$u(k) = 0.333u(k - 1) + 4e(k) - 3.808e(k - 1)$$

which can be implemented on the digital processor to give the dynamics specified.

If this compensator is implemented, we see that in fact there is an attendant improvement in the speed of the step response as indicated in Fig. 5.12; however, we also note that there is a significant loss in the ability of the system to minimize the error to a step input. It is also interesting to note the character of the control effort at the output of the digital-to-analog converter in each of the cases above. It is clear that to control the steady-state error, an integrator is needed in the forward loop to drive the error to zero, but addition of an integrator inherently slows down the loop dynamics. We explore this situation in the next example.

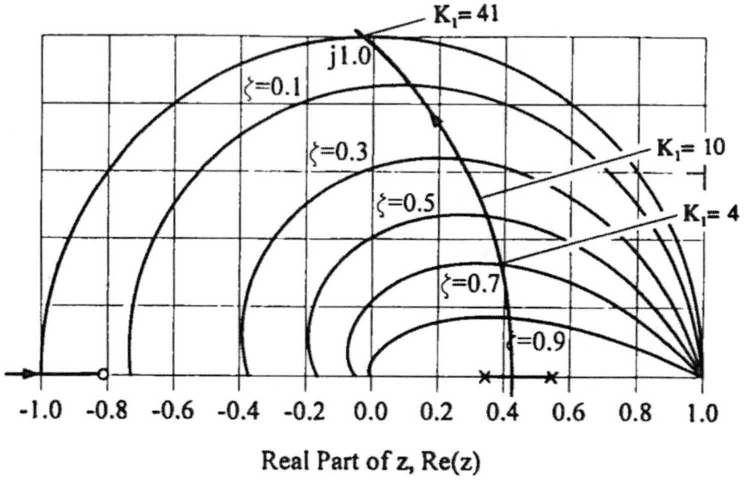


**Figure 5.11.** Root locus for compensated thermal system.

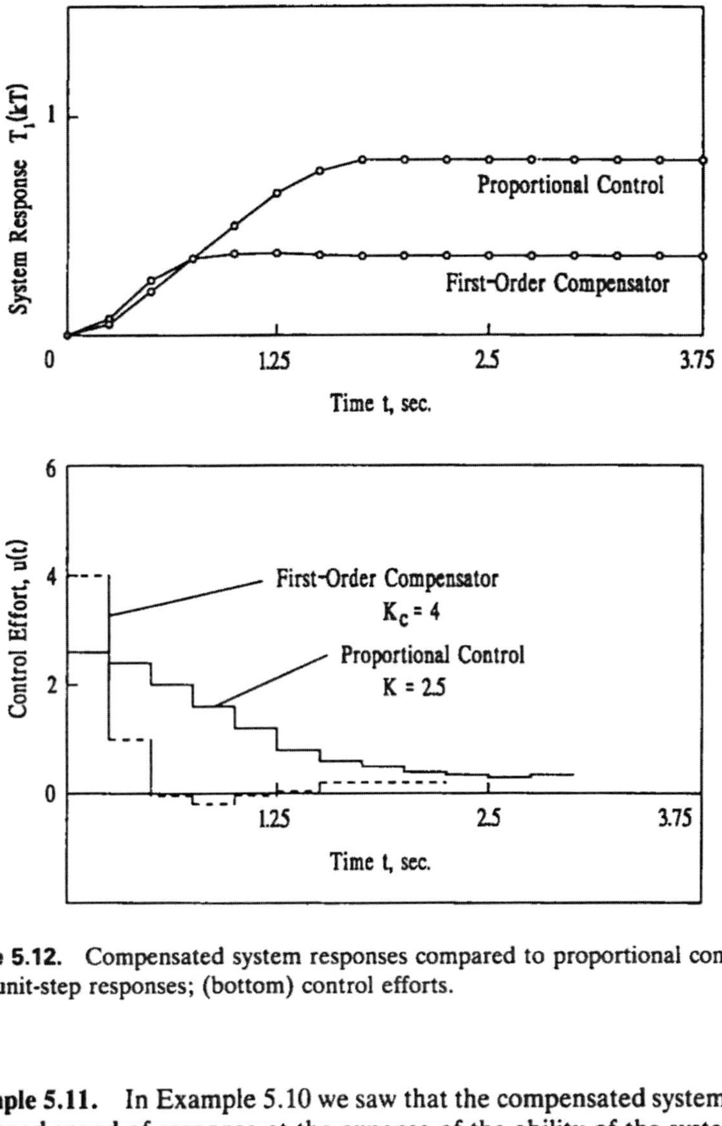


**Figure 5.12.** Compensated system responses compared to proportional control: (top) unit-step responses; (bottom) control efforts.

**Example 5.11.** In Example 5.10 we saw that the compensated system had improved speed of response at the expense of the ability of the system to control excessive steady-state errors. We know that incorporation of an integrator into the forward loop would handle the steady-state error problem successfully, so let us now attempt to cancel the other plant pole and insert compensator poles at $z = 1$ and $z = 0.2$. Hopefully, the latter pole will pull the root locus to the left sufficiently to increase the system speed.

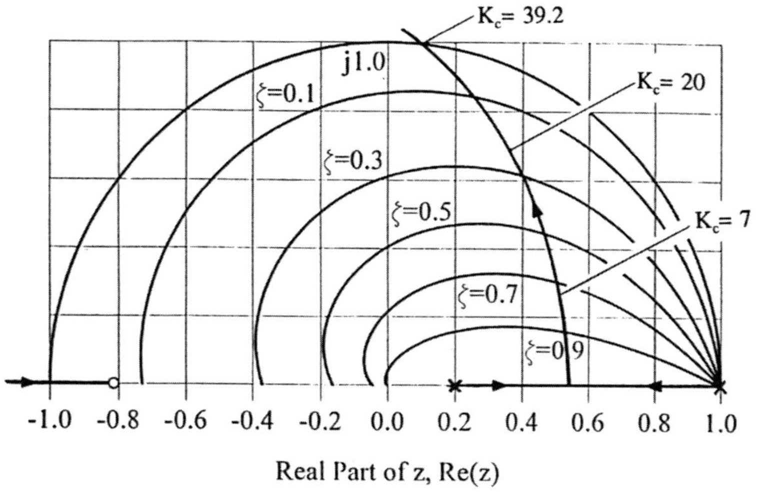


**Figure 5.13.** Root locus for compensated thermal system with integral control.

This is all done with a compensator of the form

$$D(z) = \frac{K_c(z - 0.528)(z - 0.952)}{(z - 1)(z - 0.2)}$$

The root locus for variation in the parameter $K_c$ is shown in Fig. 5.13. To give a damping ratio of about 0.7, the compensator gain $K_c$ should be chosen to be about 7.0. For this gain the difference equation for the control algorithm is

$$\begin{aligned} u(k) &= 1.2u(k - 1) - 0.2u(k - 2) \\ &\quad + 7[e(k) - 1.48e(k - 1) + 0.5026e(k - 2)] \end{aligned}$$

Step responses for this compensator and for the proportional controller of Example 3.5 are shown in Fig. 5.14. In each case the gain has been selected to give a damping ratio of $\zeta = 0.7$. In regard to quick rise time and zero steady-state error, this final design is in both ways superior to either of the other designs.

Let us examine the results of Examples 5.10 and 5.11 from the frequency response point of view by investigating the open-loop frequency response (Bode plot) of each system. These frequency response functions given by $G(z)D(z)$ evaluated at $z = e^{j\omega T}$ for various values of $\omega$ yield a Bode plot for each system. The open-loop magnitude and phase character of each system have been evaluated by digital computer and are given, respectively, in Figs. 5.15 and 5.16. If we evaluate the gain and phase

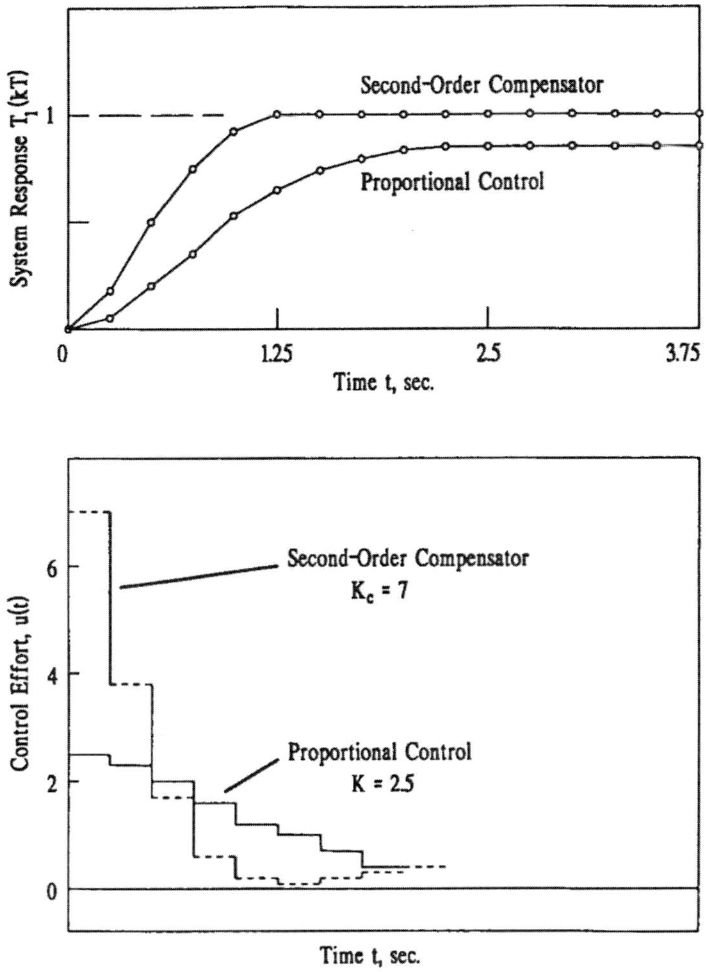


**Figure 5.14.** Responses of thermal system with second-order compensator: (top) unit-step response; (bottom) control efforts.

margin for each system, we get the data of Table 5.3. Also, we found that the system, which had the digital integrator (pole at $z = 1$) had quick response time in addition to zero steady-state error to a step reference input. We see that the first two systems had excessive gain and phase margin for a high-performance control system. If we examine the Bode plot further, we can see the reason for the poor performance of the first

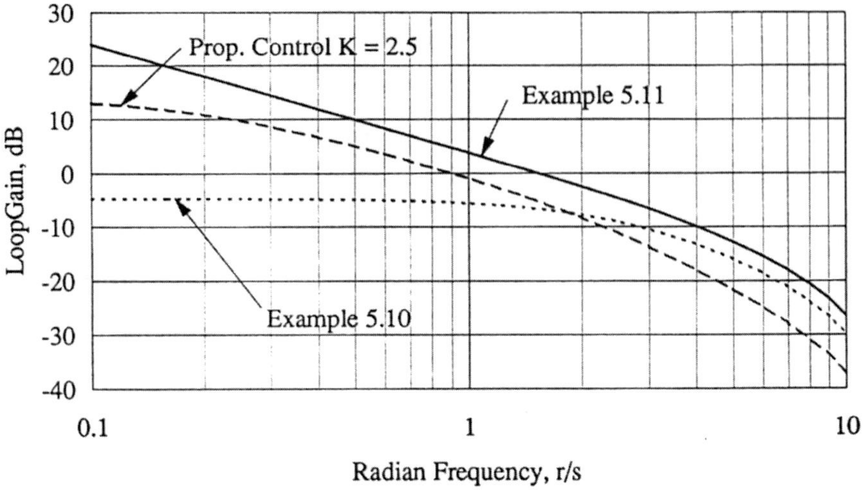


**Figure 5.15.** Magnitude character of the open-loop frequency response.

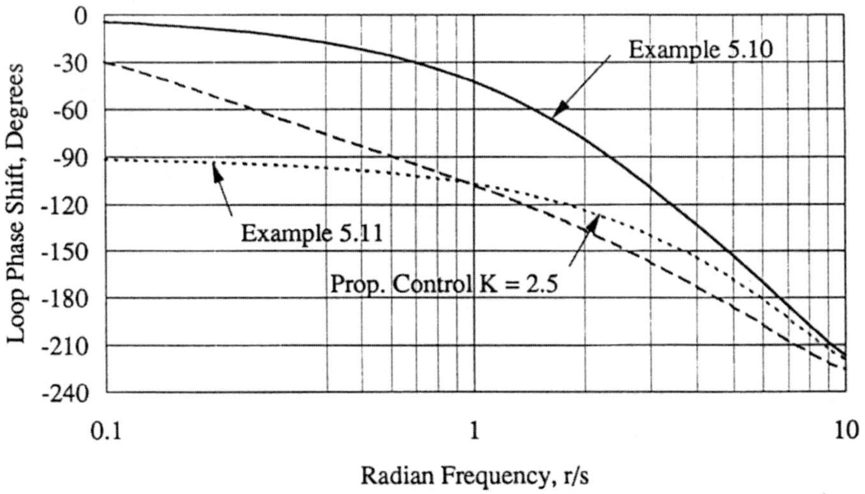


**Figure 5.16.** Phase character of the open-loop frequency response.

**Table 5.3.** Gain and Phase Margin for Each of Three Digital Control Schemes

| Compensator $D(z)$ | Gain margin (dB) | Phase margin (deg) |
|---|---|---|
| Proportional $K = 2.5$ | 9.78 | 76.16 |
| $\dfrac{4(z - 0.952)}{z - 0.333}$ | 21.41 | $\infty$ |
| $\dfrac{7(z - 0.528)(z - 0.952)}{(z - 1)(z - 0.2)}$ | 14.97 | 63.87 |

compensator and what is desirable in the second. The term that designers use is "stiffness," and this is reflected by high loop gain at the very low frequencies as that contributed by the digital integrator pole ($z = 1$). This "stiffness" property is commonly very desirable from a disturbance rejection point of view. The reason the first compensator allowed the high steady-state error was due to the extremely small low-frequency loop gain.

## 5.11 SUMMARY

In this chapter we have explored several methods of digital filter synthesis. These methods are all based on the assumption that the filter or compensator has been originally specified as a continuous-time filter or compensator either as an $s$-domain transfer function or the equivalent impulse response function or differential equation. The various methods are techniques for approximation of these various descriptions in the form of a discrete-time filtering algorithm. Finally, some examples of the problems associated with design of digital compensators are considered, as well as what is necessary to give an acceptable design.

## PROBLEMS

5.1. Given a second-order bandpass filter with the transfer function given below, find the approximate discrete-data filter transfer function $H(z)$ by the method of backward rectangular integration for $\omega_0 T = 0.1$ and $Q = 5$.

$$H(s) = \frac{(\omega_0/Q)s}{s^2 + (\omega_0/Q)s + \omega_0^2}$$

5.2. For the filter of Problem 5.1, construct the approximate digital transfer function $H(z)$ by the matched pole–zero technique. Choose $\omega_0 T = 0.2$.

5.3. In continuous-time control systems compensators are often employed to improve system performance. A commonly used compensator is the lead–lag or lag–lead filter with a transfer function

$$D(s) = \frac{s + \alpha}{s + \beta}$$

Whether the filter is lead–lag or lag–lead depends on the ratio of $\alpha$ to $\beta$. Find the approximately equivalent discrete-time filter transfer function $H(z)$ for a general sampling interval $T$ by the zero-order-hold method.

5.4. Consider the lowpass filter with transfer function

$$H(s) = \frac{\omega_0^2}{s^2 + (\omega_0/Q)s + \omega_0^2}$$

(a) Find the impulse invariant approximate discrete-time filter and the associated difference equation.

(b) Find the stability condition for the filter.

5.5. Find the approximately equivalent filter to that of Problem 5.4 by the method of bilinear transformation.

5.6. A typical lag compensator in a continuous-time system has a transfer function of the form

$$D(s) = \frac{U(s)}{E(s)} = \frac{s + 1}{10s + 1}$$

(a) Considering the critical frequencies from a Bode point of view, choose a sampling period for a discrete approximation of this filter.

(b) Using the bilinear transformation method and the value of $T$ from part (a), find the $z$-domain transfer function for an "equivalent" digital compensator.

(c) Write out the difference equation that would be implemented on the digital computer as a control algorithm.

5.7. We wish to find a discrete-time approximation to the continuous-time bandpass filter with transfer function

$$H(s) = \frac{2\zeta\omega_0 s}{s^2 + 2\zeta\omega_0 s + \omega_0^2}$$

where $\omega_0$ is the resonant or peak frequency and the damping ratio is related to $Q$ by

$$Q = \frac{1}{2\zeta}$$

If we employ the bilinear transformation for arbitrary sampling interval $T$, after some simplification we get

$$H(z) = \frac{4\zeta\omega_0T(z^2 - 1)}{z^2[4 + 4\zeta\omega_0T + (\omega_0T)^2] + z[-8 + 2(\omega_0T)^2] + [4 - 4\zeta\omega_0T + (\omega_0T)^2]}$$

If we are interested in evaluation of the frequency response, we simply let $z = e^{j\omega T}$. Since we are interested in rather general results, define dimensionless frequency as $\Omega = \omega/\omega_0$ and the ratio of resonant frequency to sampling frequency $\beta = \omega_0/\omega_s$, so $\omega_0T = 2\pi\beta$. For these definitions

$$H(z) = \frac{8\pi\zeta\beta(z^2 - 1)}{z^2(4 + 8\pi\zeta\beta + 4\pi^2\beta^2) + z(8\pi^2\beta^2 - 8) + 4(1 - 2\pi\zeta\beta + \pi^2\beta^2)}$$

Let us choose $\zeta = 0.05$ ($Q = 10$). Note also in dimensionless terms that $z$ on the unit circle becomes

$$z = e^{j2\pi\Omega\beta} \qquad \text{where } \Omega = \omega/\omega_0$$

Find the following for $\omega_0/\omega_s = \beta = \frac{1}{4}, \frac{1}{6}, \frac{1}{10}$:

(a) Resonant peak location $\Omega_p = \omega_p/\omega_0$.

(b) Frequency response height at the resonant peak.

(c) Plot the frequency response for the three values of $\beta$ given for $\Omega$ between zero and 5.

(d) Find the effective value of $Q$ from

$$Q = \frac{\omega_p}{\omega_2 - \omega_1}$$

where $\omega_1$ and $\omega_2$ are the respective 3-dB points of the actual frequency response function.

5.8. A continuous-time PI controller has a transfer function

$$D(s) = \frac{K_ps + K_i}{s} = \frac{U(s)}{E(s)}$$

Using the bilinear transformation, find an equivalent discrete-time controller transfer function and the associated difference equation algorithm.

5.9. A notch filter has a continuous-time transfer function

$$H(s) = \frac{s^2 + 1^2}{s^2 + 0.2s + 1^2}$$

(a) Choose a sampling interval ($T$) to give a good discrete-time approximation to the filter.
(b) Design an "equivalent" digital filter by the method of bilinear transformation. (Do *not* prewarp your design.)

5.10. The problem is to design a fourth-order Butterworth bandpass filter with a passband centered at 100 rad/s and bandwidth of 20 rad/s. To accomplish this design it is appropriate to start with a Butterworth low-pass prototype

$$H_{Lp}(s) = \frac{1}{s^2 + \sqrt{2}\,s + 1}$$

Then to build a bandpass filter it is appropriate to apply the bandpass transformation scaled to unity cutoff radian frequency:

$$s \rightarrow \frac{s^2 + \omega_0^2}{\mathrm{BW}s}$$

where $\omega_0 = 100$ rad/s and BW = 20 rad/s. The resulting $s$-domain filter is

$$H(s) = \frac{400s^2}{s^4 + 28.28s^3 + 20{,}400s^2 + 2.828 \times 10^5 s + 10^2}$$

Since the bandpass frequency $f_0 = \omega_0/2\pi = 15.9$ Hz, it is probably appropriate to sample at 100 Hz or $T = 0.01$ s.
(a) Synthesize the "equivalent" digital filter using the matched pole–zero technique.
(b) Synthesize the "equivalent" digital filter using the bilinear transform method—you need not prewarp the design.
(c) Evaluate the poles and zeros of each filter and compare.
(d) Evaluate each filter frequency response in the neighborhood of 100 rad/s for each filter and compare.
(e) Evaluate the frequency response of the continuous-time filter and compare with the results of part (d) assuming that the original Butterworth bandpass filter is the standard.
(f) Make any comments indicating a summary of this process.

## REFERENCES

Blandford, D. K., 1988. *The Digital Filter Analyzer*, Addison-Wesley, Reading, MA.

Franklin, G. F., J. D. Powell, and M. L. Workman, 1990. *Digital Control of Dynamic Systems*, 2nd Ed., Addison-Wesley, Reading, MA.

Katz, P., 1981. *Digital Control Using Microprocessors*, Prentice Hall International, Englewood Cliffs, NJ.

Ludeman, L. C., 1986. *Fundamentals of Digital Signal Processing*, Harper & Row, New York.

Phillips, C. L., and H. T. Nagle, 1990. *Digital Control System Analysis and Design*, 2nd Ed., Prentice Hall, Englewood Cliffs, NJ.

Strum, R. D., and D. E. Kirk, 1988. *Discrete Systems and Digital Signal Processing*, Addison-Wesley, Reading, MA.

Tustin, A., 1947. A method of analyzing the behavior of linear systems in terms of time series, *Journal of the Institution of Electrical Engineers*, *94*, pt. IIA: 130–142.

Van Valkenberg, M. E., 1982. *Analog Filter Design*, Holt, Rinehart and Winston, Fort Worth, TX.

# 6
# State-Variable Representation in Digital Control Systems

## 6.1 INTRODUCTION

In Chapters 3, 4 and 5 we investigated the conventional design of digital control systems. Two design techniques discussed were (1) treating the problem as a continuous-time control system problem, design controllers or compensators in the $s$-domain, and then convert these controllers to digital filters which are implemented in the digital processor; and (2) design the controllers and compensators directly in the $z$-domain to yield the exact discrete-time control laws or algorithms. There are problems with each of these design philosophies; the first suffers because the final digital compensator form is arrived at by a trial-and-error process, while the latter often suffers from the inexperience of the designer working in the $z$-domain. The remaining chapters of this book describe an alternative approach to those commonly referred to as conventional or "classical" design procedures. These alternative techniques comprise the large body of knowledge commonly referred to as *modern control theory*.

In the late 1950s Kalman introduced the foundations of a body of knowledge now commonly known as modern control theory, which employs matrix-vector differential equations and uses the calculus of variations to generate optimal control policies for these systems (Kalman, 1963). One important result is that for a linear system with a quadratic performance index the optimal control policy is a simple linear combination of the matrix-vector differential equation variables or states. In the early 1960s, Kalman (1960) introduced the concept of optimal state estimation when the system has random disturbances and the measurements of system outputs are contaminated by additive noise. This estimation technique was shown to be a generalization of the least-squares optimal filtering technique developed by Wiener in the late 1930s (Wiener, 1949). The modern stochastic control picture was completed when the linear quadratic stochastic optimal control problem was solved by Joseph and Tou (1961) and Gunckel and Franklin (1963).

These modern control ideas were generally developed for continuous-time control systems, but most of the ideas are applicable to discrete-time control systems. We shall, in this chapter, investigate the state-variable representation of both continuous-time and discrete-time systems.

## 6.2 CONTINUOUS-TIME STATE-VARIABLE PROBLEM

Systems describable by an $s$-domain transfer function of order $n$ are equivalently represented by a single $n$th-order differential equation or equivalently by a set of $n$ first-order differential equations. There is no unique way to represent a given transfer function in this manner; however, there are a number of well-known methods to establish this representation from a known transfer function (Brogan). The general form of the system dynamics is represented by the matrix-vector differential equation

$$\dot{\mathbf{x}}(t) = \mathbf{F}\mathbf{x}(t) + \mathbf{G}\mathbf{u}(t) \tag{6.2.1}$$

where $\mathbf{x}(t)$ is an $n$ vector of state variables, $\mathbf{u}(t)$ is an $m$ vector of control efforts, and $\mathbf{F}$ and $\mathbf{G}$ are constant $n \times n$ and $n \times m$ parameter matrices, respectively. The system outputs are given by a linear combination of the states, or what is often referred to as the measurement equation

$$\mathbf{y}(t) = \mathbf{C}\mathbf{x}(t) \tag{6.2.2}$$

where $\mathbf{y}(t)$ is an $r$ vector of outputs and $\mathbf{C}$ is a constant $r \times n$ matrix. Two examples of this will serve to illustrate the method.

**Example 6.1.** Consider the inertial plant which is described in Example 3.2 by the transfer function

$$G(s) = \frac{Y(s)}{U(s)} = \frac{1}{s^2}$$

The equivalent differential equation is

$$\ddot{y} = u(t)$$

Now define the two required state variables as

$$x_1 = y$$

and

$$x_2 = \dot{y} = \dot{x}_1$$

so the differential equations governing the system are

$$\dot{x}_1 = x_2$$
$$\dot{x}_2 = u(t)$$

or in matrix form,

$$\begin{bmatrix} \dot{x}_1 \\ \dot{x}_2 \end{bmatrix} = \begin{bmatrix} 0 & 1 \\ 0 & 0 \end{bmatrix} \begin{bmatrix} x_1 \\ x_2 \end{bmatrix} + \begin{bmatrix} 0 \\ 1 \end{bmatrix} u(t)$$

and the measurement equation is

$$y(t) = [1 \quad 0] \begin{bmatrix} x_1 \\ x_2 \end{bmatrix}$$

**Example 6.2.** Consider now the problem of the thermal system of Example 3.3. Let us define the temperatures as the state variables, so the state equations in matrix form are

$$\frac{d}{dt} \begin{bmatrix} x_1 \\ x_2 \end{bmatrix} = \begin{bmatrix} -2 & 2 \\ 0.5 & -0.75 \end{bmatrix} \begin{bmatrix} x_1 \\ x_2 \end{bmatrix} + \begin{bmatrix} 0 & 0 \\ 0.25 & 0.5 \end{bmatrix} \begin{bmatrix} T_0(t) \\ u(t) \end{bmatrix}$$

and if temperature $x_1(t)$ is the measured quantity on which to base control, the output equation is

$$y(t) = [1 \quad 0] \begin{bmatrix} x_1 \\ x_2 \end{bmatrix}$$

Often the process of selection of system state variables in physical problems is not so straightforward as in the preceding two examples. The technique for selection of state variables is explained in detail in the excellent text of Brogan (1985).

## 6.3 SOLUTION OF THE STATE EQUATION

Let us consider first the homogeneous form of (6.2.1) where $\mathbf{u}(t) = 0$, or

$$\dot{\mathbf{x}} = \mathbf{F}\mathbf{x} \tag{6.3.1}$$

Take the Laplace transform of this equation to yield

$$s\mathbf{x}(s) - \mathbf{x}(0) = \mathbf{F}\mathbf{x}(s) \tag{6.3.2}$$

Rearranging gives

$$[s\mathbf{I} - \mathbf{F}]\mathbf{x}(s) = \mathbf{x}(0) \tag{6.3.3}$$

and solving for $\mathbf{x}(s)$, we get

$$\mathbf{x}(s) = [s\mathbf{I} - \mathbf{F}]^{-1}\mathbf{x}(0) \tag{6.3.4}$$

Let us now invert the Laplace transform to yield

$$\mathbf{x}(t) = \mathcal{L}^{-1}\{[s\mathbf{I} - \mathbf{F}]^{-1}\}\mathbf{x}(0) \tag{6.3.5}$$

Since the left side of (6.3.5) is a function of the temporal variable $t$, so must be the matrix on the right side, which is commonly called the *state transition matrix* and given the symbol $\mathbf{\Phi}(t)$, so

$$\mathbf{\Phi}(t) = \mathcal{L}^{-1}\{[s\mathbf{I} - \mathbf{F}]^{-1}\} \tag{6.3.6}$$

Then (6.3.5) may be written as

$$\mathbf{x}(t) = \mathbf{\Phi}(t)\mathbf{x}(0) \tag{6.3.7}$$

We have not yet discussed how one gets the state transition matrix, but now let us consider the forced system

$$\dot{\mathbf{x}} = \mathbf{F}\mathbf{x} + \mathbf{G}\mathbf{u} \tag{6.3.8}$$

Let us, as before, take the Laplace transform to yield

$$s\mathbf{x}(s) - \mathbf{x}(0) = \mathbf{F}\mathbf{x}(s) + \mathbf{G}\mathbf{u}(s) \tag{6.3.9}$$

and now solving for the $s$-domain solution $\mathbf{x}(s)$,

$$\mathbf{x}(s) = [s\mathbf{I} - \mathbf{F}]^{-1}\mathbf{x}(0) + [s\mathbf{I} - \mathbf{F}]^{-1}\mathbf{G}\mathbf{u}(s) \tag{6.3.10}$$

Note that the first term is the same as given for the homogeneous problem and that the second term is the product of two $s$-domain matrices, so this implies convolution of time-domain functions, and noting the definition in expression (6.3.6), we get the conceptual solution to be

$$\mathbf{x}(t) = \mathbf{\Phi}(t)\mathbf{x}(0) + \int_0^t \mathbf{\Phi}(t - \tau)\mathbf{G}\mathbf{u}(\tau)\, d\tau \tag{6.3.11}$$

In this discussion we have considered the solution at some time $t$, given initial conditions at time $t = 0$. Now let us write the expression for an arbitrary starting time $t_0$, or

$$\mathbf{x}(t) = \mathbf{\Phi}(t - t_0)\mathbf{x}(t_0) + \int_{t_0}^t \mathbf{\Phi}(t - \tau)\mathbf{G}\mathbf{u}(\tau)\, d\tau \tag{6.3.12}$$

Let us now consider the system shown in Fig. 6.1, where a continuous-time plant in state-variable form is driven by a zero-order hold and the

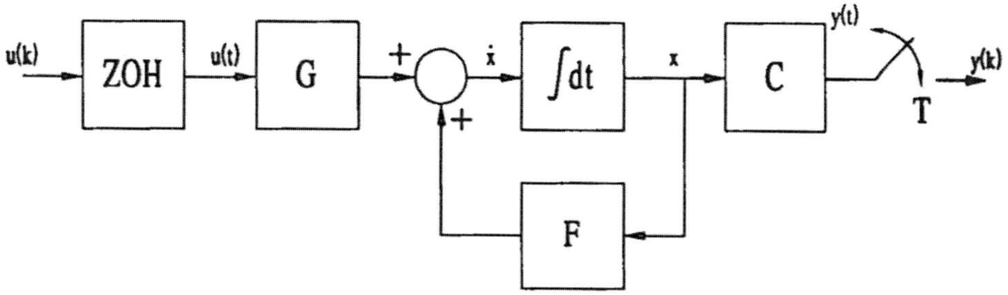


**Figure 6.1.** Continuous plant driven by a zero-order hold with sampled output.

output is sampled in the same manner as was explored in Chapter 3. If the output relation is

$$\mathbf{y}(t) = \mathbf{C}\mathbf{x}(t) \tag{6.3.13}$$

the same relation must hold at the sample instants or

$$\mathbf{y}(k) = \mathbf{C}\mathbf{x}(k) \tag{6.3.14}$$

so it is sufficient to find the states at the sampling instants. Consider the case where $t = (k + 1)T$ and $t_0 = kT$, and note that the operation of the zero-order hold is to create a vector $\mathbf{u}(t)$ according to the relation

$$\mathbf{u}(t) = \mathbf{u}(kT) \qquad kT < t < (k + 1)T \tag{6.3.15}$$

so from relations (6.3.12), we get

$$\mathbf{x}[(k + 1)T] = \mathbf{\Phi}(T)\mathbf{x}(kT) + \int_{kT}^{(k+1)T} \mathbf{\Phi}((k + 1)T - \tau)\mathbf{G}\mathbf{u}(kT)\, d\tau \tag{6.3.16}$$

But $\mathbf{u}(kT)$ is not a function of $\tau$, so the vector $\mathbf{u}(kT)$ may be extracted from the integral to the right, and we may use the shorthand notation that $\mathbf{x}(kT) = \mathbf{x}(k)$ and $\mathbf{u}(kT) = \mathbf{u}(k)$, to give

$$\mathbf{x}(k + 1) = \mathbf{\Phi}(T)\mathbf{x}(k) + \int_{kT}^{(k+1)T} \mathbf{\Phi}((k + 1)T - \tau)\, d\tau\, \mathbf{G}\mathbf{u}(k) \tag{6.3.17}$$

Let us simplify the integral of the second term by letting $(k + 1)T - \tau = \lambda$; then $d\tau = -d\lambda$, and the lower limit on $\lambda$ becomes $T$ and the upper limit becomes zero, so

$$\int_{kT}^{(k+1)T} \mathbf{\Phi}((k + 1)T - \tau)\, d\tau = -\int_{T}^{0} \mathbf{\Phi}(\lambda)\, d\lambda \tag{6.3.18}$$

or reversing the limits, we can write (6.3.17) as

$$\mathbf{x}(k + 1) = \mathbf{\Phi}(T)\mathbf{x}(k) + \int_T^0 \mathbf{\Phi}(\lambda)\, d\lambda\, \mathbf{G}\mathbf{u}(k) \tag{6.3.19}$$

Now define the following constant matrices for a fixed sampling interval $T$, or

$$\mathbf{A} = \mathbf{\Phi}(T) = \{\mathcal{L}^{-1}[(s\mathbf{I} - \mathbf{F})^{-1}]\}|_{t=T} \tag{6.3.20}$$

and

$$\mathbf{B} = \int_0^T \mathbf{\Phi}(\lambda)\, d\lambda\, \mathbf{G} \tag{6.3.21}$$

so expression (6.3.19) becomes the simple matrix-vector difference equation

$$\mathbf{x}(k + 1) = \mathbf{A}\mathbf{x}(k) + \mathbf{B}\mathbf{u}(k) \tag{6.3.22}$$

with an output equation

$$\mathbf{y}(k) = \mathbf{C}\mathbf{x}(k) \tag{6.3.23}$$

and this is the discrete-time state-variable representation of the continuous-time plant driven by a zero-order hold and followed by an output sampler. A block diagram for this discrete-time system is shown in Fig. 6.2.

We have not yet discussed methods of evaluation of $\mathbf{\Phi}(T)$, and there are many, as discussed in the excellent paper by Moeller and van Loan (1978). We shall initially explore the process dictated by relation (6.3.6) but will find that to be computationally difficult for systems of even moderate order, and hence an alternative must be sought.

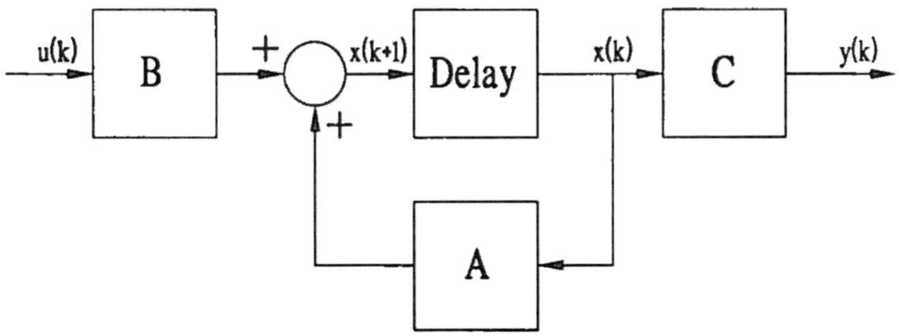


**Figure 6.2.** Block diagram for discrete-time state-variable system.

**Example 6.3.** For the problem of Example 6.1, find the discrete-state representation if this system is driven by a zero-order hold and followed by an output sampler. First find the state transition matrix $\mathbf{\Phi}(t)$ using relation (6.3.6):

$$\begin{aligned}\mathbf{\Phi}(t) &= \mathcal{L}^{-1}[(s\mathbf{I} - \mathbf{F})^{-1}] \\ &= \mathcal{L}^{-1}\begin{bmatrix} s & -1 \\ 0 & s \end{bmatrix}^{-1} \\ &= \mathcal{L}^{-1}s^{-2}\begin{bmatrix} s & 1 \\ 0 & s \end{bmatrix} = \mathcal{L}^{-1}\begin{bmatrix} \dfrac{1}{s} & \dfrac{1}{s^2} \\ 0 & \dfrac{1}{s} \end{bmatrix}\end{aligned}$$

or

$$\mathbf{\Phi}(t) = \begin{bmatrix} 1 & t \\ 0 & 1 \end{bmatrix}$$

so from relation (6.3.20) the **A** matrix is for sampling interval $T$,

$$\mathbf{A} = \mathbf{\Phi}(T) = \begin{bmatrix} 1 & T \\ 0 & 1 \end{bmatrix}$$

Now let us calculate the **B** matrix from relation (6.3.21):

$$\mathbf{B} = \int_0^T \begin{bmatrix} 1 & \lambda \\ 0 & 1 \end{bmatrix} d\lambda\, \mathbf{G} = \begin{bmatrix} \lambda & \dfrac{\lambda^2}{2} \\ 0 & \lambda \end{bmatrix}_0^T \mathbf{G}$$

and evaluating at the limits and substituting the **G** matrix yields

$$\mathbf{B} = \begin{bmatrix} T & \dfrac{T^2}{2} \\ 0 & T \end{bmatrix}\begin{bmatrix} 0 \\ 1 \end{bmatrix}$$

so the matrix **B** is

$$\mathbf{B} = \begin{bmatrix} \dfrac{T^2}{2} \\ T \end{bmatrix}$$

Then the discrete-state representation for this system is

$$\begin{bmatrix} x_1(k+1) \\ x_2(k+1) \end{bmatrix} = \begin{bmatrix} 1 & T \\ 0 & 1 \end{bmatrix}\begin{bmatrix} x_1(k) \\ x_2(k) \end{bmatrix} + \begin{bmatrix} \dfrac{T^2}{2} \\ T \end{bmatrix} u(k)$$

with an output relation

$$y(k) = [1 \quad 0]\begin{bmatrix} x_1(k) \\ x_2(k) \end{bmatrix}$$

**Example 6.4.** Let us now find the state transition matrix **A** and the **B** matrix for the thermal system considered in Example 6.2, where **F** and **G** matrices, are respectively, for no temperature disturbance $T_0(t) = 0$.

$$\mathbf{F} = \begin{bmatrix} -2 & 2 \\ 0.5 & -0.75 \end{bmatrix} \qquad \mathbf{G} = \begin{bmatrix} 0 \\ 0.5 \end{bmatrix}$$

Then $(s\mathbf{I} - \mathbf{F})$ is

$$(s\mathbf{I} - \mathbf{F}) = \begin{bmatrix} s + 2 & -2 \\ -0.5 & s + 0.75 \end{bmatrix}$$

and the inverse matrix is

$$(s\mathbf{I} - \mathbf{F})^{-1} = \frac{1}{s^2 + 2.75s + 0.5}\begin{bmatrix} s + 0.75 & 2 \\ 0.5 & s + 2 \end{bmatrix}$$

Making the indicated partial fraction expansions after noting that the denominator roots are at $s = -0.1957$ and $s = -2.554$ gives

$$(s\mathbf{I} - \mathbf{F})^{-1} = \begin{bmatrix} \dfrac{0.235}{s + 0.1957} + \dfrac{0.765}{s + 2.554} & \dfrac{0.85}{s + 0.1957} - \dfrac{0.85}{s + 2.554} \\[2ex] \dfrac{0.2125}{s + 0.1957} - \dfrac{0.2125}{s + 2.554} & \dfrac{0.765}{s + 0.1957} + \dfrac{0.235}{s + 2.554} \end{bmatrix}$$

Inversion of Laplace transforms indicated yields the state transition matrix $\mathbf{\Phi}(t)$:

$$\mathbf{\Phi}(t) = \begin{bmatrix} 0.235e^{-0.1957t} + 0.765e^{-2.554t} & 0.85(e^{-0.1957t} - e^{-2.554t}) \\ 0.2125(e^{-0.1957t} - e^{-2.554t}) & 0.765e^{-0.1957t} + 0.235e^{-2.554t} \end{bmatrix}$$

If we evaluate this matrix at $t = T = 0.25$ s, we get

$$\mathbf{\Phi}(T) = \mathbf{A} = \begin{bmatrix} 0.6277 & 0.3606 \\ 0.09016 & 0.8526 \end{bmatrix}$$

Now if we employ relation (6.3.21) to get the **B** matrix, the result is

$$\mathbf{B} = 0.5\int_0^T \begin{bmatrix} 0.85(e^{-0.1957\lambda} - e^{-2.554\lambda}) \\ 0.765e^{-0.1957\lambda} + 0.235e^{02.554\lambda} \end{bmatrix} d\lambda$$

and performing the integration and evaluating at the limits yields

$$\mathbf{B} = -\begin{bmatrix} 2.17(e^{-0.1957T} - 1) - 0.166(e^{-2.554T} - 1) \\ 1.95(e^{-0.1957T} - 1) + 0.046(e^{-2.554T} - 1) \end{bmatrix}$$

and evaluating at $T = 0.25$ s yields

$$\mathbf{B} = \begin{bmatrix} 0.2516 \\ 0.1150 \end{bmatrix}$$

The discrete-time state equations for this system are then

$$\begin{bmatrix} x_1(k+1) \\ x_2(k+1) \end{bmatrix} = \begin{bmatrix} 0.6227 & 0.3606 \\ 0.09016 & 0.8526 \end{bmatrix} \begin{bmatrix} x_1(k) \\ x_2(k) \end{bmatrix} + \begin{bmatrix} 0.02516 \\ 0.1150 \end{bmatrix} u(k)$$

## 6.4 MATRIX EXPONENTIAL SERIES APPROACH

We have seen the state transition matrix could be evaluated by Laplace transforms as in expression (6.3.6). We may verify by differentiation that the solution to

$$\dot{\mathbf{x}} = \mathbf{F}\mathbf{x} \tag{6.4.1}$$

can be written as

$$\mathbf{x}(t) = e^{\mathbf{F}t}\mathbf{x}(0) \tag{6.4.2}$$

where the exponential matrix is defined by

$$e^{\mathbf{F}t} = \mathbf{I} + \mathbf{F}t + \tfrac{1}{2}\mathbf{F}^2t^2 + \tfrac{1}{6}\mathbf{F}^3t^3 + \cdots \tag{6.4.3}$$

It is clear from comparison of relations (6.3.7) and (6.4.3) that $e^{\mathbf{F}t}$ is also the state transition matrix $\mathbf{\Phi}(t)$, or

$$\mathbf{\Phi}(t) = e^{\mathbf{F}t} \tag{6.4.4}$$

Since the **A** matrix in the discrete system representation is $\mathbf{\Phi}(T)$, then from (6.4.3) we get

$$\mathbf{A} = \sum_{i=0}^{\infty} \frac{F^i T^i}{i!} \tag{6.4.5}$$

and substitution of the series (6.4.3) into the integral relation (6.3.21) for the **B** matrix after integration gives, term by term,

$$\mathbf{B} = \sum_{i=0}^{\infty} \frac{\mathbf{F}^i T^{i+1}}{(i+1)!}\mathbf{G} \tag{6.4.6}$$

If we are given a continuous-time plant in the form of **F** and **G** matrices and we are able to select a sampling interval $T$, we may computerize the evaluation of truncated versions of relations (6.4.5) and (6.4.6) to give the discrete-time representation of the system. The **C** matrix is the same as that in the continuous-time representation.

The choice of the sampling interval $T$ should be such that it is smaller than the smallest time constant or natural period; otherwise, significant system dynamics might be missed. Once $T$ is chosen, the question arises as to how many terms should be taken in the matrix exponential series. There are several ways to look at this, but the technique of Paynter (see Takahashi et al., 1970) seems to be popular. Let $q = |F_{ij}T|_{\max}$ where $F_{ij}$ is an element of the **F** matrix and let $p$ be the number of terms required in the series. The Paynter technique suggests that $p$ and $q$ should be related by

$$\frac{1}{p!}(nq)^p e^{nq} = 0.001 \tag{6.4.7}$$

where $n$ is the order of the system. If $q$ is calculated, then $p$ must be found by trial and error, which is not an overwhelming problem with an electronic hand calculator.

**Example 6.5.** Given the inertial system of Examples 6.1 and 6.3 with **F** and **G** matrices

$$\mathbf{F} = \begin{bmatrix} 0 & 1 \\ 0 & 0 \end{bmatrix} \qquad \mathbf{G} = \begin{bmatrix} 0 \\ 1 \end{bmatrix}$$

find the **A** and **B** matrices by the method of matrix exponential series. First calculate the powers of the **F** matrix

$$\mathbf{F}^2 = \begin{bmatrix} 0 & 1 \\ 0 & 0 \end{bmatrix}\begin{bmatrix} 0 & 1 \\ 0 & 0 \end{bmatrix} = \begin{bmatrix} 0 & 0 \\ 0 & 0 \end{bmatrix}$$

Similarly, all higher-order powers of **F** are zero. The **A** matrix is given exactly by two terms of the series:

$$\mathbf{A} = \mathbf{I} + \mathbf{F}T = \begin{bmatrix} 1 & T \\ 0 & 1 \end{bmatrix}$$

Similarly, **B** is given by

$$T\mathbf{G} + \frac{1}{2}T^2\mathbf{F}\mathbf{G} = \begin{bmatrix} \dfrac{T^2}{2} \\ T \end{bmatrix}$$

This was a very fortunate case, in that higher powers of the F matrix were zero. Seldom in a real problem will this be the case, and one will need to truncate the series and assume that enough terms are retained to give reasonable approximation of the closed form of the series.

**Example 6.6.** Given the problem of Example 6.4 with a sampling interval of $T = 0.25$ s, the criterion of Paynter suggests that

$$q = \max|F_{ij}T| = (2)(0.25) = 0.5$$

Then the number of terms $p$ must satisfy

$$\frac{1}{p!}(1)^p e^1 = 0.001$$

where $n$ is the order of the system, in this case 2. This expression may be solved directly for $p!$ to yield

$$p! = 1000e^1$$

A trial-and-error solution indicates that seven terms must be retained in the series to meet the Paynter criterion.

There are often numerical difficulties associated with the direct evaluation of relations (6.4.5) and (6.4.6) becausc $(\mathbf{F}T)^i$ gets large for large $i$ as docs $i$, which makes the quotient reasonable, but calculation of the numerator and denominator of these relations separately yields a numerical problem which in many cases can be alleviated by the nested form that follows. Define a matrix $\psi(T)$ such that

$$\psi(T) = T\left(\mathbf{I} + \frac{1}{2}\mathbf{F}T\left(\mathbf{I} + \frac{1}{3}\mathbf{F}T\left(\mathbf{I} + \cdots + \frac{1}{p-1}\mathbf{F}T\left(\mathbf{I} + \frac{1}{p}\mathbf{F}T\right)\right)\right)\right) \qquad (6.4.8)$$

Careful examination of this matrix will indicate that the $\mathbf{B}$ matrix may be computed as

$$\mathbf{B} = \psi(T)\mathbf{G} \qquad (6.4.9)$$

and further that the $\mathbf{A}$ matrix may be given by

$$\mathbf{A} = \mathbf{I} + \mathbf{F}\psi(T) \qquad (6.4.10)$$

This technique is computationally efficient in that only the one series for $\psi(T)$ must be evaluated and it also avoids large numbers as encountered in the direct evaluation of (6.4.5) and (6.4.6).

## 6.5 SOLUTION OF THE DISCRETE STATE EQUATION

Consider the discrete-time system described by the time-invariant vector difference equation

$$\mathbf{x}(k + 1) = \mathbf{A}\mathbf{x}(k) + \mathbf{B}\mathbf{u}(k) \tag{6.5.1}$$

If the initial condition vector $\mathbf{x}(0)$ and the forcing sequence $\mathbf{u}(k)$ are known, we may write the solution for $\mathbf{x}(1)$ as

$$\mathbf{x}(1) = \mathbf{A}\mathbf{x}(0) + \mathbf{B}\mathbf{u}(0) \tag{6.5.2}$$

Then the solution for $\mathbf{x}(2)$ is

$$\mathbf{x}(2) = \mathbf{A}\mathbf{x}(1) + \mathbf{B}\mathbf{u}(1) \tag{6.5.3}$$

and we may substitute $\mathbf{x}(1)$ from (6.5.2) to yield

$$\mathbf{x}(2) = \mathbf{A}^2\mathbf{x}(0) + \mathbf{A}\mathbf{B}\mathbf{u}(0) + \mathbf{B}\mathbf{u}(1) \tag{6.5.4}$$

If we continue this process, the result for the state vector at $t = kT$ is

$$\mathbf{x}(k) = \mathbf{A}^k\mathbf{x}(0) + \sum_{i=0}^{k-1} \mathbf{A}^{k-i-1}\mathbf{B}\mathbf{u}(i) \tag{6.5.5}$$

It is interesting to note that the second term is a discrete-time convolution. An alternative solution to (6.5.1) will now be sought via the method of $z$-transformation. If we take the $z$-transform of expression (6.5.1), we get

$$z\mathbf{x}(z) - z\mathbf{x}(0) = \mathbf{A}\mathbf{x}(z) + \mathbf{B}\mathbf{u}(z) \tag{6.5.6}$$

and solving for $\mathbf{x}(z)$ yields

$$\mathbf{x}(z) = (z\mathbf{I} - \mathbf{A})^{-1}z\mathbf{x}(0) + (z\mathbf{I} - \mathbf{A})^{-1}\mathbf{B}\mathbf{u}(z) \tag{6.5.7}$$

If we consider only the unforced problem where the control sequence $\mathbf{u}(k) = 0$ or alternatively, $\mathbf{u}(z) = 0$, then

$$\mathbf{x}(z) = (z\mathbf{I} - \mathbf{A})^{-1}z\mathbf{x}(0) \tag{6.5.8}$$

and inverting the transforms give

$$\mathbf{x}(k) = \mathfrak{Z}^{-1}[(z\mathbf{I} - \mathbf{A})^{-1}z]\mathbf{x}(0) \tag{6.5.9}$$

Comparison of (6.5.9) and the first term of (6.5.5) indicates that

$$\mathbf{A}^k = \mathfrak{Z}^{-1}[(z\mathbf{I} - \mathbf{A})^{-1}z] \tag{6.5.10}$$

If we now examine the matrix inverse present in (6.5.10), we note that the inverse may be written as the adjoint matrix divided by the determinant,

or

$$(z\mathbf{I} - \mathbf{A})^{-1} = \frac{\operatorname{adj}(z\mathbf{I} - \mathbf{A})}{\det(z\mathbf{I} - \mathbf{A})} \tag{6.5.11}$$

where $\det(\cdot)$ is the determinant of the matrix in the parentheses. Note that this determinant is an $n$th-degree polynomial in $z$, the coefficients of which are functions of the elements of the **A** matrix. Since this polynomial appears in the denominator of each element of the inverse matrix, then in the $z$-transform inversion process, the roots of this polynomial govern the dynamic nature of the natural response sequences. This determinant is then the discrete-time system characteristic equation, and the roots of this equation are the system poles. The poles are the values of $z$ for which this polynomial vanishes, or

$$\det(z\mathbf{I} - \mathbf{A}) = 0 \tag{6.5.12}$$

**Example 6.7.** Consider the inertial plant of Example 6.1 for which the **A** matrix was

$$\mathbf{A} = \begin{bmatrix} 1 & T \\ 0 & 1 \end{bmatrix}$$

The characteristic equation is then given by

$$\det(z\mathbf{I} - \mathbf{A}) = \det\begin{bmatrix} z - 1 & -T \\ 0 & z - 1 \end{bmatrix} = 0$$

and evaluating the determinant gives the characteristic equation

$$z^2 - 2z + 1 = 0$$

or in factored form,

$$(z - 1)^2 = 0$$

Clearly, both poles of this inertial plant driven by the zero-order hold lie at $z = 1$, which corresponds to the result obtained in Example 3.2. Note that for this problem the locations of the poles are independent of the sampling interval $T$, while in most problems this is usually not the case. Another example should serve to clarify the ideas just given.

**Example 6.8.** Consider now the case of the thermal control system of Example 6.4 for a sampling interval of $T = 0.25$ s; it yielded an **A** matrix of

$$\mathbf{A} = \begin{bmatrix} 0.6277 & 0.3606 \\ 0.09016 & 0.8526 \end{bmatrix}$$

The characteristic determinant is

$$\det\begin{bmatrix} z - 0.6277 & -0.3606 \\ -0.09016 & z - 0.8526 \end{bmatrix} = 0$$

and upon multiplication of the indicated terms the characteristic polynomial is

$$z^2 - (0.6277 + 0.8526)z + (0.6277)(0.8526) - (-0.09016)(-0.3606) = 0$$

and evaluating the constants,

$$z^2 - 1.480z + 0.5026 = 0$$

The roots of this equation are given by the quadratic formula to be

$$\begin{aligned} z_{1,2} &= 0.74 \pm \sqrt{0.045} \\ &= 0.74 \pm 0.212 \\ &= 0.952,\ 0.528 \end{aligned}$$

These are the same pole locations as those given by the transfer function analysis in Example 3.3

## 6.6 TRANSFER FUNCTIONS FROM STATE-VARIABLE DESCRIPTIONS

In this section we are interested in the response of the system

$$\mathbf{x}(k + 1) = \mathbf{A}\mathbf{x}(k) + \mathbf{B}\mathbf{u}(k) \tag{6.6.1}$$

with outputs that are linear combinations of the states, or

$$\mathbf{y}(k) = \mathbf{C}\mathbf{x}(k) \tag{6.6.2}$$

when subjected to a sampled sinusoid as an input. We shall assume that the system of (6.6.1) is stable (i.e., all the poles lie inside the unit circle) such that after some period of time, a steady-state response vector $\mathbf{x}(k)$ will be a sampled sinusoid of the same frequency as the driving sinusoid $\mathbf{u}(k)$. Let us take the $z$-transform of relations (6.6.1) and (6.6.2) ignoring initial values to yield

$$z\mathbf{x}(z) = \mathbf{A}\mathbf{x}(z) + \mathbf{B}\mathbf{u}(z) \tag{6.6.3}$$

and

$$\mathbf{y}(z) = \mathbf{C}\mathbf{x}(z) \tag{6.6.4}$$

Solving (6.6.3) for $\mathbf{x}(z)$ yields

$$\mathbf{x}(z) = [z\mathbf{I} - \mathbf{A}]^{-1}\mathbf{B}\mathbf{u}(z) \tag{6.6.5}$$

and substitution of (6.6.5) into (6.6.4) gives

$$\mathbf{y}(z) = \mathbf{C}[z\mathbf{I} - \mathbf{A}]^{-1}\mathbf{B}\mathbf{u}(z) \tag{6.6.6}$$

Expression (6.6.6) represents a set of linear relations between the control efforts $\mathbf{u}(z)$ and the outputs $\mathbf{y}(z)$, so we shall define the $r \times m$ $z$-domain transfer function matrix as

$$\mathbf{G}(z) = \mathbf{C}[z\mathbf{I} - \mathbf{A}]^{-1}\mathbf{B} \tag{6.6.7}$$

If the input $u(z)$ is scalar, then $\mathbf{B} = \mathbf{b}$, a vector, and if the output $y(z)$ is also scalar, then $\mathbf{C} = \mathbf{c}^T$, and the transfer function matrix reduces to a scalar, or

$$G(z) = \mathbf{c}^T[z\mathbf{I} - \mathbf{A}]^{-1}\mathbf{b} \tag{6.6.8}$$

As developed in Chapter 2, the concept of a frequency response function can be given by letting $z = e^{j\omega T}$, where $T$ is the sampling interval and $\omega$ is the radian frequency of interest. The frequency response function matrix is given from (6.6.7) to be

$$\mathbf{G}(e^{j\omega T}) = \mathbf{C}[e^{j\omega T}\mathbf{I} - \mathbf{A}]^{-1}\mathbf{B} \tag{6.6.9}$$

and for a single-input/single-output system it is

$$G(e^{j\omega T}) = \mathbf{c}^T[e^{j\omega T}\mathbf{I} - \mathbf{A}]^{-1}\mathbf{b} \tag{6.6.10}$$

**Example 6.9.** Find the discrete-time transfer function for the inertial plant considered in Examples 6.1, 6.3, and 6.6, where

$$\mathbf{A} = \begin{bmatrix} 1 & T \\ 0 & 1 \end{bmatrix} \qquad \mathbf{b} = \begin{bmatrix} \dfrac{T^2}{2} \\ T \end{bmatrix}$$

with the displacement as the output, or

$$\mathbf{c}^T = [1 \quad 0]$$

The matrix $[z\mathbf{I} - \mathbf{A}]$ is

$$[z\mathbf{I} - \mathbf{A}] = \begin{bmatrix} z - 1 & -T \\ 0 & z - 1 \end{bmatrix}$$

and the inverse is

$$[z\mathbf{I} - \mathbf{A}]^{-1} = \begin{bmatrix} \dfrac{1}{z-1} & \dfrac{T}{(z-1)^2} \\ 0 & \dfrac{1}{z-1} \end{bmatrix}$$

The first matrix product of (6.6.8) is

$$\mathbf{c}^T[(z\mathbf{I} - \mathbf{A}]^{-1} = [1 \quad 0]\begin{bmatrix} \dfrac{1}{z-1} & \dfrac{T}{(z-1)^2} \\ 0 & \dfrac{1}{z-1} \end{bmatrix} = \begin{bmatrix} \dfrac{1}{z-1} & \dfrac{T}{(z-1)^2} \end{bmatrix}$$

and from (6.6.8) the transfer function is

$$G(z) = \mathbf{c}^T[z\mathbf{I} - \mathbf{A}]^{-1}\mathbf{b} = \begin{bmatrix} \dfrac{1}{z-1} & \dfrac{T}{(z-1)^2} \end{bmatrix}\begin{bmatrix} \dfrac{T^2}{2} \\ T \end{bmatrix}$$

or

$$G(z) = \frac{T^2}{2(z-1)} + \frac{T^2}{(z-1)^2}$$

and finding a common denominator, we have

$$G(z) = \frac{(T^2/2)(z-1) + T^2}{(z-1)^2} = \frac{T^2}{2}\frac{z+1}{(z-1)^2}$$

This is the same result as in Example 3.2.

## 6.7 CONTROLLABILITY

The concept of controllability is one of the important topics to come out of modern control theory. The problem is that of investigating whether it is possible to drive a system from some arbitrary initial condition $\mathbf{x}(0)$ to some specific final state $\mathbf{x}(n)$ in $n$ sampling intervals. The answer to this question is either yes or no, and the question as to what driving sequence $\mathbf{u}(k)$ is needed to accomplish this task remains another problem. Let us now consider the usual system

$$\mathbf{x}(k+1) = \mathbf{A}\mathbf{x}(k) + \mathbf{B}\mathbf{u}(k) \tag{6.7.1}$$

where $\mathbf{x}(k)$ is an $n$-dimensional state vector and $\mathbf{u}(k)$ is an $m$-dimensional control effort vector, and the matrices $\mathbf{A}$ and $\mathbf{B}$ are of the appropriate dimension. From relation (6.5.5) we can write the state at the $k$th time step as

$$\mathbf{x}(k) = \mathbf{A}^k\mathbf{x}(0) + \sum_{i=0}^{k-1} \mathbf{A}^{k-i-1}\mathbf{B}\mathbf{u}(i) \tag{6.7.2}$$

and for $n$ steps away from the origin,

$$\mathbf{x}(n) = \mathbf{A}^n\mathbf{x}(0) + \sum_{i=0}^{n-1} \mathbf{A}^{n-i-1}\mathbf{B}\mathbf{u}(i) \tag{6.7.3}$$

This can be rewritten as

$$\mathbf{x}(n) - \mathbf{A}^n\mathbf{x}(0) = \mathbf{A}^{n-1}\mathbf{B}\mathbf{u}(0) + \mathbf{A}^{n-2}\mathbf{B}\mathbf{u}(1) + \cdots + \mathbf{A}\mathbf{B}\mathbf{u}(n-2) + \mathbf{B}\mathbf{u}(n-1) \tag{6.7.4}$$

and rewriting in matrix form yields

$$\mathbf{x}(n) - \mathbf{A}^n\mathbf{x}(0) = [\mathbf{A}^{n-1}B \quad \mathbf{A}^{n-2}\mathbf{B} \quad \cdots \quad \mathbf{B}]\begin{bmatrix} \mathbf{u}(0) \\ \mathbf{u}(1) \\ \vdots \\ \mathbf{u}(n-1) \end{bmatrix} \tag{6.7.5}$$

This set of equations is a set of $n$ linear equations in $mn$ unknowns with the quantities on the left side being known. For there to exist a solution to these linear equations, the rank of the matrix of coefficients must be the same as the number of unknowns, or

$$\operatorname{rank}[\mathbf{A}^{n-1}\mathbf{B} \quad \mathbf{A}^{n-2}\mathbf{B} \quad \cdots \quad \mathbf{A}\mathbf{B} \quad \mathbf{B}] = n \tag{6.7.6}$$

This does not mean that this matrix need be of full rank unless the control effort sequence is scalar ($m = 1$). If $m > 1$, the uniqueness of the control sequence vector cannot be assured, which implies that there is possibly more than one sequence that will drive the system between the specified states. The matrix of (6.7.6) is called the *controllability matrix*.

The question immediately arises as to whether admission of additional steps in time will allow us to get from the initial to the final state if it was not possible to do so in $n$ steps. For one additional step the problem is formulated exactly as previously, but with one more step, and the condition for a solution vector to exist is

$$\operatorname{rank}[\mathbf{A}^{n}\mathbf{B} \quad \mathbf{A}^{n-1}\mathbf{B} \quad \cdots \quad \mathbf{A}\mathbf{B} \quad \mathbf{B}] = n \tag{6.7.7}$$

This at first looks promising, but we know from the Cayley–Hamilton theorem (proved in Appendix B) that $\mathbf{A}^n$ may be expanded as a linear combination of the lower-order powers of **A** and hence the columns of $\mathbf{A}^n\mathbf{B}$ are not linearly independent of the other columns that are in the matrix of (6.7.6). The conclusion is that allowance of another time step to achieve the desired final state does not improve the chances of being able to do so. Either the task can be done in $n$ steps or it cannot be done at all. If we allow even more steps to accomplish the task, the result will be the same, so the criterion given by expression (6.7.6) is the correct one and considerably simpler than that given in Appendix B.

Controllability is a structural property of the system which determines whether *any* control sequence could drive the system between the specific states. If the **A** and **B** matrices are such that the system is uncontrollable, there is no sense in seeking a controlling sequence to give a particular final state. The only cure for this problem is to modify the system structure to make control efforts more effective. A simple redefinition of state variables will not be sufficient to make the system controllable.

**Example 6.10.** Consider a linear discrete-time system governed by the matrix-vector difference equation

$$\begin{bmatrix} x_1(k+1) \\ x_2(k+1) \end{bmatrix} = \begin{bmatrix} 1 & -0.2 \\ 0.4 & 0.4 \end{bmatrix} \begin{bmatrix} x_1(k) \\ x_2(k) \end{bmatrix} + \begin{bmatrix} 1 \\ 1 \end{bmatrix} u(k)$$

Examine the controllability of this equation. The controllability matrix is, for this case,

$$[\mathbf{Ab} \quad \mathbf{b}]$$

The matrix product is

$$\mathbf{Ab} = \begin{bmatrix} 1 & -0.2 \\ 0.4 & 0.4 \end{bmatrix} \begin{bmatrix} 1 \\ 1 \end{bmatrix} = \begin{bmatrix} 0.8 \\ 0.8 \end{bmatrix}$$

And the controllability matrix is

$$[\mathbf{Ab} \quad \mathbf{b}] = \begin{bmatrix} 0.8 & 1 \\ 0.8 & 1 \end{bmatrix}$$

It is clear that the second column is not linearly independent of the first column and the rank of the controllability matrix is 1 and the system is uncontrollable. This fact was not obvious by examination of the original state equations. This problem is reexamined in Appendix B from a modal coordinate point of view.

## 6.8 OBSERVABILITY

The concept of observability or reconstructability is another of the fundamental ideas of modern control theory. The question to be answered is whether, based on measured system outputs at several times, the state variables at a single time may be deduced from the measurements and knowledge of the system-forcing function (control effort). It is conceivable that the structure of the system and/or the measurements taken are such that the measurements do not contain all the information about the system states.

The usual technique for system control is to generate control efforts (strategies) based on measurements of system outputs. If the measurements are missing basic information on actual system response, erroneous control efforts could be generated. The system of interest is given by the first-order matrix-vector difference equation

$$\mathbf{x}(k + 1) = \mathbf{A}\mathbf{x}(k) + \mathbf{B}\mathbf{u}(k) \tag{6.8.1}$$

with an output-measurement equation

$$\mathbf{y}(k) = \mathbf{C}\mathbf{x}(k) \tag{6.8.2}$$

and the question to be answered is: Can we find the initial state of the system $\mathbf{x}(0)$ given the sequence of measurements $\mathbf{y}(0), \mathbf{y}(1), \ldots, \mathbf{y}(n - 1)$? The response at the first time interval is, from (6.8.1),

$$\mathbf{x}(1) = \mathbf{A}\mathbf{x}(0) + \mathbf{B}\mathbf{u}(0) \tag{6.8.3}$$

and the corresponding output

$$\mathbf{y}(1) = \mathbf{C}\mathbf{x}(1) = \mathbf{C}\mathbf{A}\mathbf{x}(0) + \mathbf{C}\mathbf{B}\mathbf{u}(0) \tag{6.8.4}$$

and in general we have shown in (6.5.5) that

$$\mathbf{y}(k) = \mathbf{C}\mathbf{A}^k\mathbf{x}(0) + \sum_{i=0}^{k-1} \mathbf{C}\mathbf{A}^{k-i-1}\mathbf{B}\mathbf{u}(i) \qquad k = 1, 2, \ldots \tag{6.8.5}$$

Expression (6.8.4) could be rewritten as

$$[\mathbf{y}(1) - \mathbf{C}\mathbf{B}\mathbf{u}(0)] = \mathbf{C}\mathbf{A}\mathbf{x}(0) \tag{6.8.6}$$

and similarly for $\mathbf{y}(2)$,

$$[\mathbf{y}(2) - \mathbf{C}\mathbf{A}\mathbf{B}\mathbf{u}(0) - \mathbf{C}\mathbf{B}\mathbf{u}(1)] = \mathbf{C}\mathbf{A}^2\mathbf{x}(0) \tag{6.8.7}$$

and for the $(n - 1)$st point,

$$\left[\mathbf{y}(n - 1) - \mathbf{C}\sum_{i=0}^{n-2} \mathbf{A}^{n-2-i}\mathbf{B}\mathbf{u}(i)\right] = \mathbf{C}\mathbf{A}^{n-1}\mathbf{x}(0) \tag{6.8.8}$$

Let the left side of (6.8.6) be denoted by $\bar{\mathbf{y}}(1)$ and of (6.8.7) by $\bar{\mathbf{y}}(2)$, so we can write the output relations as

$$\begin{bmatrix} \mathbf{y}(0) \\ \bar{\mathbf{y}}(1) \\ \bar{\mathbf{y}}(2) \\ \vdots \\ \bar{\mathbf{y}}(n-1) \end{bmatrix} = \begin{bmatrix} \mathbf{C} \\ \mathbf{CA} \\ \mathbf{CA}^2 \\ \vdots \\ \mathbf{CA}^{n-1} \end{bmatrix} \mathbf{x}(0) \qquad (6.8.9)$$

Now if the output vector $\mathbf{y}(k)$ is of dimension $r$, the matrix on the right side is an $nr \times n$ matrix, and this is a set of $nr$ equations in $n$ unknowns. For a solution to exist when there are more equations than unknown, the rank of the associated matrix of coefficients must be maximal or of rank $n$ or, written mathematically,

$$\text{rank} \begin{bmatrix} \mathbf{C} \\ \mathbf{CA} \\ \mathbf{CA}^2 \\ \vdots \\ \mathbf{CA}^{n-1} \end{bmatrix} = n \qquad (6.8.10)$$

If the rank of this matrix is $n$, a solution vector $\mathbf{x}(0)$ can be found, and it will be unique. The matrix of (6.8.10) is the *observability matrix.*

One is tempted to ask: If the rank of the matrix in (6.8.10) is less than $n$, would taking one more data point, say $\mathbf{y}(n)$, aid in the solution of the initial state? This process would make the final entry in the matrix of (6.8.10) $\mathbf{CA}^n$. The problem is that $\mathbf{A}^n$ can be expanded as a linear combination of the lower-order powers of $\mathbf{A}$, and hence the additional rows contributed to the matrix are not linearly independent from the rows of the matrix of (6.8.10) and thus cannot effect an increase in rank.

**Example 6.11.** Consider the inertial plant of Example 6.3, where the state equation was

$$\begin{bmatrix} x_1(k+1) \\ x_2(k+1) \end{bmatrix} = \begin{bmatrix} 1 & T \\ 0 & 1 \end{bmatrix} \begin{bmatrix} x_1(k) \\ x_2(k) \end{bmatrix} + \begin{bmatrix} \frac{T^2}{2} \\ T \end{bmatrix} u(k)$$

Now let us measure the velocity rather than displacement, so the output equation is

$$y(k) = [0 \quad 1] \begin{bmatrix} x_1(k) \\ x_2(k) \end{bmatrix}$$

This is a second-order system with a scalar output, so the observability matrix is

$$\begin{bmatrix} \mathbf{c}^T \\ \mathbf{c}^T\mathbf{A} \end{bmatrix} = \begin{bmatrix} 0 & 1 \\ 0 & 1 \end{bmatrix}$$

The rank of this matrix is 1, and the system is unobservable. The physical reason for this is that the initial condition on $x_1(k)$ is independent of $x_2(k)$; hence there are no measurements of $x_2(k)$ that will allow the inference of $x_1(0)$.

## 6.9 STATE-VARIABLE REPRESENTATION OF DISCRETE SINGLE-INPUT/SINGLE-OUTPUT SYSTEMS

Thus far in this chapter we have discussed the discrete-time state-variable representation of continuous-time systems driven by a zero-order hold and followed by an output sampler. In this section we discuss the state-variable representation of discrete-data systems, often compensators or digital filters which are initially specified in the form of either a discrete-time transfer function $H(z)$ or an $n$th-order difference equation. There are an infinite number of ways in which the state variables may be selected. A few of the more common representations will be discussed here, but this presentation is by no means comprehensive, and any reader interested in other forms is referred to the literature.

We shall be concerned with systems specified as a $z$-domain transfer function of the form

$$H(z) = \frac{d_n z^n + d_{n-1}z^{n-1} + \cdots + d_0}{z^n + a_{n-1}z^{n-1} + \cdots + a_0} = \frac{y(z)}{u(z)} \tag{6.9.1}$$

or an equivalent difference equation

$$y_{k+n} + a_{n-1}y_{k+n-1} + \cdots + a_1 y_{k+1} + a_0 y_k = d_n u_{k+n} + d_{n-1}u_{k+n-1} + \cdots + d_0 u_k \tag{6.9.2}$$

We would like to represent this system in state variable form, which written in matrix form is

$$\mathbf{x}(k + 1) = \mathbf{A}\mathbf{x}(k) + \mathbf{b}u(k) \tag{6.9.3}$$

with output equation

$$y(k) = \mathbf{c}^T\mathbf{x}(k) + du(k) \tag{6.9.4}$$

where the elements of $\mathbf{A}$, $\mathbf{b}$, $\mathbf{c}^T$, and $d$ depend on the coefficients $d_i$ and

$a_j$ in the originally specified transfer function. The exact nature of this dependence is a function of how the state variables are chosen. If a transfer function has equal-order denominator and numerator, then in all cases long division should be performed to yield a transfer function of the form

$$H(z) = \frac{y(z)}{u(z)} = d_n + \frac{b_{n-1}z^{n-1} + \cdots + b_0}{z^n + a_{n-1}z^{n-1} + \cdots + a_0} \tag{6.9.5}$$

The $d_n$ term represents a direct feedforward of the input sequence to the output sequence.

### Parallel Realization (Partial Fraction Method)

If the poles of the transfer function (6.9.5) are known, expression (6.9.5) can be written as

$$H(z) = d_n + \frac{b_{n-1}z^{n-1} + \cdots + b_0}{(z - p_1)(z - p_2)\cdots(z - p_n)} \tag{6.9.6}$$

where we shall assume that all the $p_i$ are distinct. Now let us make a partial fraction expansion of the second term of (6.9.6) to yield

$$H(z) = d_n + \frac{A_1}{z - p_1} + \cdots + \frac{A_n}{z - p_n} \tag{6.9.7}$$

where the coefficients are given by the Heaviside method as

$$A_i = \lim_{z \to p_i} [(z - p_i)H(z)] \qquad i = 1, 2, \ldots, n \tag{6.9.8}$$

Let one of the terms of (6.9.7) be a transfer function between the input sequence and the $i$th-state variable

$$\frac{x_i(z)}{u(z)} = \frac{1}{z - p_i} \qquad i = 1, 2, \ldots, n \tag{6.9.9}$$

This transfer function can be realized as illustrated in Fig. 6.3. With this and expression (6.9.7), the entire system can be characterized by the parallel structure of Fig. 6.4. It is interesting to note that a set of state variables has been defined in (6.9.9), so the state equations are given by inverting (6.9.9), or

$$x_i(k + 1) = p_i x_i(k) + u(k) \qquad i = 1, 2, \ldots, n \tag{6.9.10}$$

Then from Fig. 6.4 or expression (6.9.7), the output sequence is given by

$$y(k) = d_n u(k) + A_1 x_1(k) + \cdots + A_n x_n(k) \tag{6.9.11}$$

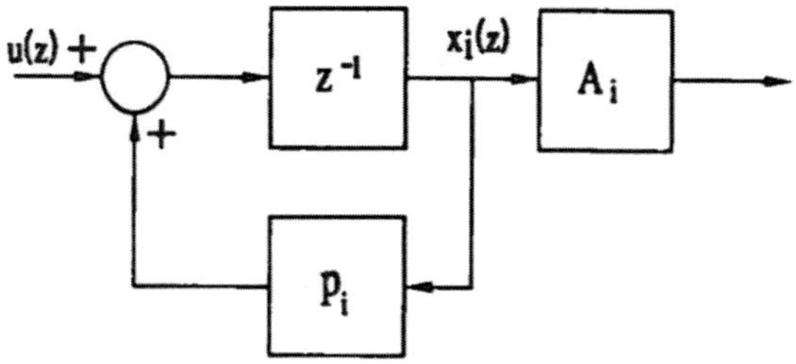


**Figure 6.3.** Realization of the simple first-order transfer function.

The matrix forms of equations (6.9.10) and (6.9.11) are

$$\begin{bmatrix} x_1(k+1) \\ x_2(k+1) \\ \vdots \\ x_n(k+1) \end{bmatrix} = \begin{bmatrix} p_1 & 0 & \cdots & & 0 \\ 0 & p_2 & & & \vdots \\ \vdots & & \ddots & & \\ & & & p_{n-1} & 0 \\ 0 & & & 0 & p_n \end{bmatrix} \begin{bmatrix} x_1(k) \\ x_2(k) \\ \vdots \\ x_n(k) \end{bmatrix} + \begin{bmatrix} 1 \\ 1 \\ \vdots \\ 1 \end{bmatrix} u(k) \tag{6.9.12}$$

and

$$y(k) = [A_1 \quad A_2 \quad \cdots \quad A_n] \begin{bmatrix} x_1(k) \\ x_2(k) \\ \vdots \\ x_n(k) \end{bmatrix} + d_0 u(k) \tag{6.9.13}$$

An example problem will be useful in illustration of this realization.

**Example 6.12.** Find the parallel realization of the following transfer function:

$$H(z) = \frac{z^2 + 2z + 1}{z^2 + 5z + 6}$$

Long division of the transfer function yields

$$H(z) = 1 + \frac{-3z - 5}{z^2 + 5z + 6}$$

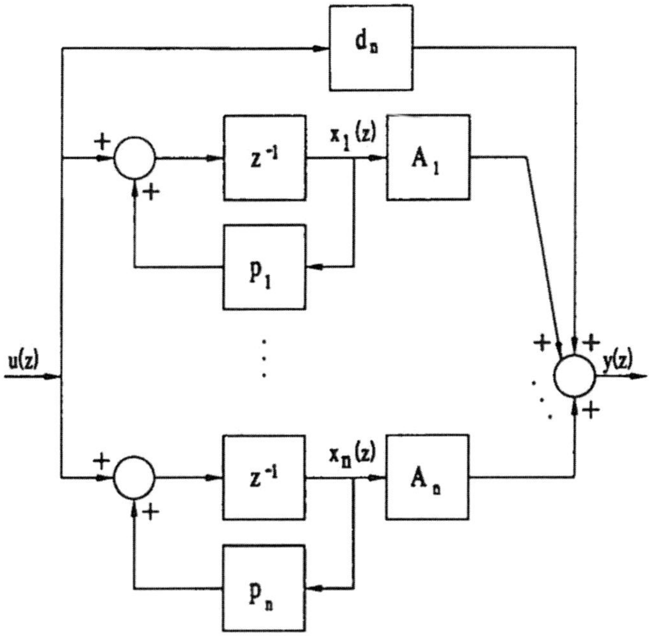


**Figure 6.4.** Parallel realization of a transfer function.

Let us make a partial fraction expansion of the second term to yield

$$H(z) = 1 + \frac{1}{z + 2} + \frac{-4}{z + 3}$$

The state-variable form will be, from relations (6.9.12) and (6.9.13),

$$\begin{bmatrix} x_1(k + 1) \\ x_2(k + 1) \end{bmatrix} = \begin{bmatrix} -2 & 0 \\ 0 & -3 \end{bmatrix} \begin{bmatrix} x_1(k) \\ x_2(k) \end{bmatrix} + \begin{bmatrix} 1 \\ 1 \end{bmatrix} u(k)$$

with an output relation

$$y(k) = [1 \quad -4] \begin{bmatrix} x_1(k) \\ x_2(k) \end{bmatrix} + u(k)$$

The block diagram of this realization is illustrated in Fig. 6.5. Transfer functions that possess pairs of complex poles should have the pairs of poles treated as quadratics which yield some second-order subsystems in the parallel structure. Also in the case of repeated poles, the structure will have to be such that it follows the usual forms for partial fraction expansion of functions with higher-order poles.

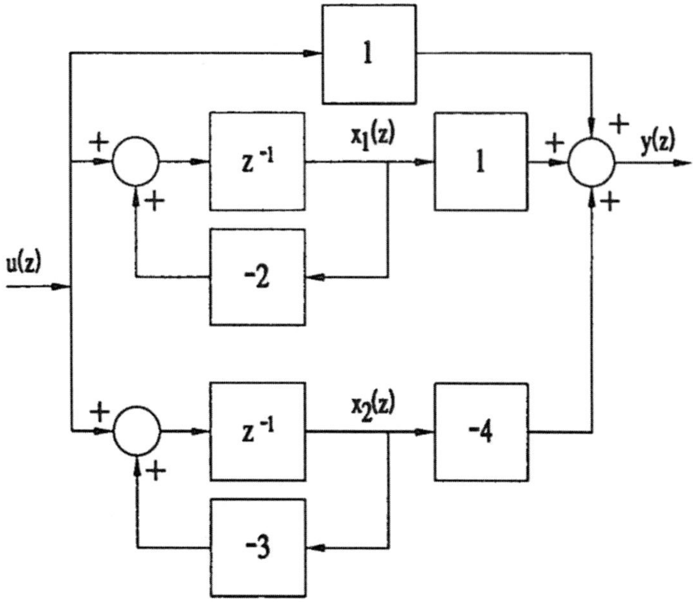


**Figure 6.5.** Parallel realization of Example 6.12.

## Cascade Realization (Iterative Programming)

In this section we consider again the transfer function of (6.9.5), but now we shall assume that all the poles and zeros of the function are known such that it may be written in factored form as

$$H(z) = \frac{\alpha(z - z_1)(z - z_2) \cdots (z - z_m)}{(z - p_1)(z - p_2) \cdots (z - p_n)} \qquad m \leq n \tag{6.9.14}$$

where we have assumed that the numerator is of lower order than the denominator. If this is not the case, carry out long division to yield the form of (6.9.5). The second term of (6.9.5) can be realized by the method that follows. The first term can then be realized as a parallel construction of feedforward of the input sequence. Upon factoring (6.9.14), we get

$$H(z) = \frac{\alpha_1}{z - p_1} \cdots \frac{\alpha_{n-m}}{z - p_{n-m}} \cdot \frac{\alpha_{n-m+1}(z - z_1)}{z - p_{n-m+1}} \cdots \frac{\alpha_n(z - z_m)}{z - p_n} \tag{6.9.15}$$

This expression can be realized by a cascade of $n$ blocks each with a transfer function of one of the terms of (6.9.15), as illustrated in Fig. 6.6.

Let us now define the state variables as indicated in Fig. 6.6, so

$$
\begin{aligned}
x_1(z) &= \frac{\alpha_1}{z - p_1} u(z) \\
x_2(z) &= \frac{\alpha_2}{z - p_2} x_1(z) \\
&\vdots \\
x_{n-m}(z) &= \frac{\alpha_{n-m}}{z - p_{n-m}} x_{n-m-1}(z) \\
&\vdots \\
x_n(z) &= \frac{\alpha_n(z - z_m)}{z - p_n} x_{n-1}(z)
\end{aligned}
\tag{6.9.16}
$$

When these relations are inverted, the following set of difference equations results:

$$
\begin{aligned}
x_1(k+1) &= p_1 x_1(k) + \alpha_1 u(k) \\
x_2(k+1) &= p_2 x_2(k) + \alpha_2 x_1(k) \\
&\vdots \\
x_{n-m}(k+1) &= p_{n-m} x_{n-m}(k) + \alpha_{n-m} x_{n-m-1}(k) \\
x_{n-m+1}(k+1) &= p_{n-m+1} x_{n-m+1}(k) + \alpha_{n-m+1}[x_{n-m}(k+1) - z_1 z_{n-m}(k)] \\
&\vdots \\
x_n(k+1) &= p_n x_n(k) + \alpha_n[x_{n-1}(k+1) - z_m x_{n-1}(k)]
\end{aligned}
\tag{6.9.17}
$$

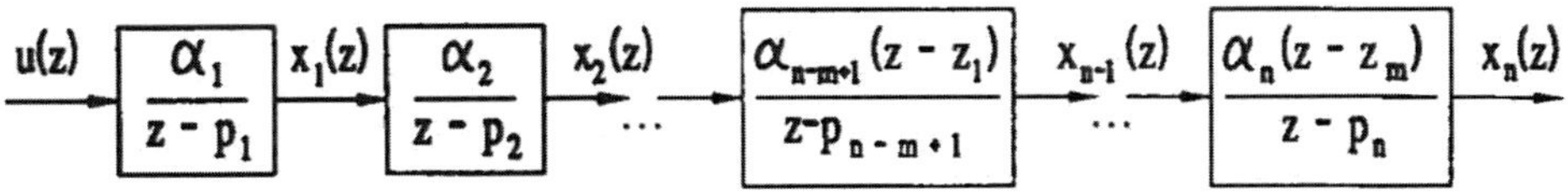


**Figure 6.6.** Cascade realization of a transfer function.

with an output equation

$$y(k) = x_n(k) \tag{6.9.18}$$

It is interesting to note that expressions (6.9.17) in matrix form are not of the form

$$\mathbf{x}(k + 1) = \mathbf{A}\mathbf{x}(k) + \mathbf{b}u(k) \tag{6.9.19}$$

because the right side of the last $m$ of equations (6.9.17) involve the advanced version of the input sequence to each respective block. The first $n - m$ of relations (6.9.17) are clearly of the correct form, but the difficulty lies in the remainder. The problem can be solved conceptually by forward-substituting the $(n - m)$st relation into the next relation and then substituting each succeeding relation into the next one until all dependence of the right-hand sides on the index $(k + 1)$ has been eliminated. The process is too complex to be generalized and is best illustrated by an example.

**Example 6.13.** Consider the following transfer function, which is the same as that in Example 6.12:

$$H(z) = \frac{z^2 + 2z + 1}{z^2 + 5z + 6} = \frac{z^2 + 2z + 1}{(z + 2)(z + 3)}$$

Long division yields

$$H(z) = 1 + \frac{-3z - 5}{(z + 2)(z + 3)}$$

We shall now realize the second term as $\hat{H}(z)$ to give

$$\hat{H}(z) = \frac{-1}{z + 2}\,\frac{3z + 5}{z + 3}$$

So the system structure is as illustrated in Fig. 6.7. Now the state equations

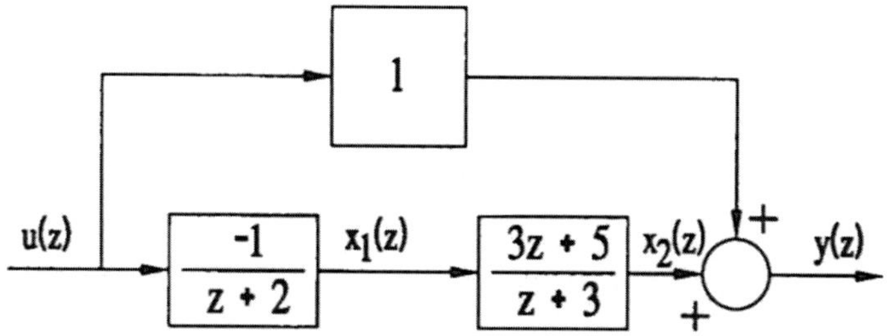


**Figure 6.7.** Cascade realization of Example 6.12.

are

$$x_1(k+1) = -2x_1(k) - u(k) \tag{a}$$

$$x_2(k+1) = -3x_2(k) + 3x_1(k+1) + 5x_1(k) \tag{b}$$

and we may eliminate the $x_1(k+1)$ term from (b) by substituting (a) to yield a second state equation of the proper form

$$x_2(k+1) = -3x_2(k) - x_1(k) - 3u(k)$$

so the matrix form of the state-variable representation is

$$\begin{bmatrix} x_1(k+1) \\ x_2(k+1) \end{bmatrix} = \begin{bmatrix} -2 & 0 \\ -1 & -3 \end{bmatrix} \begin{bmatrix} x_1(k) \\ x_2(k) \end{bmatrix} + \begin{bmatrix} -1 \\ -3 \end{bmatrix} u(k)$$

with an output given by

$$y(k) = [0 \quad 1] \begin{bmatrix} x_1(k) \\ x_2(k) \end{bmatrix} + u(k)$$

## Controllable Canonical Form

The controllable canonical form is sometimes also referred to as direct programming, although this should not be confused with what will be referred to as the direct form. We shall be concerned with transfer functions with numerators of lower order than denominators (otherwise, perform long division and form a feedforward of the input sequence). The transfer function of interest is given, after possible long division, to be

$$H(z) = \frac{b_{n-1}z^{n-1} + \cdots + b_0}{z^n + a_{n-1}z^{n-1} + \cdots + a_0} + d_n = \frac{y(z)}{u(z)} \tag{6.9.20}$$

Let us define an intermediate variable $w(z)$ such that

$$\frac{w(z)}{u(z)} = \frac{1}{z^n + a_{n-1}z^{n-1} + \cdots + a_0} \tag{6.9.21}$$

and thus

$$y(z) = (b_{n-1}z^{n-1} + \cdots + b_0)w(z) + d_n u(z) \tag{6.9.22}$$

Inverting (6.9.21) we get an $n$th-order difference equation

$$w(k+n) + a_{n-1}w(k+n-1) + \cdots + a_0 w(k) = u(k) \tag{6.9.23}$$

Inversion of (6.9.22) yields an output equation

$$y(k) = b_{n-1}w(k+n-1) + \cdots + b_0 w(k) + d_n u(k)$$

Define the state variables as

$$\begin{aligned} x_1(k) &= w(k) \\ x_2(k) &= w(k+1) = x_1(k+1) \\ x_3(k) &= w(k+2) = x_2(k+1) \\ &\vdots \\ x_n(k) &= w(k+n-1) = x_{n-1}(k+1) \end{aligned} \tag{6.9.24}$$

Substitution of relations (6.9.24) into expression (6.9.23) gives

$$x_n(k+1) = -a_{n-1}x_n(k) - a_{n-2}x_{n-1}(k) - \cdots - a_0x_1(k) + u(k) \tag{6.9.25}$$

with the other $(n-1)$ state equations defined by (6.9.24). The output relation becomes

$$y(k) = b_{n-1}x_n(k) + b_{n-2}x_{n-1}(k) + \cdots + b_0x_1(k) + d_nu(k) \tag{6.9.26}$$

The matrix form of this representation is

$$\begin{bmatrix} x_1(k+1) \\ x_2(k+1) \\ \vdots \\ \\ x_n(k+1) \end{bmatrix} = \begin{bmatrix} 0 & 1 & 0 & \cdots & 0 \\ 0 & 0 & 1 & \cdots & 0 \\ \vdots & & & & \\ 0 & & & \cdots & 1 \\ -a_0 & -a_1 & -a_2 & \cdots & -a_{n-1} \end{bmatrix} \begin{bmatrix} x_1(k) \\ x_2(k) \\ \vdots \\ \\ x_n(k) \end{bmatrix} + \begin{bmatrix} 0 \\ 0 \\ \vdots \\ \\ 1 \end{bmatrix} u(k) \tag{6.9.27}$$

with an output of

$$y(k) = [b_0 \quad \cdots \quad b_{n-1}] \begin{bmatrix} x_1(k) \\ x_2(k) \\ \vdots \\ x_n(k) \end{bmatrix} + d_nu(k) \tag{6.9.28}$$

The block diagram for this realization is shown in Fig. 6.8. An example will help to clarify this realization.

**Example 6.14.** Consider the transfer function of Example 6.12 when long division has been performed; it is of the form

$$H(z) = \frac{-3z - 5}{z^2 + 5z + 6} + 1$$

The state representation is given by inspection

$$x_1(k + 1) = x_2(k)$$

and from the denominator

$$x_2(k + 1) = -5x_2(k) - 6x_1(k) + u(k)$$

or in matrix form,

$$\begin{bmatrix} x_1(k + 1) \\ x_2(k + 1) \end{bmatrix} = \begin{bmatrix} 0 & 1 \\ -6 & -5 \end{bmatrix} \begin{bmatrix} x_1(k) \\ x_2(k) \end{bmatrix} + \begin{bmatrix} 0 \\ 1 \end{bmatrix} u(k)$$

$$y(k) = [-5 \quad -3] \begin{bmatrix} x_1(k) \\ x_2(k) \end{bmatrix} + u(k)$$

The block diagram for this example is shown in Fig. 6.9. Note that if the

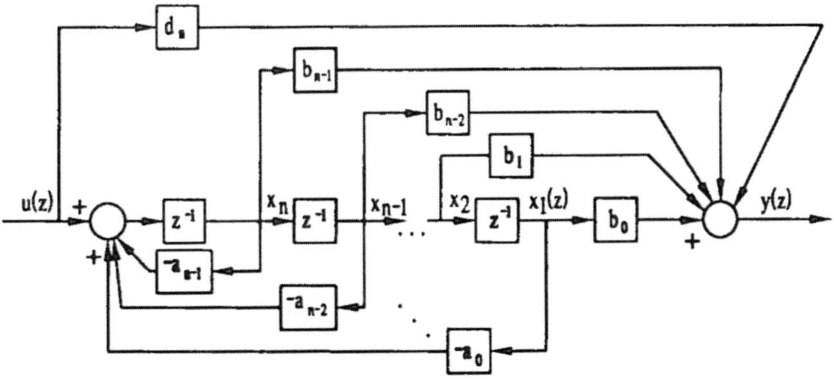


**Figure 6.8.** Controllable canonical form for state representation of a discrete-time transfer function.

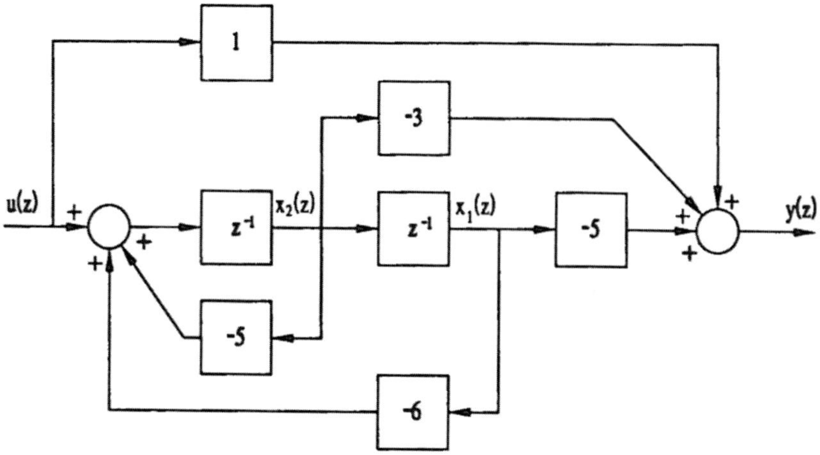


**Figure 6.9.** Controllable canonical realization of Example 6.12.

numerator polynomial is of lower order than the denominator, the long division will not be necessary; thus the $u(k)$ term will not appear explicitly in the $y(k)$ or output relation.

### Observable Canonical Form

The observable canonical form is sometimes referred to in the literature as the direct form. We start this discussion with the following form of the transfer function:

$$H(z) = \frac{d_n + d_{n-1}z^{-1} + \cdots + d_0 z^{-n}}{1 + a_{n-1}z^{-1} + \cdots + a_0 z^{-n}} \tag{6.9.29}$$

Let us define the state variables as follows:

$$\begin{aligned} x_1(k+1) &= -a_{n-1}y(k) + d_{n-1}u(k) + x_2(k) && \text{(a)} \\ x_2(k+1) &= -a_{n-2}y(k) + d_{n-2}u(k) + x_3(k) && \text{(b)} \\ &\vdots && \vdots \\ x_n(k+1) &= -a_0 y(k) + d_0 u(k) && (n) \end{aligned} \tag{6.9.30}$$

where

$$y(k) = x_1(k) + d_n u(k) \tag{6.9.31}$$

Expression (6.9.31) could be substituted into expressions (6.9.30) to give the following state equations:

$$\begin{aligned} x_1(k+1) &= -a_{n-1}x_1(k) + x_2(k) + (d_{n-1} - a_{n-1}d_n)u(k) \\ x_2(k+1) &= -a_{n-2}x_1(k) + x_3(k) + (d_{n-2} - a_{n-2}d_n)u(k) \\ &\ \vdots \\ x_n(k+1) &= -a_0x_1(k) + (d_0 - a_0d_n)u(k) \end{aligned} \tag{6.9.32}$$

To verify the validity of this representation, let us substitute relations (6.9.30) successively into (6.9.31). Substitute the first relation of (6.9.30) into (6.9.31) to give

$$\begin{aligned} y(k) = {} & -a_{n-1}u(k-1) + d_{n-1}u(k-1) \\ & + x_2(k-1) + d_nu(k) \end{aligned} \tag{6.9.33}$$

Then substitute the second to give

$$\begin{aligned} y(k) = {} & -a_{n-1}y(k-1) + d_{n-1}u(k-1) - a_{n-2}y(k-2) \\ & + d_{n-2}u(k-2) + x_3(k-2) + d_nu(k) \end{aligned} \tag{6.9.34}$$

and after substitution of the last of the state equations, the resulting form is

$$y(k) = -\sum_{i=1}^{n} a_{n-i}y(k-i) + \sum_{i=0}^{n} d_{n-i}u(k-i) \tag{6.9.35}$$

which is a difference equation which if $z$-transformed will give the transfer function of (6.9.29). The matrix form of the state-variable representation of (6.9.32) is

$$\begin{bmatrix} x_1(k+1) \\ x_2(k+1) \\ \vdots \\ x_n(k+1) \end{bmatrix} = \begin{bmatrix} -a_{n-1} & 1 & 0 & \cdots & 0 \\ -a_{n-2} & 0 & 1 & & \\ \vdots & & & & 1 \\ -a_0 & 0 & & \cdots & 0 \end{bmatrix} \begin{bmatrix} x_1(k) \\ x_2(k) \\ \vdots \\ x_n(k) \end{bmatrix} + \begin{bmatrix} d_{n-1} - a_{n-1}d_n \\ d_{n-2} - a_{n-2}d_n \\ \vdots \\ d_0 - a_0d_n \end{bmatrix} u(k) \tag{6.9.36}$$

with an output expression

$$y(k) = [1 \quad 0 \quad \cdots \quad 0]\begin{bmatrix} x_1(k) \\ x_2(k) \\ \vdots \\ x_n(k) \end{bmatrix} + d_n u(k) \tag{6.9.37}$$

If the numerator is of lower order than the denominator ($d_0 = 0$), the $b$ vector simplifies considerably. The block diagram representation of the system is as shown in Fig. 6.10. Again note that the $d_0$ block represents a direct feedforward of the input.

We shall now illustrate this technique with the example used to discuss all the previous forms.

**Example 6.15.** Consider the transfer function of the previous examples:

$$H(z) = \frac{z^2 + 2z + 1}{z^2 + 5z + 6} = \frac{1 + 2z^{-1} + z^{-2}}{1 + 5z^{-1} + 6z^{-2}}$$

The state-variable form can be given by inspection to be

$$\begin{bmatrix} x_1(k+1) \\ x_2(k+1) \end{bmatrix} = \begin{bmatrix} -5 & 1 \\ -6 & 0 \end{bmatrix}\begin{bmatrix} x_1(k) \\ x_2(k) \end{bmatrix} + \begin{bmatrix} -3 \\ -5 \end{bmatrix} u(k)$$

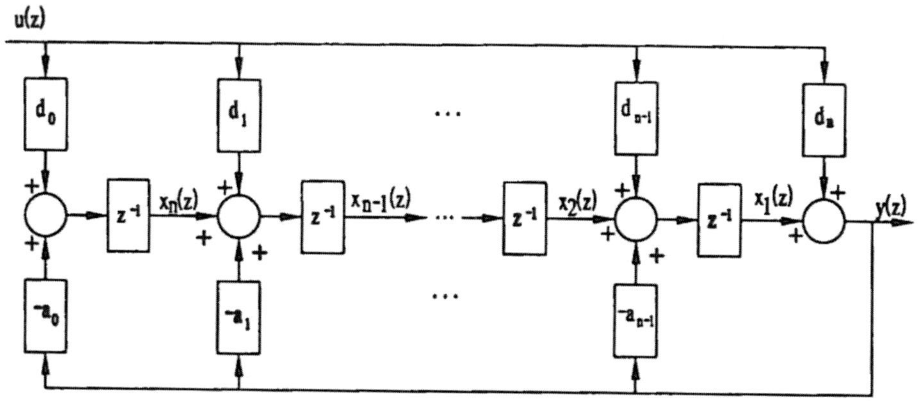


**Figure 6.10.** Block diagram of the observable canonical form.

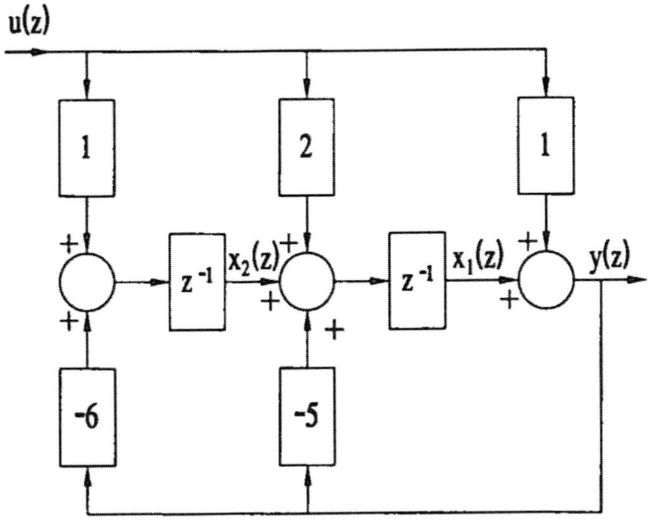


**Figure 6.11.** Observable canonical realization of Example 6.13.

with the output relation

$$y(k) = x_1(k) + u(k)$$

The block diagram for this system is shown in Fig. 6.11.

## 6.10 STATE-VARIABLE REPRESENTATION OF COMPOSITE CONTROL SYSTEMS

In this section we discuss briefly those systems that have purely continuous-time portions and purely discrete-time portions. The continuous-time portions are assumed to be driven by zero-order holds, and these outputs coupled to the digital computer will be coupled through ideal samplers. The reference inputs will be assumed to enter only the discrete-time portion of the system. Those states associated with the continuous dynamics, the zero-order hold, and the sampler will be denoted as $\mathbf{x}_c(k)$, while those associated with the discrete states are denoted as $\mathbf{x}_d(k)$ and the vector reference input sequence will be denoted as $\mathbf{r}(k)$. The entire system can be represented by the following set of state equations. The continuous-time dynamics with sampler and zero-order hold are governed by

$$\mathbf{x}_c(k + 1) = \mathbf{A}_{11}x_c(k) + \mathbf{B}_1\mathbf{u}(k) \tag{6.10.1}$$

with an output or measurement equation

$$\mathbf{y}_c(k) = \mathbf{C}_1\mathbf{x}_c(k) \tag{6.10.2}$$

where we have assumed that a feedforward of the input sequence was not required. The discrete compensator dynamics are then in state form,

$$\mathbf{x}_d(k + 1) = \mathbf{A}_{22}\mathbf{x}_d(k) + \mathbf{B}_2[\mathbf{r}(k) - \mathbf{y}_c(k)] \tag{6.10.3}$$

with an output equation that gives the control effort

$$\mathbf{u}(k) = \mathbf{C}_2\mathbf{x}_d(k) + \mathbf{D}_2[\mathbf{r}(k) - \mathbf{y}_c(k)] \tag{6.10.4}$$

These four equations are reflected in Fig. 6.12. All of these can be written as a composite state equation by substituting (6.10.2) into (6.10.3) and (6.10.4) and substitution of (6.10.4) into (6.10.1) to give

$$\begin{bmatrix} \mathbf{x}_c(k + 1) \\ \mathbf{x}_d(k + 1) \end{bmatrix} = \begin{bmatrix} \mathbf{A}_{11} - \mathbf{B}_1\mathbf{D}_2\mathbf{C}_1 & \mathbf{B}_1\mathbf{C}_2 \\ -\mathbf{B}_2\mathbf{C}_1 & \mathbf{A}_{22} \end{bmatrix}\begin{bmatrix} \mathbf{x}_c(k) \\ \mathbf{x}_d(k) \end{bmatrix} + \begin{bmatrix} \mathbf{B}_1\mathbf{D}_2 \\ \mathbf{B}_2 \end{bmatrix}\mathbf{r}(k) \tag{6.10.5}$$

Expression (6.10.5) will be useful in simulation studies and for evaluation of the closed-loop frequency response. This would allow us to check the

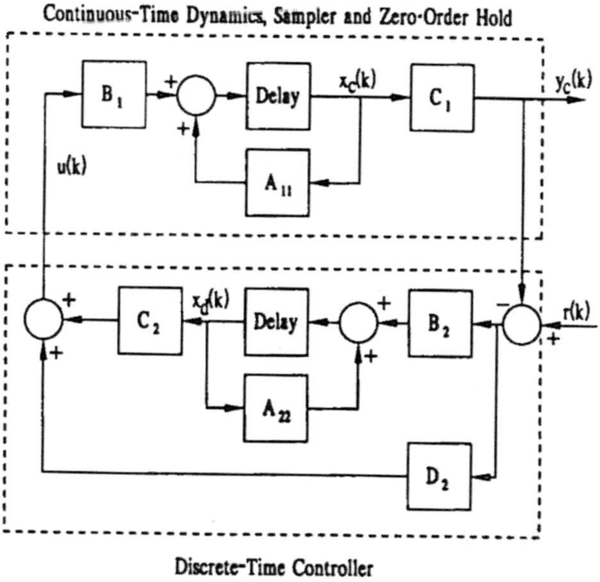


**Figure 6.12.** Composite control system block diagram.

ability of the system to track reference input sequence and whether frequency-domain or time-domain specifications have been met

**Example 6.16.** Consider the thermal control system with temperature $x_1(k)$ as the discrete-time output which was considered in Example 5.11. Find the state-variable representation of the composite system. The compensator chosen was of the form

$$D(z) = \frac{7(z - 0.528)(z - 0.952)}{(z - 1)(z - 0.2)} = \frac{U(z)}{E(z)}$$

The state representation of the continuous-time plant with zero-order hold and output sampler was given in Example 6.4 to be

$$\begin{bmatrix} x_{1c}(k+1) \\ x_{2c}(k+1) \end{bmatrix} = \begin{bmatrix} 0.6277 & 0.3606 \\ 0.09016 & 0.8526 \end{bmatrix} \begin{bmatrix} x_{1c}(k+1) \\ x_{2c}(k+1) \end{bmatrix} + \begin{bmatrix} 0.02516 \\ 0.1147 \end{bmatrix} u(k)$$

with output

$$y(k) = [1 \quad 0] \begin{bmatrix} x_{1c}(k) \\ x_{2c}(k) \end{bmatrix}$$

The compensator transfer function can be long-divided to yield

$$D(z) = 7 + \frac{-1.96z + 2.119}{z^2 - 1.2z + 0.2} = \frac{U(z)}{R(z) - Y(z)}$$

The second term can be represented by the controllable canonical form as

$$\begin{bmatrix} x_{1d}(k+1) \\ x_{2d}(k+1) \end{bmatrix} = \begin{bmatrix} 0 & 1 \\ -0.2 & 1.2 \end{bmatrix} \begin{bmatrix} x_{1d}(k) \\ x_{2d}(k) \end{bmatrix} + \begin{bmatrix} 0 \\ 1 \end{bmatrix} [r(k) - y(k)]$$

The output is given by the numerator and the constant term after long division as

$$u(k) = [2.119 \quad -1.96] \begin{bmatrix} x_{1d}(k) \\ x_{2d}(k) \end{bmatrix} + 7[r(k) - y(k)]$$

Inserting all these matrices into expression (6.10.5), the composite system model is

$$\begin{bmatrix} x_{1c}(k+1) \\ x_{2c}(k+1) \\ x_{1d}(k+1) \\ x_{2d}(k+1) \end{bmatrix}$$

$$
= \begin{bmatrix} 0.4513 & 0.3606 & 0.0531 & -0.04931 \\ -0.715 & 0.8526 & 0.24305 & -0.2248 \\ 0 & 0 & 0 & 1 \\ -1 & 0 & -0.2 & 1.2 \end{bmatrix} \begin{bmatrix} x_{1c}(k) \\ x_{2c}(k) \\ x_{1d}(k) \\ x_{2d}(k) \end{bmatrix} + \begin{bmatrix} 0.17612 \\ 0.8029 \\ 0 \\ 1 \end{bmatrix} r(k)
$$

The unit-step response of this composite control system can now be evaluated easily by a digital computer to yield the response shown in Fig. 6.13.

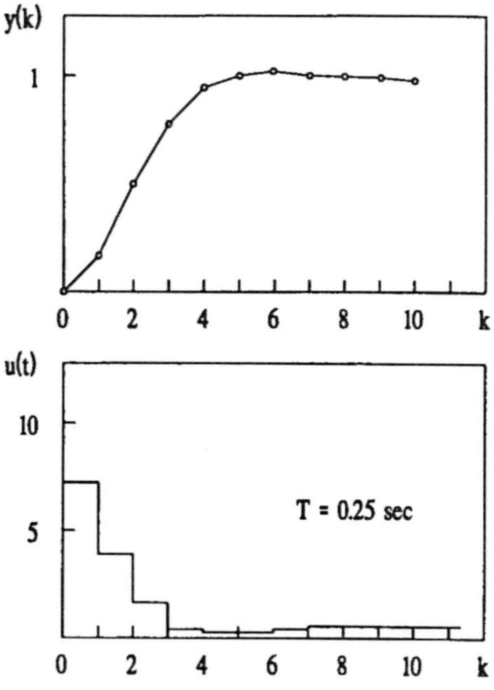


**Figure 6.13.** Step response and control effort for compensated thermal system.

## 6.11 SUMMARY

In this chapter we have introduced the concept of state-variable representation for continuous-time control system elements and the discrete-time state representation was developed for these systems driven by digital-to-analog converter and with sampled system outputs. In the former problem the representation is in the form of *n* coupled first-order differential equations, while the latter representation is in the form of *n* coupled first-order difference equations. Solution techniques have been developed for both systems, and their relations to transfer functions and frequency response have been discussed.

The important concepts of controllability and observability have been discussed and the necessary and sufficient conditions developed. Various state-variable representations of discrete-time systems originally specified as a transfer function are developed and the state-variable representation of systems that have some discrete-time elements and some continuous-time elements is developed in the context of using it as a simulation tool.

The material in this chapter is preliminary in nature and should serve to introduce the topics in the remainder of the book.

## PROBLEMS

6.1. For the continuous-time dynamic plant shown, accomplish the following tasks:

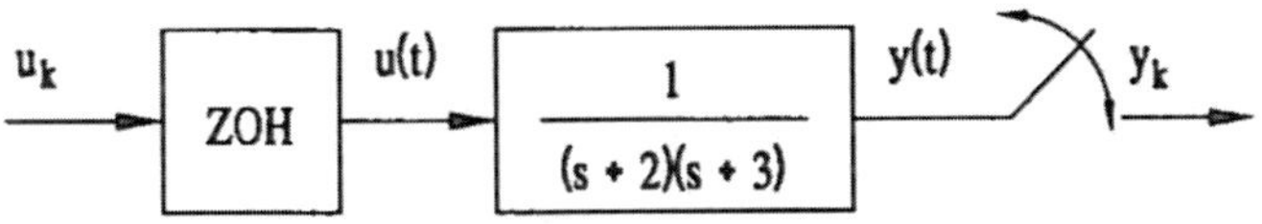


(a) Find the continuous-time state-variable description of each system, identifying **F**, **G**, and **C**. Use the following state variables:

$$x_1 = y$$
$$x_2 = \dot{y}$$

(b) Find the discrete-time models of the systems above. This discrete-time model should be between $u_k$ and $y_k$. Find the **A**, **B**, and **C** matrices.

6.2. The positioning system described by the transfer function

$$G(s) = \frac{1}{s(s+1)} = \frac{Y(s)}{U(s)}$$

is also describable by state equations by letting $x_1 = y$ and $x_2 = \dot{y} = \dot{x}_1$.

(a) Show that the continuous-time state equations are

$$\frac{d}{dt}\begin{bmatrix} x_1 \\ x_2 \end{bmatrix} = \begin{bmatrix} 0 & 1 \\ 0 & -1 \end{bmatrix}\begin{bmatrix} x_1 \\ x_2 \end{bmatrix} + \begin{bmatrix} 0 \\ 1 \end{bmatrix} u(t)$$

with

$$y(t) = [1 \quad 0]\begin{bmatrix} x_1 \\ x_2 \end{bmatrix}$$

(b) Show that the state transition matrix is

$$\mathbf{\Phi}(t) = \begin{bmatrix} 1 & 1 - e^{-t} \\ 0 & e^{-t} \end{bmatrix}$$

(c) Show that the **A** matrix for a sampling interval of $T = 1$ s is

$$\mathbf{A} = \mathbf{\Phi}(1) = \begin{bmatrix} 1 & 0.632 \\ 0 & 0.368 \end{bmatrix}$$

(d) Show that the discrete **b** matrix is

$$\mathbf{b} = \begin{bmatrix} 0.368 \\ 0.632 \end{bmatrix}$$

6.3. With the **A** and **b** matrices given in Problem 6.2, find the discrete-time transfer function $G(z)$ from the relation

$$G(z) = \mathbf{c}^T[z\mathbf{I} - \mathbf{A}]^{-1}\mathbf{b}$$

6.4. Draw the block diagram for the continuous-time system of Problem 6.1 using integrators, multipliers, and summers only.

6.5. Draw the block diagram for the discrete-time system of Problem 6.1 using delays, multipliers, and summers only.

6.6. If we consider the fluid flow system illustrated below and write the differential equations for the mass balance, we get

$$A_1 \frac{dx_1}{dt} = \frac{1}{R_1}(x_2 - x_1) + u(t)$$

and

$$A_2 \frac{dx_2}{dt} = \frac{1}{R_1}(x_1 - x_2) - \frac{1}{R_2} x_1$$

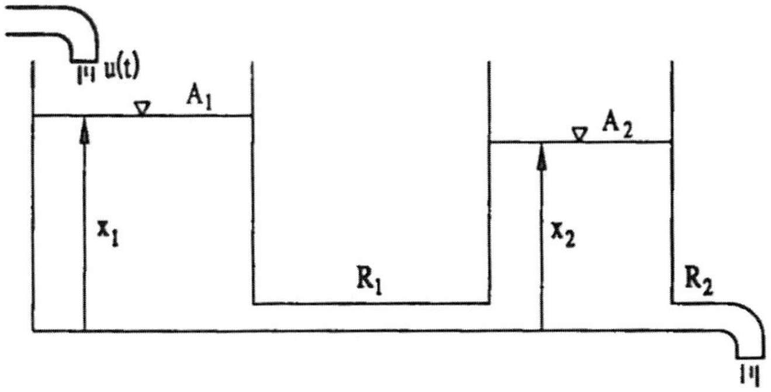


(a) For the parameters $A_1 = A_2 = 1$, $R_1 = 0.5$, and $R_2 = 0.333$, show that the system matrices are

$$\mathbf{F} = \begin{bmatrix} -2 & 2 \\ 2 & -5 \end{bmatrix} \qquad \mathbf{g} = \begin{bmatrix} 1 \\ 0 \end{bmatrix}$$

(b) Find the state-transition matrix $\mathbf{\Phi}(t)$.

(c) Show that for a sampling interval of $T = 0.1$ s, the discrete-time system matrices are

$$\mathbf{A} = \begin{bmatrix} 0.8336 & 0.1424 \\ 0.1424 & 0.6200 \end{bmatrix} \qquad \mathbf{b} = \begin{bmatrix} 0.0912 \\ 0.0080 \end{bmatrix}$$

6.7. Investigate the controllability of a system with the following **A** and **b** matrices:

$$\mathbf{A} = \begin{bmatrix} 0.5 & 0.6 \\ 0.2 & 0.4 \end{bmatrix} \qquad \mathbf{b} = \begin{bmatrix} 2 \\ 1 \end{bmatrix}$$

6.8. Investigate the controllability of a system with the same **A** matrix as Problem 6.7 and with

$$\mathbf{b} = \begin{bmatrix} 1 \\ 1 \end{bmatrix}$$

6.9. Investigate the observability of the system of Problem 6.7 with

$$\mathbf{c}^T = [1 \quad 1]$$

6.10. Investigate the observability of the system of Problem 6.7 with

$$\mathbf{c}^T = [1 \quad -2]$$

6.11. Given the first-order system shown, (a) find the continuous-time state-variable representation of the plant alone; (b) find the discrete-time state equation for $T = 0.1$ s.

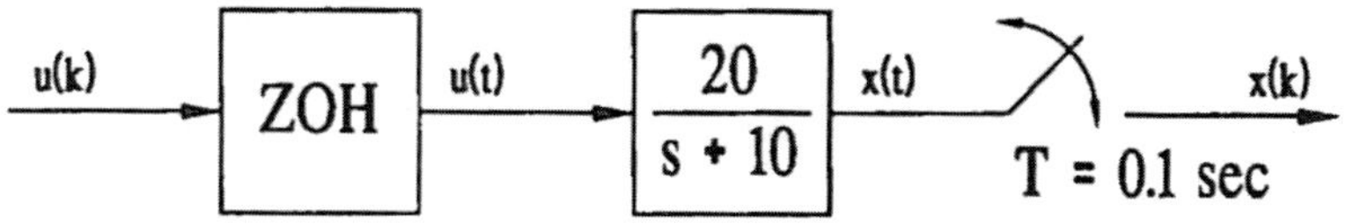


6.12. The position control of a sprung mass was discussed in Problem 3.17 and was governed by the following differential equation:

$$\ddot{y} + 4\dot{y} + 400y = 2u(t)$$

(a) Find the state-variable representation if the displacement and velocity of the mass are taken as the state variables, or

$$\mathbf{x}(t) = \begin{bmatrix} y \\ \dot{y} \end{bmatrix}$$

(b) Sketch the signal-flow diagram for this system. If the system is driven by a zero-order hold and followed by a sampler sampling the displacement $x_1(t)$, find the discrete-time state-variable description for $T = 0.05$ s. This may be done using Laplace transforms or the matrix exponential series. (Check $a_{11} = 0.569$ and $a_{12} = 0.0381$.)

6.13. Find the state-variable representation in the controllable canonical form for the two-tank system with transfer function found in Problem 3.12. (Note that the states are now no longer the respective tank levels.)

$$G(z) = \frac{H_2(z)}{U(z)} = 0.008 \frac{z + 0.790}{z^2 - 1.4536z + 0.4966}$$

6.14. Find the state-variable representation in the observable canonical form for the transfer function of Problem 6.13.

6.15. In Example 5.11 a compensator transfer function of the form

$$D(z) = \frac{U(z)}{E(z)} = \frac{7(z - 0.528)(z - 0.952)}{(z - 1)(z - 0.2)}$$

was found to compensate the thermal system adequately. Put this transfer function in controllable canonical state-variable form.

## REFERENCES


Astrom, K. J., and B. Whittenmark, 1984. *Computer Controlled System: Theory and Design*, Prentice Hall, Englewood Cliffs, NJ.

Brogan, W. L., 1985. *Modern Control Theory*, 2nd Ed., Prentice Hall, Englewood Cliffs, NJ.

Chen, C. T., 1984. *Linear Systems: Analysis and Design*, Holt, Rinehart and Winston, New York.

De Russo, P. M., R. F. Roy, and C. M. Close, 1965. *State Variables for Engineers*, Wiley, New York.

Franklin, G. F., J. D. Powell, and M. L. Workman, 1990. *Digital Control of Dynamic Systems*, 2nd Ed., Addison-Wesley, Reading, MA.

Friedland, B., 1986. *Control System Design: An Introduction to State Space methods*, McGraw-Hill, New York.

Gunckel, T. F., and G. F. Franklin, 1963. A general solution for linear sampled-data control systems, *Trans. ASME: Journal of Basic Engineering*, *85D*: 197–203.

Joseph, P. D., and J. T. Tou, 1961. On linear control theory, *Trans. AIEE*, *80*(2): 193–196.

Kailath, T., 1980. *Linear Systems*, Prentice Hall, Englewood Cliffs, NJ.

Kalman, R. E., 1960. A new approach to linear filtering and prediction problems, *Trans. ASME: Journal of Basic Engineering*, *82D*: 35–45.

Kalman R. E., 1963. The theory of optimal control and the calculus of variations, in *Mathematical Optimization Techniques*, R. Bellman, Ed., University of California Press, Berkeley, CA.

Moler, C., and C. van Loan, 1978. Nineteen dubious ways to compute the exponential of a matrix, *SIAM Review*, *20*(2): 83–86.

Phillips, C. L., and H. T. Nagle, Jr., 1990. *Digital Control System Analysis and Design*, 2nd Ed., Prentice Hall, Englewood Cliffs, NJ.

Takahashi, Y., M. F. Rabins, and D. M. Auslander, 1970. *Control and Dynamic Systems*, Addison-Wesley, Reading, MA.

Wiener, N., 1949. *The Interpolation and Smoothing of Stationary Time Series*, MIT Press, Cambridge, MA.

# 7 Quantization and Error Effects

## 7.1 INTRODUCTION

To this point in the book the numbers involved in digital control algorithms (both digital filter coefficients and the variables) have been treated as if they could be represented to as many decimal places as needed. In reality, since all digital processors deal with numbers in binary form, they must, by necessity, be represented by a binary word composed of a finite number of binary digits or bits. The finiteness of the word length in the representations of these numbers, which in general require an infinite number of bits, means that nearly all the numbers represented in the processor will be in error to some degree.

In this chapter we consider the effects of these errors on digital filters or compensators and then the effect on the closed-loop digital control system as a whole. The errors may be classified into three types, which are treated separately here. Type 1 errors are analog-to-digital (A/D) conversion errors where the samples of an analog signal are to be represented by a finite-length binary word. These errors are commonly referred to as quantization errors and can be handled statistically as random "noise" sequences injected into the control loop. In the past decade, which has seen the introduction of 16- and 32-bit processors, these errors have become less significant than they were in the time of 8-bit processors. Type 2 errors are multiplication errors where the product of two finite-length binary words must be represented by another word of the same length. This type of error can be treated by the same statistical techniques as type 1 errors and, as we shall show, they are also a function of the way in which the structure of a digital filter is realized. Type 3 errors are referred to as *parameter storage errors* because control algorithm coefficients must be stored as finite-length words. We know that both pole and zero locations and frequency response characteristics of analog and digital filters are very sensitive to changes in the coefficient values in the transfer function and hence the need for this consideration. The effect of these errors can be treated by investigating mathematically the sensitivity of pole locations to changes in coefficient values.

## 7.2 QUANTIZATION ERRORS (TYPE 1 ERRORS)

In Chapter 3 we treated the analog-to-digital converter as if it gave us an exact representation of the size of the input voltage to the analog-to-digital converter. In this section we shall discuss the analog-to-digital (A/D) converter, which can be thought of as the combination of three elements: an ideal sampler, a quantizer, and a binary coder illustrated in Fig. 7.1. The ideal sampler produces an evenly spaced temporal sequence of numbers as discussed in Chapter 3, while the quantizer represents these numbers by a set of finitely spaced levels of 0, $\pm q$, $\pm 2q$, and so on. The binary coder codes these quantized samples into a binary form for processing by the digital processor.

The input–output character for a rounding quantizer is shown in Fig. 7.2a, and that for a truncating quantizer is as shown in Fig. 7.2b. Let us now consider the rounding quantizer as given in Fig. 7.2a and a coder that will code the output levels into 4-bit binary numbers. Two schemes for this coding operation are given in Table 7.1. We see that 15 levels of quantization result if sign-magnitude coding is employed, while 16 levels result if 2's complement is employed. In general, if the coding is to be in $n$ bits, 2's-complement coding will result in $2^n$ levels, while sign-magnitude coding

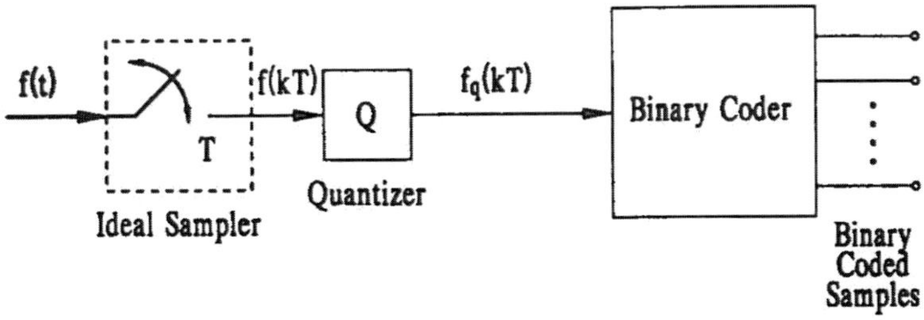


**Figure 7.1.** Analog-to-digital converter conceptualization.

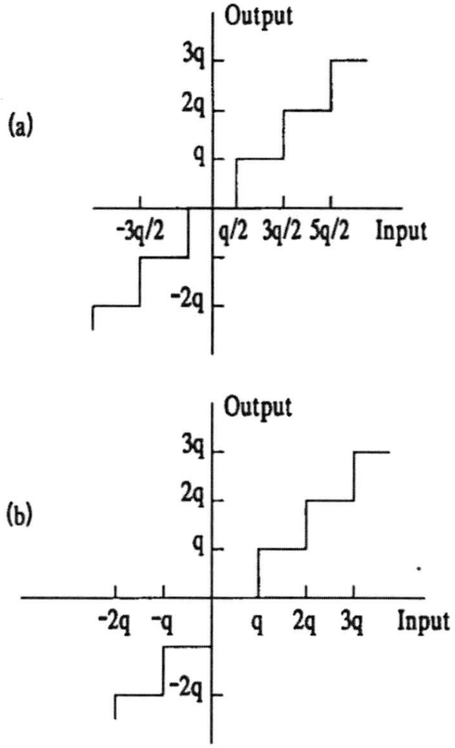


**Figure 7.2.** Two possible analog-to-digital converter input–output characteristics: (a) rounding A/D; (b) truncating A/D.

will yield $2^n - 1$ levels. The preferred technique today is to use 2's-complement representation for the negative numbers because most modern digital processors perform the digital arithmetic in 2's-complement form. There are other coding schemes which have built-in redundancy that yield error-detecting codes.

Let us consider a signal input to a rounding A/D converter with the character of Fig. 7.2a. The input signal and the output sample sequence are shown in Fig. 7.3 along with the error sequence, which is the difference between the samples and the original function at the sampling instants.

If the input to analog-to-digital converter has a full-scale voltage of $V_{fs}$ and the output is coded in $n$ bits, there are $2^n$ levels of the input that can be represented, so the quantization level $q$ is

$$q = \frac{V_{fs}}{2^n}$$

**Table 7.1.** Several Schemes of Binary Coding of Quantizer Outputs

| Output | Sign-magnitude code | 2's-Complement code |
|---|---|---|
| $7q$ | 0111 | 0111 |
| $6q$ | 0110 | 0110 |
| $5q$ | 0101 | 0101 |
| $4q$ | 0100 | 0100 |
| $3q$ | 0011 | 0011 |
| $2q$ | 0010 | 0010 |
| $q$ | 0001 | 0001 |
| 0 | 0000 | 0000 |
| $-q$ | 1001 | 1111 |
| $-2q$ | 1010 | 1110 |
| $-3q$ | 1011 | 1101 |
| $-4q$ | 1100 | 1100 |
| $-5q$ | 1101 | 1011 |
| $-6q$ | 1110 | 1010 |
| $-7q$ | 1111 | 1001 |
| $-8q$ | — | 1000 |

**Example 7.1.** If we have an 8-bit A/D converter that covers the input range of $\pm 10$ V, the quantization level is

$$q = \frac{20}{2^8} 0.078125 \text{ V} = 78 \text{ mV}$$

From this example it is clear that these errors are small, but if only a 4-bit A/D were used, the error would be 16 times higher and most surely significant. Whether or not these errors are significant is determined by the particular application and the degree of precision required in the control task.

If the input signal traverses a reasonable portion of the full scale of the analog-to-digital converter range, it is reasonable to assume that the error sequence $e(k)$ is a sequence of independent uniformly distributed random variables on the interval $(-q/2, q/2)$. Examination of the typical error sequence of Fig. 7.3 illustrates that this is probably a good assumption. These assumptions were first discussed from a theoretical point of view by Widrow (1956) and later in the text by Gold and Rader (1969). In addition to the nature of the probabilistic structure of the error sequence, we shall assume that this structure is time-invariant or stationary. Let us now investigate some of the probabilistic aspects of these assumptions. We know

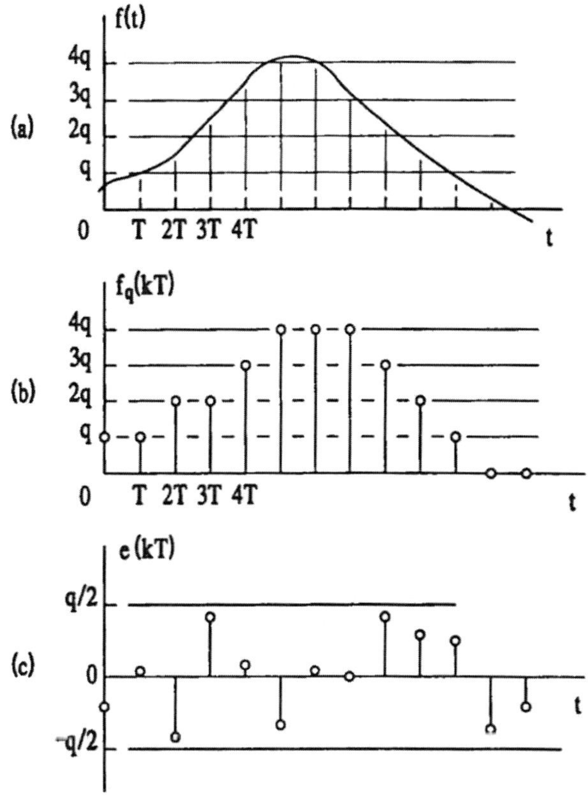


**Figure 7.3.** (a) Continuous function to be sampled in quantized form; (b) quantized samples; (c) quantization error sequence.

that the probability density function of the random variable $e(k)$ is a function $p_e(x)$ such that

$$p_e(x)\,dx = \text{Prob}[x \leq e(k) < x + dx] \tag{7.2.1}$$

The uniform density function for the random variable $e(k)$ on the interval $(-q/2, q/2)$ is shown in Fig. 7.4. The statistical average or expected value of any function of the random variable $f(e(k))$ is given by

$$E[f(e(k))] = \int_{-\infty}^{\infty} f(x)p_e(x)\,dx \tag{7.2.2}$$

where $p_e(x)$ is the probability density function of the error sequence. When

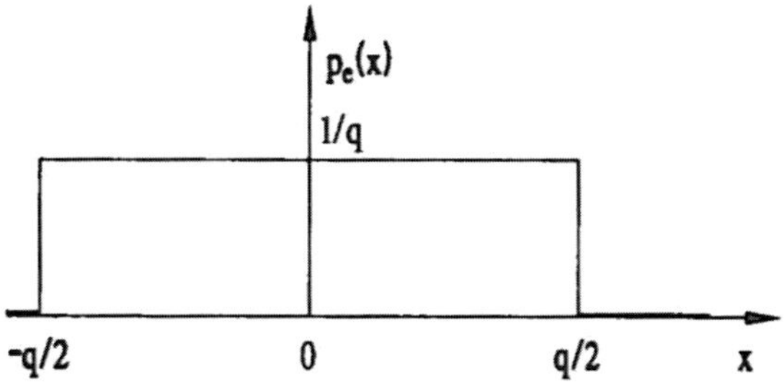


**Figure 7.4.** Density function for uniformly distributed quantization errors.

$f(e(k))$ is $e(k)$ itself, the expected value of the random variable itself is

$$E[e(k)] = \int_{-\infty}^{\infty} x p_e(x)\, dx \tag{7.2.3}$$

and for this particular density function

$$E[e(k)] = \frac{1}{q}\int_{-q/2}^{q/2} x\, dx = \frac{1}{q}\frac{x^2}{x}\bigg|_{q/2}^{q/2} = 0 \tag{7.2.4}$$

This zero average value is the property given by the symmetry of the density function of Fig. 7.4 about the origin. The mean-square value of the random error sequence $e(k)$ is given by the second moment of the density function about the origin, or

$$E[e^2(k)] = \int_{-\infty}^{\infty} x^2 p_e(x)\, dx \tag{7.2.5}$$

and for the uniform density function the mean-square value is

$$E[e^2(k)] = \frac{1}{q}\int_{-q/2}^{q/2} x^2\, dx = \frac{q^2}{12} \tag{7.2.6}$$

The statistical independence of the elements of the error sequence implies that they are uncorrelated, or

$$E[e(k)e(j)] = \begin{cases} 0 & j \neq k \\ \dfrac{q^2}{12} & j = k \end{cases} \tag{7.2.7}$$

If we employ the impulse notation of Chapter 2, this is written simply as

$$E[e(k)e(j)] = \frac{q^2}{12}\,\delta(j - k) \tag{7.2.8}$$

This type of noise, which is uncorrelated from sample to sample, is sometimes referred to as discrete-time *white noise*.

The development above has been for the rounding quantizer, while a similar analysis for a truncating quantizer, as illustrated in Fig. 7.2b, will yield errors that are uniformly distributed on $(0, q)$ with a mean $q/2$ and mean-square value of $q^2/3$. Thus the variance (the mean square about the mean) is

$$E[(e(k) - E[e(k)]^2)^2] = \frac{q^2}{12} \tag{7.2.9}$$

## 7.3 RESPONSE OF A DISCRETE TRANSFER FUNCTION TO QUANTIZATION ERRORS

As stated in Section 7.2, the quantization errors appear as noise at the input to a discrete transfer function or as noise injected into the loop. Let us consider the discrete transfer function of Fig. 7.5, with $H(z)$ defined by

$$H(z) = \frac{Y(z)}{E(z)} \tag{7.3.1}$$

In Section 2.12 we have shown that the response of this discrete system can be calculated by discrete-time convolution as

$$y(k) = \sum_{m=0}^{k} h(m)e(k - m) \tag{7.3.2}$$

where the $h(m)$ sequence is the impulse response sequence given by the inverse $z$-transform of the transfer function $H(z)$. We shall assume that the $e(k)$ sequence is stationary (i.e., its statistical properties are not time dependent) and that the digital filter $H(z)$ is stable or that the poles of $H(z)$ lie within the unit circle. We shall take the upper limit on the sum to be infinity to ensure the stationarity of the output sequence $y(k)$, or

$$y(k) = \sum_{m=0}^{\infty} h(m)e(k - m) \tag{7.3.3}$$

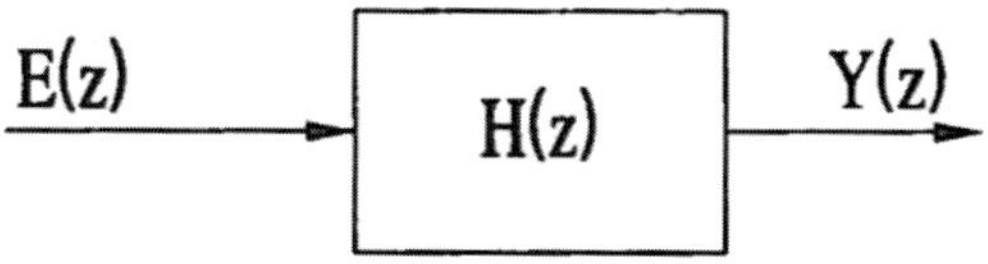


**Figure 7.5.** Discrete transfer function driven by quantization error sequence.

Let us now investigate the expected value of the output sequence

$$E[y(k)] = E\left[\sum_{m=0}^{\infty} h(m)e(k - m)\right] \tag{7.3.4}$$

Interchanging the operations of expectation and summation yields

$$E[y(k)] = \sum_{m=0}^{\infty} h(m)E[e(k - m)] \tag{7.3.5}$$

and if the error sequence is stationary, we may write

$$E[y(k)] = E[e(k)] \sum_{m=0}^{\infty} h(m) \tag{7.3.6}$$

But we know that since $H(z)$ is the $z$-transform of $h(m)$,

$$H(z) = \sum_{m=0}^{\infty} h(m)z^{-m} \tag{7.3.7}$$

Expression (7.3.6) may now be rewritten as

$$E[y(k)] = E[e(k)]H(z)|_{z=1} \tag{7.3.8}$$

but $H(z)$ evaluated at $z = 1$ is equivalent to $H(e^{j\omega T})$ evaluated at $\omega = 0$ or $H(1)$, which is the dc gain of the discrete transfer function. For the rounding quantizer, the zero mean error implies a zero mean output sequence.

By employing (7.3.3) the expression for the square of the stationary output sequence may be formed, or

$$y^2(k) = \sum_{m=0}^{\infty} \sum_{n=0}^{\infty} h(m)h(n)e(k - m)e(k - n) \tag{7.3.9}$$

The expected value of this yields the mean-square output, or

$$E[y^2(k)] = \sum_{m=0}^{\infty} \sum_{n=0}^{\infty} h(m)h(n)E[e(k - m)e(k - n)] \tag{7.3.10}$$

But from the property of (7.2.7) the expected value involved in (7.3.10) is then

$$E[e(k - m)e(k - n)] = \begin{cases} 0 & m \neq n \\ \dfrac{q^2}{12} & m = n \end{cases} \tag{7.3.11}$$

The mean-square output then becomes

$$E[y^2(k)] = \frac{q^2}{12} \sum_{m=0}^{\infty} h^2(m) \tag{7.3.12}$$

Now let us see if we can obtain a relationship similar to (7.3.12) in the $z$-domain. Consider the product $H(z)H(z^{-1})$, or, by definition of the $z$-transform, we get

$$H(z)H(z^{-1}) = \sum_{m=0}^{\infty} h(m)z^{-m} \sum_{n=0}^{\infty} h(n)z^{n} \tag{7.3.13}$$

or

$$H(z)H(z^{-1}) = \sum_{m=0}^{\infty} \sum_{n=0}^{\infty} h(m)h(n)z^{n-m} \tag{7.3.14}$$

Now consider the contour integral

$$I_{mn} = \frac{1}{2\pi j} \oint_c h(m)h(n)z^{n-m-1}\, dz \tag{7.3.15}$$

If we employ the residue theorem from complex variable theory, we can show that this integral is zero unless $m = n$, and then

$$I_{mn} = \begin{cases} h^2(m) & m = n \\ 0 & m \neq n \end{cases} \tag{7.3.16}$$

Now let us return to expression (7.3.14) and multiply by $z^{-1}$ and integrate both sides around a contour on the unit circle, which yields

$$\oint_c H(z)H(z^{-1})z^{-1}\, dz = \oint_c \sum_{m=0}^{\infty} \sum_{n=0}^{\infty} h(m)h(n)z^{n-m-1}\, dz \tag{7.3.17}$$

and employing relation (7.3.16),

$$\oint_c H(z)H(z^{-1})z^{-1}\, dz = 2\pi j \sum_{m=0}^{\infty} h^2(m) \tag{7.3.18}$$

Now let us substitute (7.3.18) into (7.3.12) to yield the desired expression

$$E[y^2(k)] = \frac{q^2}{12} \frac{1}{2\pi j} \oint_c H(z)H(z^{-1})z^{-1}\, dz \tag{7.3.19}$$

where the method of residues may be used to evaluate the contour integral. An example will serve to clarify the technique.

**Example 7.2.** Consider a filter with transfer function

$$H(z) = \frac{Y(z)}{E(z)} = \frac{Bz}{z - \alpha} \qquad 0 < \alpha < 1$$

with impulse response sequence

$$h(k) = B\alpha^k \qquad k > 0$$

Find the mean-square output from the filter due to a rounding quantizer with quantization level $q$. The mean-square value of the quantization error is given to be

$$E[e^2(k)] = \frac{q^2}{12}$$

The mean-square output is, then, from expression (7.3.12),

$$E[y^2(k)] = \frac{q^2}{12} \sum_{m=0}^{\infty} B^2\alpha^{2m}$$

The closed form of this geometric sum is easily obtainable from a set of math tables, or we may obtain the closed-form result by the method of contour integration of expression (7.3.19), or

$$E[y^2(k)] = \frac{q^2B^2}{12} \frac{1}{2\pi j} \oint_c \frac{z}{z - \alpha} \frac{z^{-1}}{z^{-1} - \alpha} z^{-1}\, dz$$

or, rewriting,

$$E[y^2(k)] = \frac{q^2B^2}{12} \frac{1}{2\pi j} \oint_c \frac{1}{z - \alpha} \frac{1}{1 - \alpha z}\, dz$$

where the contour $C$ is taken on the unit circle. Application of the residue theorem yields

$$\begin{aligned} E[y^2(k)] &= \frac{q^2B^2}{12} \frac{1}{2\pi j} 2\pi j \left[ z - \alpha \frac{1}{z - \alpha} \frac{1}{1 - \alpha z} \right]_{z=\alpha} \\ &= \frac{q^2B^2}{12} \frac{1}{1 - \alpha^2} \end{aligned}$$

**Example 7.3.** Consider the digital control system shown in Fig. 7.6, where the control action is a proportional controller with a given $K = 0.4$. Find the root-mean-squared error in the output due to quantization in the analog-to-digital converter. The first thing to accomplish is to evaluate the transfer function between the quantization noise $Q(z)$ and the output $Y(z)$. The output is

$$Y(z) = \frac{0.4}{z - 0.8} [-Q(z) - Y(z)]$$

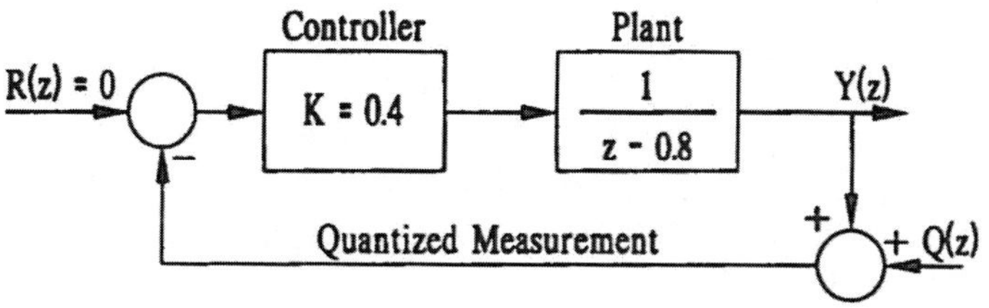


**Figure 7.6.** Discrete-time control system with quantized feedback.

Collecting like terms yields a transfer function

$$\frac{Y(z)}{Q(z)} = H(z) = \frac{-0.4}{z - 0.4}$$

Then

$$H(z^{-1}) = \frac{z}{z - 2.5}$$

and

$$z^{-1}H(z^{-1}) = \frac{1}{z - 2.5}$$

The integrand of the complex contour integral is

$$H(z)H(z^{-1})z^{-1} = \frac{-0.4}{(z - 0.4)(z - 2.5)}$$

Employing the residue theorem to evaluate the mean-square output, we get

$$E[y^2(k)] = \frac{q^2}{12}(-0.4)\left(\frac{1}{z - 2.5}\right)\bigg|_{z=0.4}$$

or evaluating all this, we obtain

$$E[y^2(k)] = \frac{q^2}{12}\left(\frac{-0.4}{-2.1}\right) = 0.1904\,\frac{q^2}{12}$$

or the root-mean-square error is

$$\sqrt{E[y^2(k)]} = 0.125q$$

**Example 7.4.** Consider the effect of quantization on the output response of the thermal control system of Example 3.6. The analog-to-digital converter is assured to be a 2's-complement 8-bit device with a range of $\pm 10$ output units, so the quantization level is

$$q = \frac{20}{2^8} = 0.0781$$

The system is shown in Fig. 7.7, where $Q(z)$ represents the quantization noise. The mean-square quantization error is

$$E[q^2(k)] = \frac{q^2}{12} = 5.08 \times 10^{-4}$$

As in Example 7.3, we could calculate the transfer function between $Q(z)$ and the output $T_1(z)$, which is

$$\frac{T_1(z)}{Q(z)} = H(z) = \frac{-0.0625(z + 0.816)}{z^2 - 1.418z + 0.5536}$$

or in factored form,

$$H(z) = \frac{-0.0625(z + 0.816)}{(z - 0.709 + j0.226)(z - 0.709 - j0.226)}$$

After some algebra we get for $z^{-1}H(z^{-1})$,

$$z^{-1}H(z^{-1}) = \frac{-0.0921(z + 1.225)}{z^2 - 2.56z + 1.806}$$

The poles of the integrand, which lie interior to the unit circle, are the closed-loop poles, or

$$z_{1,2} = 0.709 \pm j0.226$$

The integrand is of the form

$$z^{-1}H(z^{-1})H(z) = \frac{0.00576(z + 0.816)(z + 1.125)}{(z^2 - 2.56z + 1.806)(z - 0.709 + j0.226)(z - 0.709 - j0.226)}$$

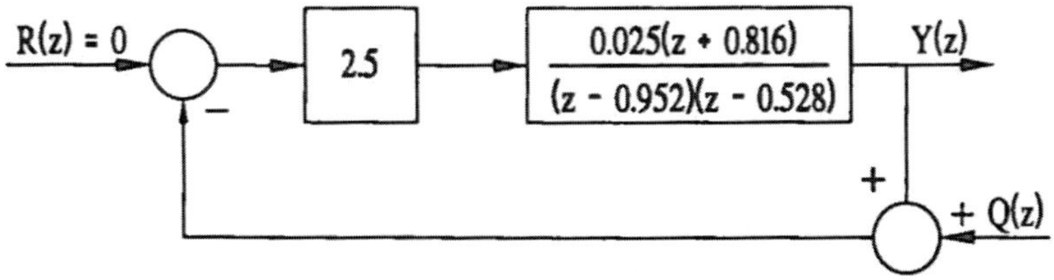


**Figure 7.7.** Thermal control system with quantized output.

Evaluating the integral by application of the residue theorem for complex variable theory gives after much tedious, complex algebra,

$$E[y^2(k)] = \frac{0.00576q^2}{12}[12.29\angle 44.35° + 12.29\angle -44.35°]$$

or the mean-square output due to the quantization noise is

$$E[y^2(k)] = \frac{q^2}{12}(0.00576)(17.5) = 0.0084q^2$$

and the root-mean-square (rms) output is

$$\sqrt{E[y^2(k)]} = 0.0917q$$

If we evaluate this for the quantization level earlier specified ($q = 0.0781$), then

$$\sqrt{E[y^2(k)]} = 0.00716$$

## 7.4 BOUND ON THE OUTPUT MAGNITUDE (BERTRAM'S BOUND)

Given the system of Fig. 7.5, we have already shown how to calculate the mean and mean-square outputs due to quantization error, but as the reader will recall, this required complex contour integration. Bertram (1958) has shown that an algebraic bound is possible and that is the topic of this section. It was shown in (7.3.2) that the system output due to an error sequence would be

$$y(k) = \sum_{m=0}^{k} h(m)e(k-m) \tag{7.4.1}$$

The absolute value of $y(k)$ is

$$|y(k)| = \left|\sum_{m=0}^{k} h(m)e(k-m)\right| \tag{7.4.2}$$

and the right side will be less than the absolute value of each of the products involved under the summation, or

$$|y(k)| \le \sum_{m=0}^{k} |h(m)e(k-m)| \tag{7.4.3}$$

This is the same as assuming that the error always has a sign the same as that of corresponding impulse response sequence value. We know for a rounding quantizer that the error is bounded from above, or

$$e(k) \leq \frac{q}{2} \tag{7.4.4}$$

Then

$$|y(k)| \leq \frac{q}{2} \sum_{m=0}^{k} |h(m)| \tag{7.4.5}$$

and in the case of the stationary output,

$$|y(k)| \leq \frac{q}{2} \sum_{m=0}^{\infty} |h(m)| \tag{7.4.6}$$

**Example 7.5.** Evaluate the bound on the error for the control system of Example 7.3, where the transfer function from quantization noise to output was

$$\frac{Y(z)}{Q(z)} = H(z) = \frac{-0.4}{z - 0.4}$$

Long division of this transfer function yields a transfer function

$$H(z) = 1 - \frac{z}{z - 0.4}$$

or an impulse response sequence of

$$h(k) = \delta(k) - (0.4)^k$$

The sum given in (7.4.6) is equal to 0.667, and hence the Bertram bound on the stationary response is

$$|y(k)| \leq 0.667 \frac{q}{2} = 0.333q$$

which is a factor of 2.66 higher than the rms error given exactly in Example 7.3. This is a considerable error, but the answer was given with relative ease.

## 7.5 MULTIPLICATION ERRORS (TYPE 2 ERRORS)

In the process of implementation of digital control algorithms, it has been found necessary to carry out multiplication and addition. A typical control algorithm might be of the form

$$u(k) = \alpha u(k - 1) + \beta e(k) \tag{7.5.1}$$

The operations of multiplication of $u(k - 1)$ by parameter $\alpha$ and $e(k)$ by parameter $\beta$ will be in error because of the way digital processors carry out the multiplication process.

Consider, for example, a 2-bit (decimal) processor in which we wish to carry out the multiplication of two numbers, say 0.16 and 0.24, and we must similarly represent their product by two significant decimal digits, or

$$0.16 \times 0.24 \cong 0.03 \tag{7.5.2}$$

where in reality the exact answer is 0.0384, which indicates an error of 22%, and this is typical. We find that the errors in this case will be uniformly distributed on the interval of the last retained digit, or in this case on the interval $(-0.01, 0)$. The floating-point arithmetic of the BASIC language commonly involved in personal computers functions in this manner where the exponent is kept track of by some additional bits.

The filter or compensator of equation (7.5.1) can be realized as illustrated in Fig. 7.8. Since the errors are incurred in the multiplication process, they can be represented by additive noise sequences $n_1(k)$ and $n_2(k)$ injected into the loop immediately following the operations of multiplication, as illustrated in Fig. 7.9. Since the filter is still a linear system superposition applies, and the output sequence $u(k)$ can be thought of as being composed of parts due to $e(k)$, the $\beta$ multiplier error $n_2(k)$, and the $\alpha$ multiplier error $n_1(k)$. One must establish a transfer function between each of these inputs and the output sequence. In this case a simple transfer function may be derived, but in filters with a more complicated topology the input–output relation in the $z$-domain may be given by use of the Mason gain formula (Mason, 1953). If we consider for the moment only the $\alpha$ multiplier error $y(k)$, a filter of the form shown in Fig. 7.10 results. The relation between

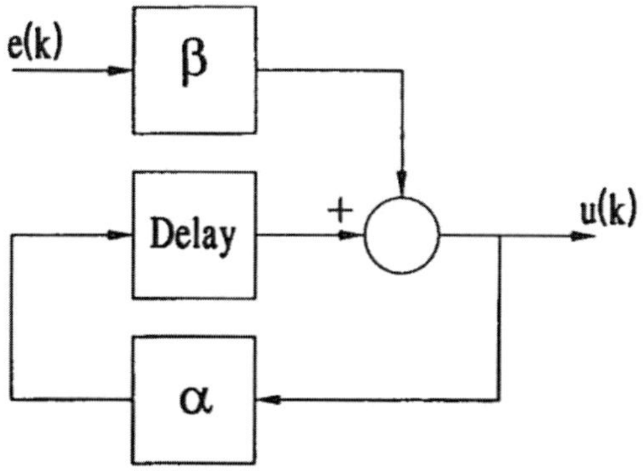


**Figure 7.8.** Digital compensator.

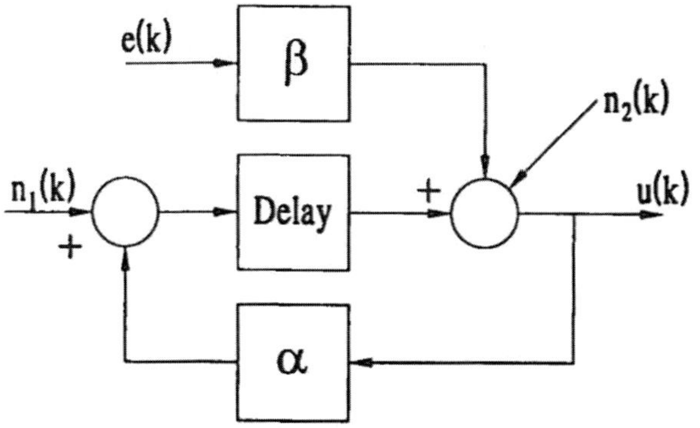


**Figure 7.9.** Multiplier error noise in a digital compensator.

the $z$-domain multiplier error and that part of the filter output due to that error is

$$\frac{U(z)}{N_1(z)} = \frac{1}{z - \alpha} \tag{7.5.3}$$

At this point it becomes clear that the method developed to describe the response of the filter to quantization error sequences is completely applicable to this case; hence no further time will be spent analyzing the situation.

To this point we have considered multiplier errors only in a simple first-order system. In that case there were no variations in the way in which the filter algorithm could be realized, but in the case of higher-order filters we know that they can be implemented in different fashions, as shown in Chapter 6, where state variable forms were discussed. The realization method will determine what coefficients must be stored and the number

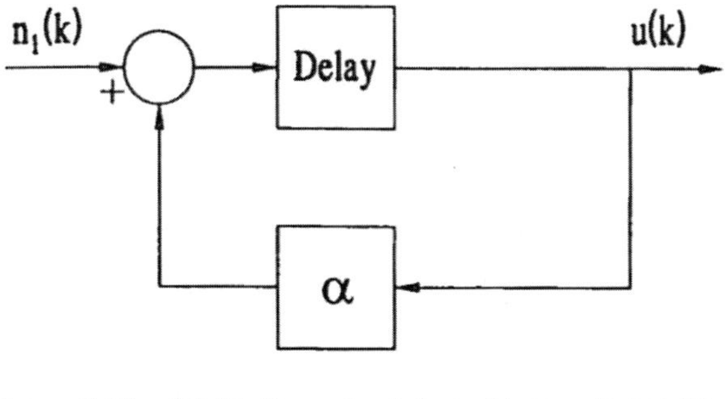


**Figure 7.10.** Multiplier noise injected into a digital filter.

of arithmetic operations to be performed per cycle. The errors induced by multiplication are heavily dependent on the realization chosen. These concepts are best illustrated by an example taken from Chapter 5.

**Example 7.6.** Consider the digital bandpass filter derived by the impulse invariant method in Example 5.8 with a transfer function

$$H(z) = \frac{Y(z)}{U(z)} = \frac{0.20z^2 - 0.1138z}{z^2 - 0.9853 + 0.8186}$$

There are four nonunity coefficients in this transfer function and we shall consider the controllable and observable canonical forms. We shall treat the errors in the coefficient multiplication as the noise sequences indicated in the following table:

| Coefficient | Multiplier |
|---|---|
| −0.1138 | $n_1(k)$ |
| 0.20 | $n_2(k)$ |
| 0.9853 | $n_3(k)$ |
| −0.8186 | $n_4(k)$ |

Let us now consider the observable canonical form, which is illustrated in Fig. 7.11. The noise sequences have been added immediately following the appropriate multiplication. The transfer function between $n_1(k)$ and $y(k)$, for example, is

$$\frac{Y(z)}{N_1(z)} = \frac{z}{z^2 - 0.9853z + 0.8186}$$

while the transfer function from $n_3(k)$ to $y(k)$ is the same or

$$\frac{Y(z)}{N_3(z)} = \frac{z}{z^2 - 0.9853z + 0.8186}$$

Now consider the controllable canonical realization as illustrated in Fig. 7.12, where the same notation has been employed to denote the multiplier noise sequences. The transfer function from $n_1(k)$ to the output $y(k)$ is

$$\frac{Y(z)}{N_1(z)} = 1$$

and the transfer function between $n_3(k)$ and the output is

$$\frac{Y(z)}{N_3(z)} = \frac{-0.1138z}{z^2 - 0.9853z + 0.8186}$$

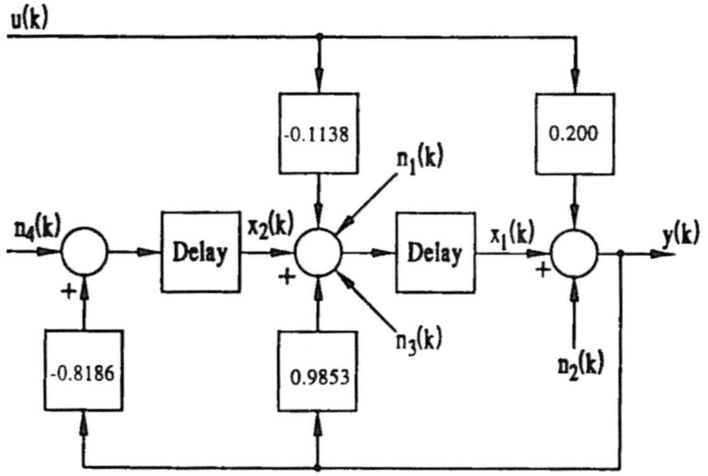


**Figure 7.11.** Observable canonical realization for bandpass filter.

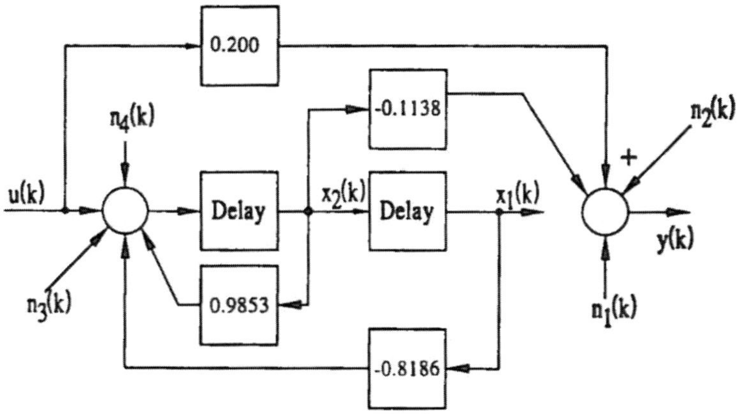


**Figure 7.12.** Controllable canonical realization for bandpass filter.

It is clear that the effects of the multiplication errors on the output in the two cases is quite different.

## 7.6 FINITE-WORD-LENGTH REPRESENTATION OF DIGITAL FILTER COEFFICIENTS (TYPE 3 ERRORS)

In Section 7.1 we discussed several types of errors. We have already discussed the first two types of errors, and in this section we address the third. When we implement a digital filter in the form of a linear difference equation on a digital processor, the coefficients of the difference equation must be represented by a digital word of a finite number of binary bits. In general, a coefficient should require an infinite number of bits, but since this is not possible, we would like to know how the performance of the filter represented with only approximate coefficients deviates from that of the originally specified filter. In fact, in the case of a high-$Q$ bandpass filter, the finite-word-length coefficients can cause the filter to be unstable. The effect on the filter dynamics may be found by investigating the changes in the filter pole and zero locations due to deviations in the coefficient values.

Since the numerators and denominators of digital filter transfer functions are polynomials in the complex variable $z$, it is sufficient for us to investigate the changes in the roots of a polynomial with respect to changes in one of the coefficients.

Let us consider a polynomial in the variable $z$ with coefficients that are functions of the parameter $q$. We are interested in the values of $z$ for which the polynomial vanishes:

$$P(z, q) = 0 \tag{7.6.1}$$

Let us denote these values of $z$ as $z_1, \ldots, z_n$, or

$$P(z_i, q) = 0 \qquad i = 1, \ldots, n \tag{7.6.2}$$

We would like to investigate how the root of the polynomial changes when the parameter changes to $q + \Delta q$. We shall denote the change in the root location as $\Delta z_i$ so that the new polynomial roots satisfy

$$P(z_i + \Delta z_i, q + \Delta q) = 0 \qquad i = 1, \ldots, n \tag{7.6.3}$$

The situation is illustrated in Fig. 7.13. Let us now expand the polynomial in a Taylor series about the nominal parameter $q$ and the nominal root $z_i$,

or

$$P(z_i + \Delta z_i, q + \Delta q) = P(z_i, q) + \left.\frac{\partial P(z, q)}{\partial z}\right|_{z=z_i} \Delta z_i$$

$$+ \left.\frac{\partial P(z, q)}{\partial q}\right|_{z=z_i} \Delta q + \text{(higher-order terms)} \qquad (7.6.4)$$

If we note (7.6.2) and (7.6.3), it is clear that the left side and the first term on the right side of (7.6.4) are both zero, so after ignoring the higher-order terms we may write

$$\left.\frac{\partial P}{\partial z}\right|_{z=z_i} \Delta z_i = -\left.\frac{\partial P}{\partial q}\right|_{z=z_i} \Delta q \qquad (7.6.5)$$

or to a first-order approximation,

$$\Delta z_i = -\left.\frac{\partial P/\partial q}{\partial P/\partial z}\right|_{z=z_i} \Delta q \qquad (7.6.6)$$

In the analysis above we have neglected the higher-order terms in the Taylor series expansion. This truncation may not be valid unless $\Delta q$ is particularly small. It is best only to solve (7.6.5) for the ratio $\Delta z_i/\Delta q$, which gives

$$\frac{\Delta z_i}{\Delta q} = -\left.\frac{\partial P/\partial q}{\partial P/\partial z}\right|_{z=z_i} \qquad (7.6.7)$$

which gives the variation in the pole location per unit variation of the

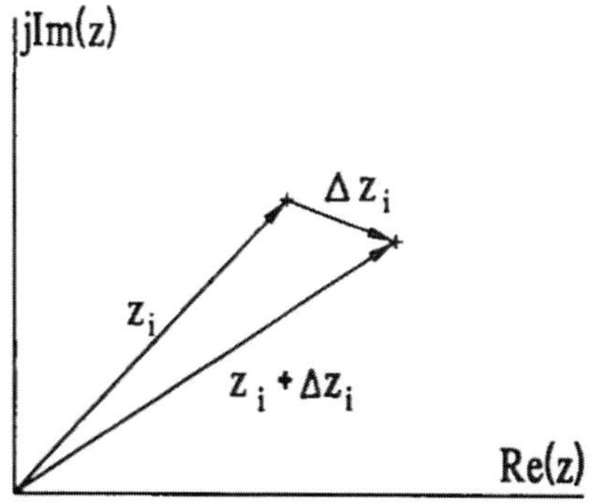


**Figure 7.13.** Characteristic root change in $z$-plane.

parameter $q$ about the nominal value in the neighborhood of the nominal root $z_i$.

**Example 7.7.** Consider the following polynomial in $z$ and examine the rate of change in the roots with respect to changes in the coefficients:

$$P(z) = z^2 + a_1 z + a_0$$

The nominal values of the parameters being $a_1 = -1.4$ and $a_0 = 0.4$ and the nominal pole locations being $z_1 = 0.6$ and $z_2 = 0.8$, we have

$$\frac{\partial P}{\partial z} = 2z + a_1$$

$$\frac{\partial P}{\partial a_1} = z$$

$$\frac{\Delta z_1}{\Delta a_1} = \left.\frac{-z}{2z + a_1}\right|_{z=0.6} = \frac{-0.6}{2(0.6) - 1.4} = 3$$

and for the second root,

$$\frac{\Delta z_2}{\Delta a_1} = \left.\frac{-z}{2z + z_1}\right|_{z=0.8} = \frac{-0.8}{2(0.8) - 1.4} = -4$$

Now evaluate the rate of change of the second pole location with respect to the second parameter:

$$\frac{\partial P}{\partial a_0} = 1$$

$$\frac{\Delta z_1}{\Delta a_0} = \left.\frac{-1}{2z + a_1}\right|_{z=0.6} = \frac{-1}{1.2 - 1.4} = 5$$

and for the second root location,

$$\frac{\Delta z_2}{\Delta a_0} = \left.\frac{-1}{2z + a_1}\right|_{z=0.8} = \frac{-1}{1.6 - 1.4} = -5$$

The numbers derived in Example 7.7 illustrate the fact that the rates of change of pole locations are not the same for either of the poles or either of the coefficients. There is one problem: These numbers are meaningless unless one knows something about the nominal pole locations and the nominal values of the coefficients. An alternative and related method that alleviates this problem is given in the following section.

## 7.7 ROOT SENSITIVITY ANALYSIS

In this section we investigate the problem outlined in Section 7.6, but we do it on a per unit basis to give the results a better perspective. Let us define the root sensitivity function $S_q^{z_i}$, which is the ratio of the per unit change in pole location to the per unit change in the same parameter $q$, or

$$S_q^{z_i} = \frac{\Delta z_i/|z_i|}{\Delta q/|q|} \tag{7.7.1}$$

or

$$S_q^{z_i} = \frac{|q|}{|z_i|}\frac{\Delta z_i}{\Delta q} \tag{7.7.2}$$

Note that the root sensitivity function as given in (7.7.2) is a complex number and the directional property is indicative of the direction of motion of the root with increasing $\Delta q$. We have already evaluated the quantity in parentheses on the right side of (7.7.2) and it is given in expression (7.6.7). If this is substituted into (7.7.2), the resulting expression for the root sensitivity function is

$$S_q^{z_i} = \frac{-|q|}{|z_i|}\left.\frac{\partial P/\partial q}{\partial P/\partial z}\right|_{z=z_i} \tag{7.7.3}$$

This expression could be further specialized for general polynomial functions with the parameter $q$ being one of the polynomial coefficients.

By means of a root sensitivity analysis, we may evaluate which coefficient of the difference equation has the highest influence on each of the root locations.

**Example 7.8.** Consider the same polynomial as that of Example 7.7. Calculate all four root sensitivities

$$P(z) = z^2 + a_1 z + a_0$$

where the nominal values are $a_1 = -1.4$ and $a_0 = 0.48$ with nominal pole locations of $z_i = 0.6$ and $z_2 = 0.8$.

The derivatives have already been evaluated in Example 7.7, and thus we calculate the sensitivities as

$$S_{a_1}^{z_1} = \frac{1.4}{0.6}(3) = 7$$

$$S_{a_1}^{z_2} = \frac{1.4}{0.8}(-4) = -7$$

$$S_{a_0}^{z_1} = \frac{0.48}{0.6}(5) = 4$$

$$S_{a_0}^{z_2} = \frac{0.48}{0.8}(-5) = -3$$

It turns out that both pole locations are quite a bit more sensitive to parameter $a_1$. In the case of foot number 1 ($z_1 = 0.6$) a 1% change in coefficient $a_1$ will cause a 7% change in the location of $z_1$ in the positive direction along the real axis of the $z$-plane.

**Example 7.9.** Consider the effect of inaccuracies in the representation of the proportional digital controller gain $K$ of the thermal control system shown in Fig. 7.14. The nominal gain will be chosen as in Example 3.6 to yield a closed-loop damping ratio of 0.7, or

$$K = 2.5$$

The closed-loop characteristic equation was given in Example 3.5 to be

$$P(z, K) = z^2 - z(1.48 - 0.25K) + (0.5026 + 0.0204K)$$

For a nominal controller gain of 2.5, the nominal closed-loop roots are

$$z_{1,2} = 0.709 \pm j0.226 = 0.744\angle \pm 17.6°$$

where $z_1$ will represent the root with the positive imaginary part. To evaluate the root sensitivity from relation (7.7.3), we need to calculate the

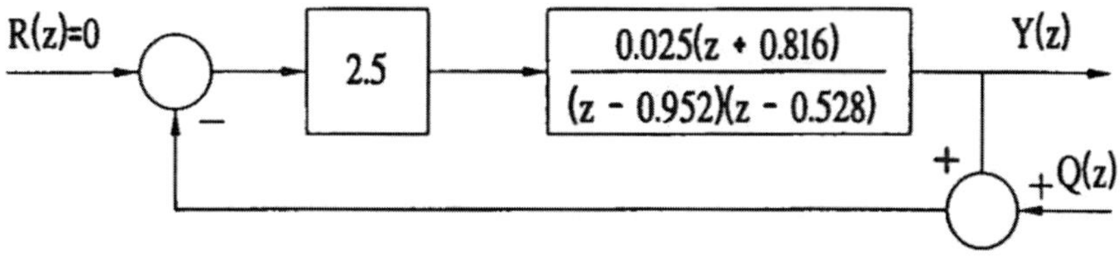


**Figure 7.14.** Thermal control system.

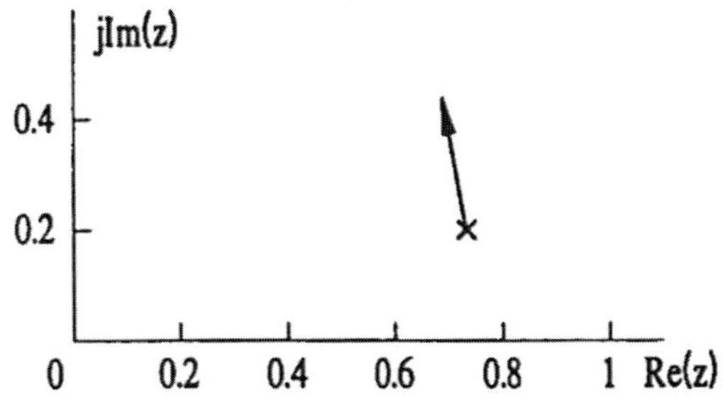


**Figure 7.15.** Root sensitivity vector direction.

following partial derivatives:

$$\frac{\partial P}{\partial K} = 0.025z + 0.0204$$

and

$$\frac{\partial P}{\partial z} = 2z - 1.48 + 0.025K$$

If we evaluate these at the nominal root location ($z_1 = 0.709 + j0.226$) and nominal gain, then

$$\frac{\partial P}{\partial K} = 0.0177 + j0.00565 + 0.0204 = 0.0385\angle 8.44^\circ$$

$$\frac{\partial P}{\partial z} = 1.418 + j0.452 - 1.48 + 0.0625 = 0.452\angle 90^\circ$$

This root sensitivity function for $z$ is, from relation (7.7.3),

$$S_K^{z_1} = \frac{-2.5}{0.744}\,\frac{0.0385\angle 8.44^\circ}{0.452\angle 90^\circ} = 0.286\angle 98.4^\circ$$

We should now spend some time interpreting this complex number as to the meaning of its magnitude and angle. In the neighborhood of a gain of 2.5, a 10% change in gain will cause a 2.86% change in the magnitude of the pole locations, and for positive changes in the gain away from 2.5, the root will move in a direction of 98.4° from the nominal pole location, as illustrated in Fig. 7.15. Positive deviations in the gain $K$ away from the nominal will yield a system with lower damping than specified, while negative deviations in the gain will yield a closed-loop system with higher-than-specified damping.

**Example 7.10.** Consider the control system shown in Fig. 7.16, where a compensator has been inserted to improve dynamic system performance. Nominally, the compensator would have a zero at $z = 0.8$ ($a = 0.8$) which would cancel the plant pole, and the compensator pole would be made as fast as possible at $z = 0.4$ ($b = 0.4$). The closed-loop pole would be located at $z = 0.2$, which would imply that the nominal gain is $K = 0.2$. The control algorithm corresponding to the compensator shown is

$$u_{k+1} = bu_k + K(e_{k+1} - ae_k)$$

It is clear that the parameters $K$, $a$, and $b$ must be stored in the digital processor in the form of finite-length digital words, and hence we would like to evaluate the sensitivity of the resultant pole to the stored coefficients in the neighborhood of the nominal values. The characteristic polynomial containing the parameters is

$$P(z) = z^2 - (0.8 + b - K)z + 0.8b - Ka = 0$$

Now let us evaluate the required partial derivatives, or

$$\frac{\partial P}{\partial a} = -K$$

$$\frac{\partial P}{\partial b} = -z + 0.8$$

$$\frac{\partial P}{\partial K} = z - a$$

and

$$\frac{\partial P}{\partial z} = 2z - 0.8 - b + K$$

Evaluating these partial derivatives at the nominal parameter values and

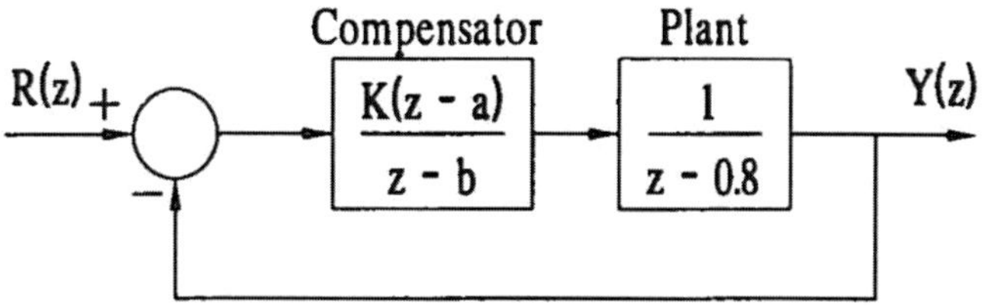


**Figure 7.16.** Compensated closed-loop system.

at the nominal closed-loop root location ($z = 0.2$), we get

$$\frac{\partial P}{\partial a} = -0.2$$

$$\frac{\partial P}{\partial b} = 0.6$$

$$\frac{\partial P}{\partial K} = -0.6$$

and

$$\frac{\partial P}{\partial z} = -0.6$$

Inserting these quantities into relation (7.7.3), we get the following sensitivities:

$$S_a^{z_1} = \frac{-0.8(-0.2)}{0.2(-0.6)} = -1.333$$

$$S_b^{z_1} = \frac{-0.4(0.6)}{0.2(-0.6)} = 2$$

$$S_K^{z_1} = \frac{-0.2(-0.6)}{0.2(-0.6)} = -1$$

The sensitivity $S_b^{z_1} = 2$ indicates that for every unit change in the parameter $b$ the pole in question will translate 2 units in the positive direction in the $z$-plane. The sensitivity $S_K^{z_1} = -1$ indicates that for every unit change in the parameter $K$ the pole at $z = 0.2$ will translate in the negative direction 1 unit in the $z$-plane.

## 7.8 SUMMARY

In this chapter we have investigated problems created in digital control systems by virtue of the finite word length of a digital word used to represent numbers which in general do not have finite-word-length representation. It was shown that the errors in samples of an arbitrary signal could be modeled by insertion of equivalent measurement noise into the control loop following the A/D converter. Under the appropriate assumptions, the root-mean-square (rms) deviations in output and/or control effort for such noise may be predicted.

Another problem is that of the way in which digital processors accomplish the operation of multiplication of two numbers. It was shown that this can be examined by treating these errors as noise injected into the control loop following each multiplication operation and employing a previously developed technique to predict rms deviations in system variables.

The third problem encountered was the representation of control algorithm (digital filter) coefficients by finite-length binary words. Errors in the coefficient representation yield systems with considerably different dynamic character than originally specified. The tool used for examining this problem was the root sensitivity function, which predicts the deviation in closed-loop characteristic root locations with variations in the control algorithm coefficients or parameters.

## PROBLEMS

7.1. An A/D converter is to be used to drive a 4-bit microprocessor for calculating the control strategy. Two's-complement coding is to be used on the A/D output and it is to convert a range of 0 to 10 V. Find the quantization level for this converter.

7.2. In an 8-bit microprocessor digital control application an A/D converter has a range of ±5 V. Assume rounding quantization and 2's-complement algebra represents the A/D output.

(a) Find the quantization level $q$.

(b) A compensator of the form

$$D(z) = \frac{z}{z^2 - 1.4z + 0.64}$$

is to be implemented on the processor. Calculate the mean-square output due to errors in the A/D conversion process.

(c) The filter in part (b) is to be represented in observable canonical (direct) form. Draw the block diagram for the filter.

(d) Let us assume that the mean-square multiplier error in the 1.4 coefficient is 0.1 and that the multiplier errors are independent from time step to time step.

7.3. The fluid-level control system considered in Problem 3.6 had a plant transfer function of

$$G(z) = \frac{0.008(z + 0.79)}{z^2 - 1.4536z + 0.4966}$$

A proportional controller with a gain $K = 7$ closed loop poles at $z_{1,2} = 0.698 \pm j0.229$, which yielded a closed-loop damping ratio of $\zeta = 0.7$. Find the root sensitivity functions $S_K^{z_1}$ and $S_K^{z_2}$.

7.4. A digital filter with the transfer function

$$H(z) = \frac{z + 1}{z^2 - 1.8z + \beta}$$

is to be implemented.

(a) Find the poles $z_1$ and $z_2$ of this filter as functions of the parameter $\beta$.

(b) Evaluate the derivative $\partial z/\partial\beta$ at $\beta = 1$ to get a feel for the sensitivity of the pole locations due to the parameter $\beta$.

(c) Evaluate the root sensitivity functions $S_\beta^{z_1}$ and $S_\beta^{z_2}$.

7.5. The fluid-level control system considered in Problem 3.6 had a plant transfer function of

$$G(z) = \frac{0.008(z + 0.79)}{z^2 - 1.4536z + 0.4966}$$

A proportional controller with gain $K = 7$ gave closed-loop roots of $z = 0.698 \pm j0.229$. The controller is to be implemented on a 4-bit microprocessor with an analog-to-digital converter with an input range of $\pm 10$ V. The system is shown below.

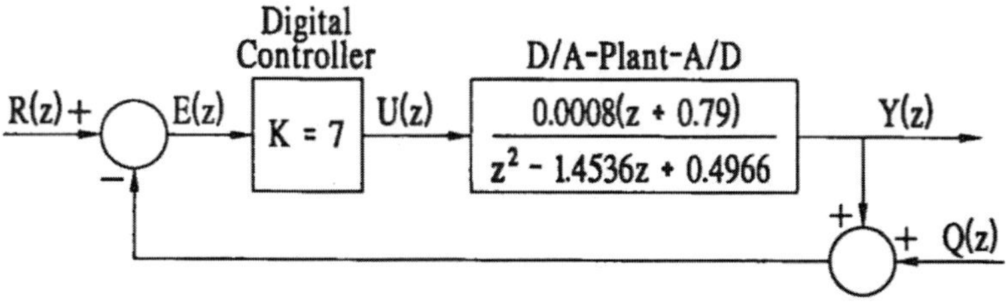


(a) Find the quantization level $q$ of the analog-to-digital converter assuming 2's-complement representation of the sample values.

(b) Find the rms value of the perturbations in the fluid level $y(t)$ due to the quantization.

(c) Find the rms value of the perturbations in the control effort $u(t)$ due to the quantization.

## REFERENCES

Bertram, J. E., 1958. The effects of quantization in sampled-feedback systems, *Trans. AIEE (Applications and Industry)*, 77: 177–181.

Budak, A., 1974. *Passive and Active Network Analysis and Synthesis*, Houghton Mifflin, Boston.

Gold, B., and C. M. Rader, 1969. *Digital Processing of Signals*, McGraw-Hill, New York.

Huelsman, L., Ed., 1970. *Active Filters: Lumped, Distributed, Integrated, Digital, and Parametric*, McGraw-Hill, New York.

Mason, S. J., 1953. Feedback theory: some properties of signal flow graphs, *Proc. IRE*, *41*(9): 1144–1156.

Moroney, P., 1983. *Issues in the Implementation of Digital Feedback Compensator*, MIT Press, Cambridge, MA.

Widrow, B., 1956. A study of rough amplitude quantization by means of Nyquist sampling theory, *IRE Trans. on Circuit Theory*, *CT3*(4): 266–276.

# 8

# State-Space Approach to Control System Design

## 8.1 INTRODUCTION

In the earlier chapters we introduce the conventional approach to control system design. With the introduction of modern control theory, completely new techniques for system design have been developed, which include optimal control system synthesis, state-estimation techniques, and stochastic estimation and control techniques. The development of the digital computer has aided in solution of the computational problems that were a product of this development.

In this chapter we have a look at control system design using complete state feedback and also using the estimated state when incomplete measurements of the state are available. Also in this chapter, an algorithm for calculating state feedback and state estimator gains will be given. State-variable designs with reference inputs present will be explored and the necessary modifications to the control law and state estimator will be developed.

## 8.2 STATE-VARIABLE FEEDBACK AND SYSTEM DESIGN

Consider the discrete-time plant described by the vector difference equation

$$\mathbf{x}_{k+1} = \mathbf{A}\mathbf{x}_k + \mathbf{B}\mathbf{u}_k \tag{8.2.1}$$

A regulator is a system that controls the output about zero in the presence of external disturbances, so control will be provided based on negative proportional state feedback, which will be of the form

$$\mathbf{u}_k = -\mathbf{F}\mathbf{x}_k \tag{8.2.2}$$

where the matrix **F** is referred to as the feedback gain matrix. This situation is illustrated in Fig. 8.1. If the control is state feedback as postulated in (8.2.2), this may be substituted into (8.2.1) and the resulting homogeneous difference equation is

$$\mathbf{x}_{k+1} = (\mathbf{A} - \mathbf{B}\mathbf{F})\mathbf{x}_k \tag{8.2.3}$$

The closed-loop eigenvalues or the eigenvalues of (8.2.3) are given by the characteristic equation

$$\alpha_c(z) = \det[z\mathbf{I} - \mathbf{A} + \mathbf{B}\mathbf{F}] = 0 \tag{8.2.4}$$

If the state is of dimension $n$ and the control effort is of dimension $m$, the feedback matrix **F** has $nm$ elements. If the system (**A**, **B**) is controllable, the choice of the elements of **F** will control the location of the closed-loop system poles. Since there are only $n$ closed-loop roots, the required $mn$ elements of **F** will not be determined uniquely. We shall assume that the elements of **F** are real and hence we restrict the closed-loop poles to real values or complex-conjugate pairs. Since there are $nm$ elements of **F** to be selected and only $n$ poles to place, we have $nm - n$ degrees of freedom in the selection of these feedback coefficients. There are many techniques for this selection process which are considered in the literature (Wonham, 1967; Davison, 1968; Heymann, 1968), and a summary of some of these processes is given by Brogan (1985).

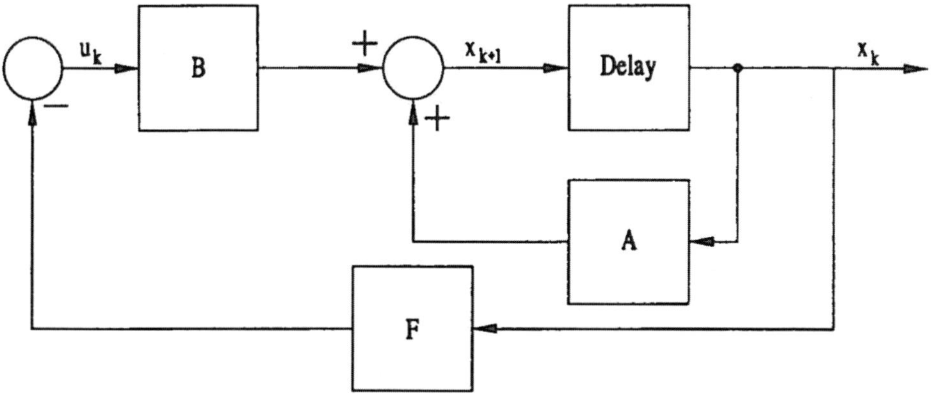


**Figure 8.1.** State-feedback-controlled linear discrete system.

If the control effort $u_k$ is a scalar sequence, the matrix **F** will be a row vector which we shall term $\mathbf{f}^T$. This vector consists of only $n$ elements, which will be uniquely determined by specifying all $n$ roots of the characteristic equation (8.2.4) with the restriction that complex roots occur in conjugate pairs. Equating the desired closed-loop characteristic equation with that of (8.2.4) yields a set of $n$ linear equations in the elements of $\mathbf{f}^T$. Several examples of this technique will now be given.

**Example 8.1.** Consider the simple inertial plant of Example 6.1, which is governed by the difference equation

$$\begin{bmatrix} x_1 \\ x_2 \end{bmatrix}_{k+1} = \begin{bmatrix} 1 & 0.1 \\ 0 & 1 \end{bmatrix} \begin{bmatrix} x_1 \\ x_2 \end{bmatrix}_k + \begin{bmatrix} 0.005 \\ 0.1 \end{bmatrix} u_k$$

If we choose state feedback, the control law is

$$u_k = -[f_1 \quad f_2] \begin{bmatrix} x_1 \\ x_2 \end{bmatrix}_k$$

where $f_1$ and $f_2$ are the feedback gain coefficients. The plant equation becomes

$$\begin{bmatrix} x_1 \\ x_2 \end{bmatrix}_{k+1} = \begin{bmatrix} 1 & 0.1 \\ 0 & 1 \end{bmatrix} \begin{bmatrix} x_1 \\ x_2 \end{bmatrix}_k - \begin{bmatrix} 0.005f_1 & 0.005f_2 \\ 0.1f_1 & 0.1f_2 \end{bmatrix} \begin{bmatrix} x_1 \\ x_2 \end{bmatrix}_k$$

so the characteristic determinant is

$$\alpha_c(z) = \det \begin{bmatrix} z - 1 + 0.005f_1 & -0.1 + 0.005f_2 \\ 0.1f_1 & z - 1 + 0.1f_2 \end{bmatrix} = 0$$

which yields

$$z^2 - (2 - 0.005f_1 - 0.1f_2)z + (1 - 0.005f_1)(1 - 0.1f_2) + 0.01f_1 - 0.0005f_1f_2 = 0$$

and after simplification,

$$z^2 - (2 - 0.005f_1 - 0.1f_2)z + 1 + 0.005f_1 - 0.1f_2 = 0 \qquad \text{(a)}$$

For any desired closed-loop root location $\lambda_1$ and $\lambda_2$ we could write

$$\alpha_c(z) = (z - \lambda_1)(z - \lambda_2) = z^2 - (\lambda_1 + \lambda_2)z + \lambda_1\lambda_2 = 0 \qquad \text{(b)}$$

Equating (a) and (b) will yield $f_1$ and $f_2$. Say that we desire a real pair of roots at $z = 0.7 \pm j0.2$; then we can write

$$\alpha_c(z) = z^2 - 1.4z + 0.53$$

We now find that by equating coefficients of powers of $z$ we get

$$2 - 0.005f_1 - 0.1f_2 = 1.4$$

and in constant terms we get

$$1 + 0.005f_1 - 0.1f_2 = 0.53$$

Solving these equations simultaneously, we get

$$f_1 = 13.0, \qquad f_2 = 5.35$$

which are reasonable values of gain for implementation in physical hardware.

**Example 8.2.** For the thermal system of Examples 3.3 and 6.2 choose the state feedback gain matrix $\mathbf{f}^T$ to place the closed-loop system poles at $z = 0.5 \pm j0.2$.

With the given characteristic roots the closed-loop characteristic equation will be

$$\alpha_c(z) = (z - 0.5 - j0.2)(z - 0.5 + j0.2) = z^2 - z + 0.29 = 0 \qquad \text{(a)}$$

The system of interest here is

$$\begin{bmatrix} x_1 \\ x_2 \end{bmatrix}_{k+1} = \begin{bmatrix} 0.6277 & 0.3597 \\ 0.0899 & 0.8526 \end{bmatrix} \begin{bmatrix} x_1 \\ x_2 \end{bmatrix}_k + \begin{bmatrix} 0.0251 \\ 0.1150 \end{bmatrix} u_k$$

The characteristic determinant from (8.2.4) is

$$\det \begin{bmatrix} z - 0.6277 + 0.0251f_1 & 0.0251f_2 - 0.3597 \\ -0.0899 + 0.115f_1 & z - 0.8526 + 0.115f_2 \end{bmatrix} = 0$$

Expansion of this determinant yields a characteristic polynomial of

$$\alpha_c(z) = z^2 - z(1.48 - 0.0251f_1 - 0.115f_2)$$
$$+ 0.5025 + 0.02f_1 - 0.069f_2 = 0 \qquad \text{(b)}$$

Equating expressions (a) and (b) dictates that the polynomial coefficients must be equal or gives the following set of simultaneous linear equations:

$$0.0251f_1 + 0.115f_2 = 0.48$$
$$0.0201f_1 - 0.0698f_2 = -0.2125$$

which must be solved for the feedback gains $f_1$ and $f_2$. The solution yields a feedback gain matrix of

$$\mathbf{f}^T = [f_1 \quad f_2] = [2.248 \quad 3.685]$$

Implementing this control law yields the feedback control system shown in Fig. 8.2. If we start the system with initial conditions $x_1(0) = 2$ and $x_2(0) = 1$, the responses will be as shown in Fig. 8.3.

In the two preceding examples we have seen that the expansion of the determinant containing the symbolic feedback gain coefficients can be a tedious task. In Section 8.8 we discuss an algorithm which is a reasonable programming task to accomplish the pole placement task.

## 8.3 FEEDBACK CONTROL WITH INCOMPLETE STATE INFORMATION

Often in a large system it is unfeasible to measure all the state variables, while sometimes we can only measure a few or a linear combination of the states. For feedback control we would like to be able to reconstruct the state variables from the measured variables.

The problem with which we are concerned is that which is described by the matrix-vector difference equation

$$\mathbf{x}_{k+1} = \mathbf{A}\mathbf{x}_k + \mathbf{B}\mathbf{u}_k \tag{8.3.1}$$

with measurements $\mathbf{y}_k$ given by a linear combination of the states, or

$$\mathbf{y}_k = \mathbf{C}\mathbf{x}_k \tag{8.3.2}$$

This situation is illustrated in Fig. 8.4.

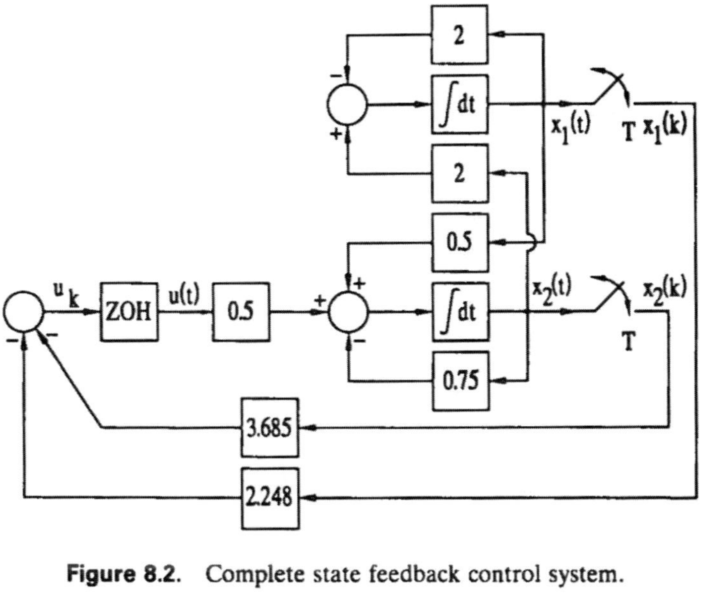


**Figure 8.2.** Complete state feedback control system.

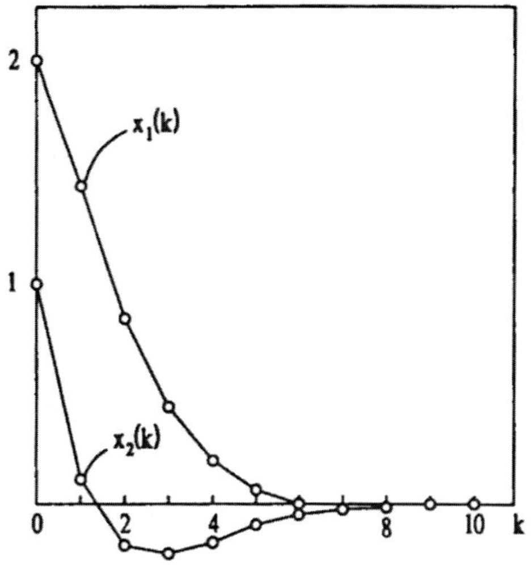


**Figure 8.3.** State-variable responses for thermal system with state feedback control.

One possible way to control this system about the zero equilibrium position would be to construct the control effort vector as a linear combination of the measured variables, or

$$\mathbf{u}_k = -\mathbf{F}\mathbf{y}_k \tag{8.3.3}$$

If the system is controllable and the rank of **C** is full, it is possible to place $r$ poles, where $r$ is the row dimension of **C**. If the number of elements in **F** is fewer than $n$, it is possible to place as many poles as there are elements and the remainder

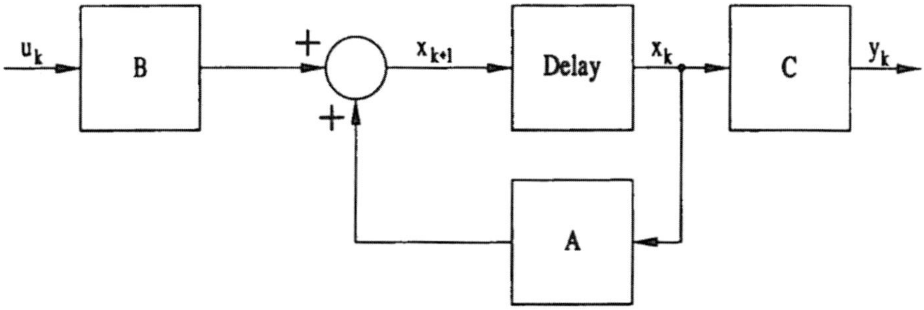


**Figure 8.4.** Discrete linear system.

will then be fixed. This problem is discussed by Davison (1970) and a summary is provided by Brogan (1985).

An alternative approach is that of estimating the state based on measurements that are a linear combination of the states and then generating the control effort based on the estimated states. We shall assume that at best only some of the states are measured directly. We want to design a state estimator or observer (Luenberger, 1964) which when given a sequence $\mathbf{y}_k$ and the input $\mathbf{u}_k$ reconstructs an estimate of the $\mathbf{x}_k$ sequence. An observer or state estimator is another dynamic system that has inputs $\mathbf{u}_k$ and $\mathbf{y}_k$, the output of which is an estimate of $\mathbf{x}_k$ which we shall call $\hat{\mathbf{x}}_k$.

## 8.4 OPEN-LOOP ESTIMATOR OR OBSERVER

The obvious way to solve the posed problem is to use a model of that portion of the system described by Eq. (8.3.1) as shown in Fig. 8.5. If we know $\mathbf{A}$ and $\mathbf{B}$ and the initial condition $\mathbf{x}_0$ with sufficient accuracy, this

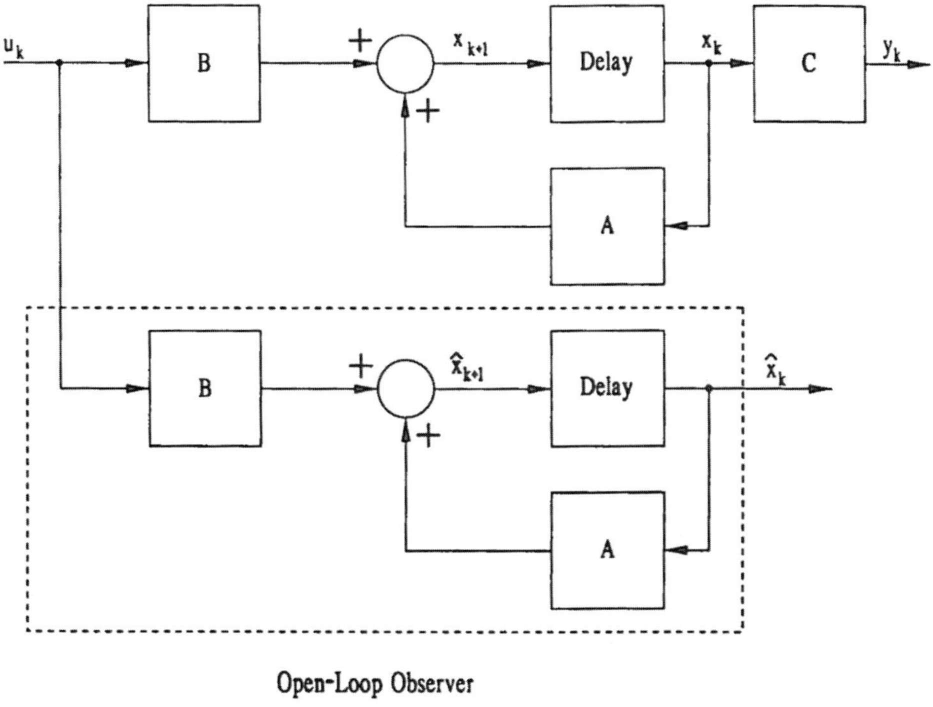


**Figure 8.5.** Open-loop observer.

model will accurately estimate the state sequence $\mathbf{x}_k$. The real problem lies in the fact that we always know $\mathbf{A}$ and $\mathbf{B}$ only approximately and seldom do we know $\mathbf{x}_0$ at all. In the case where the matrices of the model are not the same as those of the original system and/or the initial conditions differ from the real initial conditions, the estimator output will diverge from the actual state and, as a result, will be of little practical value.

In the scheme proposed above, the state was reconstructed without regard to the measurement sequence $\mathbf{y}_k$. Surely we can devise a scheme that will do a better job by employing the measurements in the estimation task.

## 8.5 ASYMPTOTIC PREDICTION ESTIMATOR OR OBSERVER

In this section we consider the system of (8.3.1) and (8.3.2) and assume that the matrix pair $(\mathbf{A}, \mathbf{C})$ is observable. We would like to build an estimator that will drive the error in the estimate to zero in the asymptotic sense. It would seem reasonable to design an estimator or observer that includes the previous estimate $\hat{\mathbf{x}}_k$ and the measurement $\mathbf{y}_k$ to estimate $\hat{\mathbf{x}}_{k+1}$. A linear estimator of such form would be

$$\hat{\mathbf{x}}_{k+1} = \mathbf{M}\hat{\mathbf{x}}_k + \mathbf{K}\mathbf{y}_k + \mathbf{z}_k \tag{8.5.1}$$

where $\mathbf{M}$ and $\mathbf{K}$ are matrices yet to be selected and $\mathbf{z}_k$ is a yet unspecified vector. The plant of interest is given by expressions (8.2.1) and (8.2.2), so we may define the state error $\tilde{\mathbf{x}}_k$ as

$$\tilde{\mathbf{x}}_k = \mathbf{x}_k - \hat{\mathbf{x}}_k \tag{8.5.2}$$

If we subtract (8.5.1) from (8.2.1), we get, after substitution of (8.5.2),

$$\tilde{\mathbf{x}}_{k+1} = \mathbf{A}\mathbf{x}_k - \mathbf{M}\hat{\mathbf{x}}_k + \mathbf{B}\mathbf{u}_k - \mathbf{K}\mathbf{y}_k - \mathbf{z}_k \tag{8.5.3}$$

We would like the errors in the estimate to be independent of the control effort sequence $\mathbf{u}_k$, so it is appropriate to choose $\mathbf{z}_k = \mathbf{B}\mathbf{u}_k$, so the error expression becomes

$$\tilde{\mathbf{x}}_{k+1} = \mathbf{A}\mathbf{x}_k - \mathbf{M}\hat{\mathbf{x}}_k - \mathbf{K}\mathbf{y}_k \tag{8.5.4}$$

but $\mathbf{y}_k = \mathbf{C}\mathbf{x}_k$, so

$$\tilde{\mathbf{x}}_{k+1} = (\mathbf{A} - \mathbf{K}\mathbf{C})\mathbf{x}_k - \mathbf{M}\hat{\mathbf{x}}_k \tag{8.5.5}$$

If we now choose $\mathbf{M} = \mathbf{A} - \mathbf{K}\mathbf{C}$, we get a simple homogeneous difference equation governing the error, or

$$\tilde{\mathbf{x}}_{k+1} = (\mathbf{A} - \mathbf{K}\mathbf{C})\tilde{\mathbf{x}}_k \tag{8.5.6}$$

Selection of the elements of the **K** matrix will locate the poles that govern the error dynamics. This problem is computationally the same as the state feedback pole placement problem discussed in Section 8.2.

Now let us return to the estimator form of expression (8.5.1) and substitute for **M** and $\mathbf{z}_k$ from the previous analysis to yield

$$\hat{\mathbf{x}}_{k+1} = (\mathbf{A} - \mathbf{KC})\hat{\mathbf{x}}_k + \mathbf{K}\mathbf{y}_k + \mathbf{B}\mathbf{u}_k \tag{8.5.7}$$

or

$$\hat{\mathbf{x}}_{k+1} = \mathbf{A}\hat{\mathbf{x}}_k + \mathbf{K}(\mathbf{y}_k - \mathbf{C}\hat{\mathbf{x}}_k) + \mathbf{B}\mathbf{u}_k \tag{8.5.8}$$

It is interesting to note that the estimator is a replica of the original plant (8.2.1) driven by the error in the estimate of $\mathbf{y}_k$. The block diagram of this estimator is given in Fig. 8.6. If the estimator is stable, the error will go to zero asymptotically. We can control the error dynamics by selecting the estimator gain matrix **K** to locate the poles anywhere we desire. The poles of the error dynamics are the eigenvalues of matrix **A** − **KC** and thus are

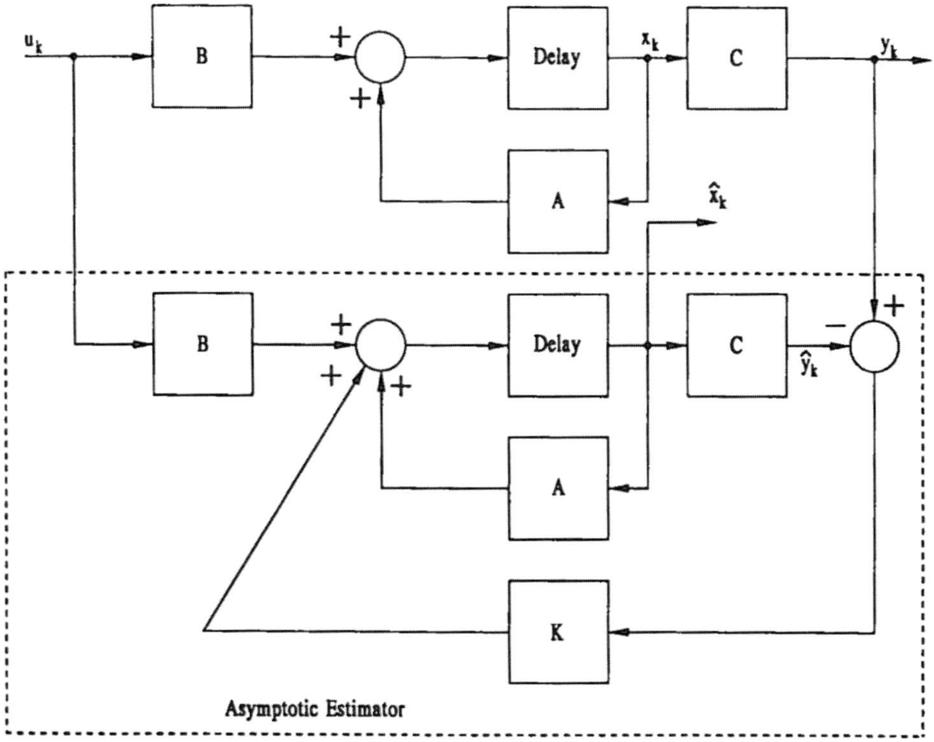


**Figure 8.6.** Asymptotic prediction estimator.

roots of

$$\det[\mathbf{I}z - \mathbf{A} + \mathbf{KC}] = 0 \tag{8.5.9}$$

It is best if the dynamics of the estimator are faster than those of the closed-loop control system. If the system is observable, there are no unique elements of **K** to give specified pole locations. If, on the other hand, the output $\mathbf{y}_k$ is scalar and hence **K** is a vector, we can find the elements of **K** uniquely to locate the poles of the estimator as desired, the only limitation being that complex poles must occur in conjugate pairs and that the system be observable.

Now let us examine a control system where an estimator is used to reconstruct the states and then the state estimate is fed back to generate the control effort sequence, or the control law is now

$$\mathbf{u}_k = -\mathbf{F}\hat{\mathbf{x}}_k \tag{8.5.10}$$

This entire scheme is shown in Fig. 8.7. For the estimator we have shown

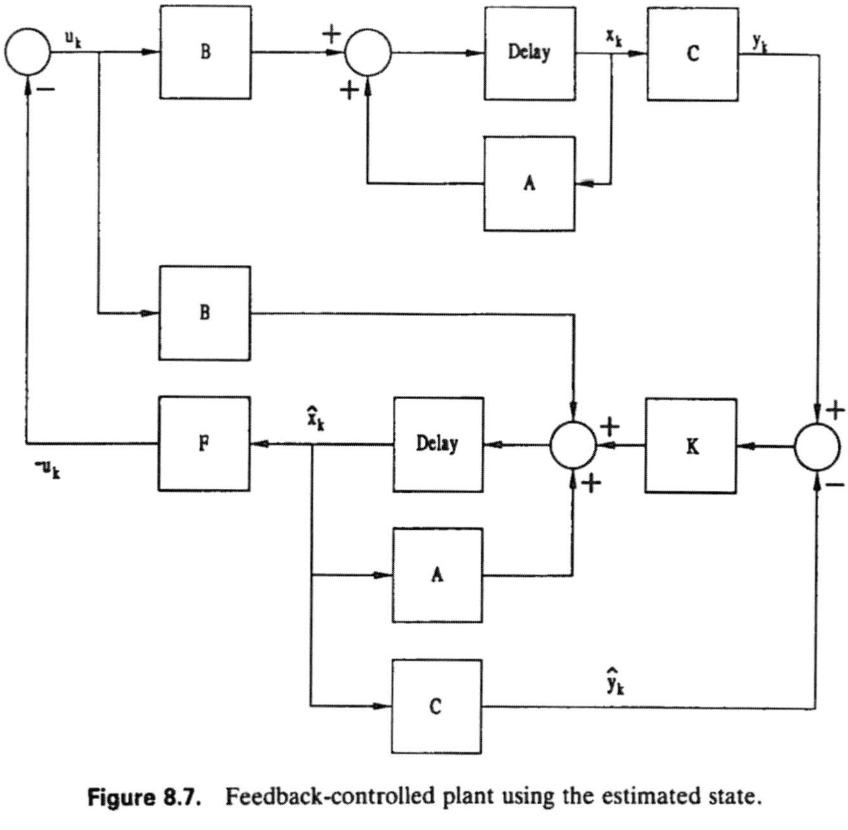


**Figure 8.7.** Feedback-controlled plant using the estimated state.

that the error equation is

$$\tilde{\mathbf{x}}_{k+1} = [\mathbf{A} - \mathbf{KC}]\tilde{\mathbf{x}}_k \tag{8.5.11}$$

and the plant equation (8.2.1) combined with (8.5.10) is

$$\mathbf{x}_{k+1} = \mathbf{Ax}_k - \mathbf{BF}\hat{\mathbf{x}}_k \tag{8.5.12}$$

or, noting relation (8.5.2),

$$\mathbf{x}_{k+1} = \mathbf{Ax}_k - \mathbf{BF}(\mathbf{x}_k - \tilde{\mathbf{x}}_k) \tag{8.5.13}$$

Writing (8.5.11) and (8.5.13) in matrix form,

$$\begin{bmatrix} \tilde{\mathbf{x}}_{k+1} \\ \mathbf{x}_{k+1} \end{bmatrix} = \begin{bmatrix} \mathbf{A} - \mathbf{KC} & \mathbf{0} \\ \mathbf{BF} & \mathbf{A} - \mathbf{BF} \end{bmatrix} \begin{bmatrix} \tilde{\mathbf{x}}_k \\ \mathbf{x}_k \end{bmatrix} \tag{8.5.14}$$

Taking the $z$-transform yields a determinant for the characteristic equation of the complete system, or

$$\det \begin{bmatrix} \mathbf{I}z - (\mathbf{A} - \mathbf{KC}) & \mathbf{0} \\ -\mathbf{BF} & \mathbf{I}z - (\mathbf{A} - \mathbf{BF}) \end{bmatrix} = 0 \tag{8.5.15}$$

The zero matrix in the upper right corner is a nice coincidence, so the determinant above may be written as the product of two determinants, or

$$\det[\mathbf{I}z - (\mathbf{A} - \mathbf{KC})]\det[\mathbf{I}z - (\mathbf{A} - \mathbf{BF})] = 0 \tag{8.5.16}$$

This fact is not as obvious as might appear initially and is proven in the book by Brogan (1985). Examination of these determinants indicates that $n$ poles of the complete system come from the second determinant and their values are controlled by selection of the elements of the feedback control matrix **F**. The poles of the observer are dictated by the first determinant in the selection of **K**. In general, the elements of matrix **F** will not be unique unless the control effort $\mathbf{u}_k$ is scalar, in which case it is easily determined by solution of linear equations, as shown in Section 8.2.

This result is called the *separation principle* in that we can design the observer and the controller separately. It is clear that all the roots of (8.5.16) will contribute to the closed-loop response of the system. After the elements of **F** are selected to give the closed-loop control system the correct pole locations, the elements of **K** should be picked so that the estimator poles are faster (roughly by a factor of 2) than the closed-loop poles.

With the control law (8.5.10) substituted into the estimator equation, we get

$$\hat{\mathbf{x}}_{k+1} = \mathbf{A}\hat{\mathbf{x}}_k - \mathbf{BF}\hat{\mathbf{x}}_k + \mathbf{K}(\mathbf{y}_k - \mathbf{C}\hat{\mathbf{x}}_k) \tag{8.5.17}$$

or

$$\hat{\mathbf{x}}_{k+1} = [\mathbf{A} - \mathbf{BF} - \mathbf{KC}]\hat{\mathbf{x}}_k + \mathbf{Ky}_k \tag{8.5.18}$$

and the control law is

$$\mathbf{u}_k = -\mathbf{F}\hat{\mathbf{x}}_k \tag{8.5.19}$$

Expression (8.5.18) is the usual form of the prediction estimator employed when implementing an observer in software. The flowchart for such a process is illustrated in Fig. 8.8. Now if $\mathbf{y}_k$ and $\mathbf{u}_k$ are both scalar sequences, we can solve for a transfer function of the observer–controller combination, which is what conventionally would be called a compensator. The compensator transfer function is the ratio of $-U(z)$ to $Y(z)$, which is

$$D(z) = \frac{-U(z)}{Y(z)} = \mathbf{F}[\mathbf{I}z - (\mathbf{A} - \mathbf{BF} - \mathbf{KC})]^{-1}\mathbf{K} \tag{8.5.20}$$

while it is clear that the compensator poles come from

$$\det[\mathbf{I}z - (\mathbf{A} - \mathbf{BF} - \mathbf{KC})] = 0$$

Clearly, these poles are different from the roots of (8.5.9) as they are for the open-loop compensator. If we examine (8.5.18), we find the reason why this was called a prediction estimator, since the state at $(k + 1)T$ was calculated based on the measurement taken at $kT$.

**Example 8.3.** Consider the inertial plant of Examples 6.1, 6.3, and 8.1 when the simple inertial plant and zero-order hold are described by

$$\begin{bmatrix} x_1 \\ x_2 \end{bmatrix}_{k+1} = \begin{bmatrix} 1 & 0.1 \\ 0 & 1 \end{bmatrix}\begin{bmatrix} x_1 \\ x_2 \end{bmatrix}_k + \begin{bmatrix} 0.005 \\ 0.1 \end{bmatrix} u_k$$

with measurements on the position variable $x_1$, or

$$y_k = [1 \quad 0]\begin{bmatrix} x_1 \\ x_2 \end{bmatrix}_k$$

If we use the prediction estimator outlined in Section 8.5 and given by expression (8.5.8), we get

$$\begin{bmatrix} \hat{x}_1 \\ \hat{x}_2 \end{bmatrix}_{k+1} = \begin{bmatrix} 1 & 0.1 \\ 0 & 1 \end{bmatrix}\begin{bmatrix} \hat{x}_1 \\ \hat{x}_2 \end{bmatrix}_k + \begin{bmatrix} 0.005 \\ 0.1 \end{bmatrix} u_k + \begin{bmatrix} k_1 \\ k_2 \end{bmatrix}[x_1(k) - \hat{x}_1(k)]$$

and the characteristic equation associated with the observer is given by expression (8.5.16) to be

$$\alpha_e(z) = \det(z\mathbf{I} - \mathbf{A} + \mathbf{KC})$$

or

$$\alpha_e(z) = \det\begin{bmatrix} z - 1 + k_1 & -0.1 \\ k_2 & z - 1 \end{bmatrix} = 0$$

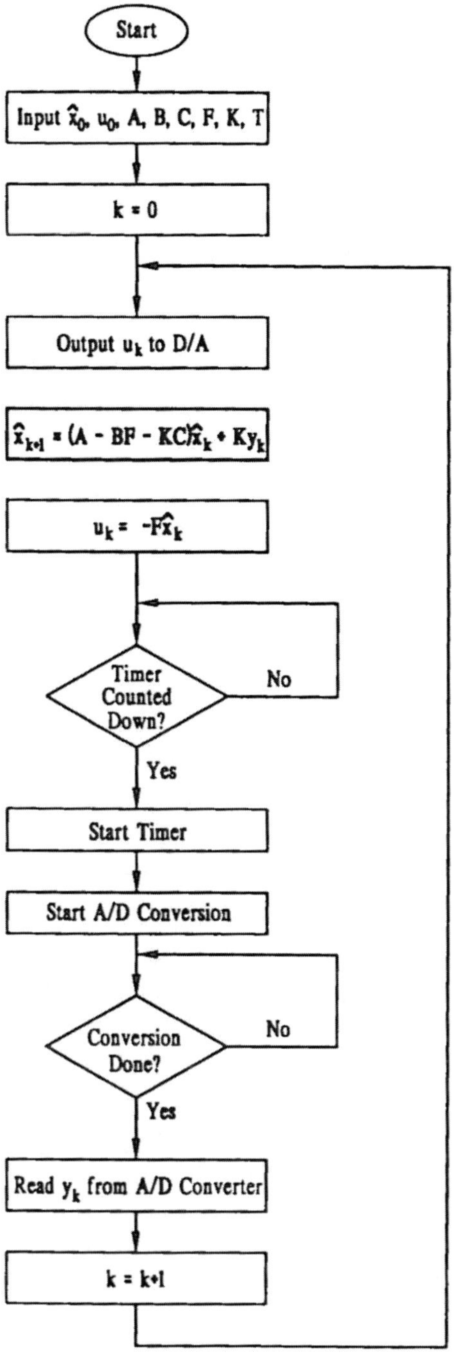


**Figure 8.8.** Flowchart for implementation of estimated state feedback with a prediction observer.

Upon expansion of the determinant the following polynomial characteristic equation results:

$$\alpha_e(z) = z^2 - z(2 - k_1) + (1 - k_1 + 0.1k_2) = 0 \qquad \text{(a)}$$

If we want the observer roots to be faster than either the open-loop roots or those given by state feedback in Example 8.1, we could locate the roots at $z = 0.5 \pm j0.3$, which would yield an observer characteristic equation of

$$\alpha_e(z) = z^2 - z + 0.34 \qquad \text{(b)}$$

If we now equate the coefficients of (a) and (b), we get simultaneous linear equations for equality of the coefficients

$$2 - k_1 = 1$$

and

$$1 - k_1 + 0.1k_2 = 0.34$$

Solving these for the observer gains yields

$$k_1 = 1$$

$$k_2 = 3.4$$

We shall use this estimator to estimate the states of the inertial plant with no control effort applied and an initial state of

$$\mathbf{x}_0 = \begin{bmatrix} 0 \\ 1 \end{bmatrix}$$

The initial state of the estimator will be chosen to be zero because often no better information exists. Figure 8.9(a) and (b) show both states and their estimates based on measurement of state one. The estimates are seen to follow the states to a reasonable degree after an initial transient error.

If we now employ the feedback gains of Example 8.1 to feed back the estimated states to create the control effort, we are given the state and state estimate histories illustrated in Fig. 8.10(a) and (b), where it is seen that the estimator follows the states to a reasonable degree. One might think that feedback involving the measured state $x_1(k)$ and the estimated unmeasured state $\hat{x}_2(k)$ might provide better performance, and this is true if there is no additive random measurement noise that will put random fluctuations into the control effort. If will be shown in a later chapter that it is better in this case to use estimated states, and in the statistical case the observer is called a Kalman filter.

If we now would like to examine the combined estimation and control problem as an equivalent digital compensator as given by expression (8.5.20),

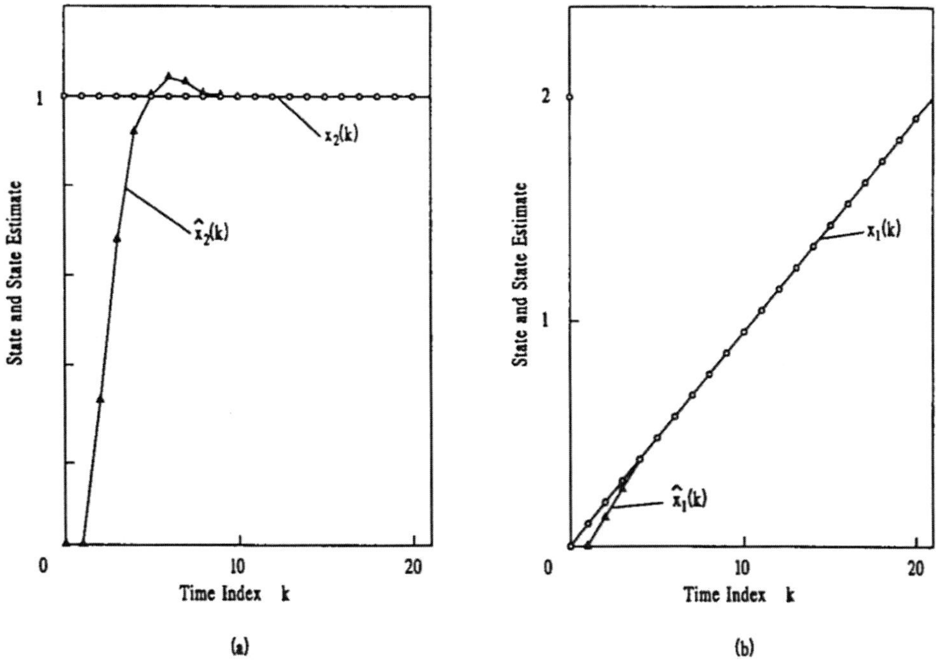


**Figure 8.9.** Responses and estimated responses for an observer with no control efforts: (a) angular displacement and its estimate; (b) angular velocity and its estimate.

then

$$D(z) = \mathbf{f}^T[\mathbf{I}z - (\mathbf{A} - \mathbf{b}\mathbf{f}^T - \mathbf{k}\mathbf{c}^T)]^{-1}\mathbf{k}$$
$$= (13 \quad 5.35)\begin{bmatrix} z - 1 + 0.065 + 1 & -0.1 + 0.02675 \\ 1.3 + 3.4 & z - 1 + 0.535 \end{bmatrix}^{-1}\begin{bmatrix} 1 \\ 3.4 \end{bmatrix}$$

or

$$D(z) = (13 \quad 5.35)\begin{bmatrix} z - 0.065 & -0.07325 \\ 4.7 & z - 0.465 \end{bmatrix}^{-1}\begin{bmatrix} 1 \\ 3.4 \end{bmatrix}$$

If one is careful with the numbers, the resulting transfer function is

$$D(z) = \frac{31.19z - 26.77}{z^2 - 0.4z + 0.3141} = \frac{-U(z)}{X_1(z)}$$

The difference equation that must be implemented on the digital computer is

$$u(k + 2) - 0.4u(k + 1) + 0.3141u(k) = -31.19x_1(k + 1) + 26.77x_1(k)$$

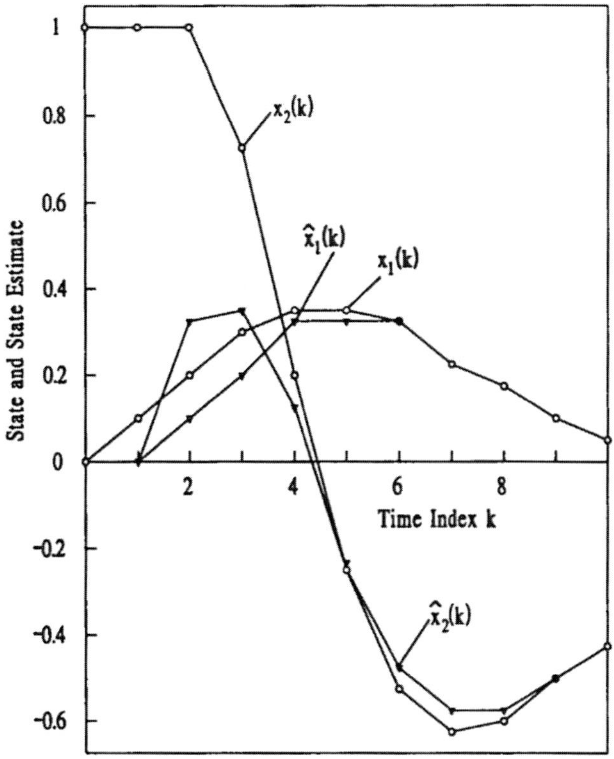


**Figure 8.10.** Responses and estimated responses for inertial system employing estimated state feedback. State variable $x_1(k)$ is the position and $x_2(k)$ is the velocity.

It is hoped that this extensive example will clarify application of the above-developed theory to real problems.

**Example 8.4.** Consider the same problem in which state feedback was employed in Example 8.2, but here let us measure only the temperature sequence $x_1(k)$. Let us use estimated feedback to provide the control. Since the separation principle holds, half of the poles may be specified by employing the feedback matrix calculated in Example 3.3. It remains for us to calculate the observer gain matrix such that the poles of the observer are faster than those of the feedback-controlled plant.

The output relation in this problem is

$$y(k) = [1 \quad 0]\begin{bmatrix} x_1(k) \\ x_2(k) \end{bmatrix}$$

so

$$\mathbf{c}^T = [1 \quad 0]$$

The observer poles may be determined from the appropriate determinant of relation (8.5.16), or

$$\det[z\mathbf{I} - \mathbf{A} + \mathbf{k}\mathbf{c}^T] = 0$$

or substitution of the values for this problem yields

$$\det\begin{bmatrix} z - 0.6277 + k_1 & -0.3597 \\ -0.0899 + k_2 & z - 0.8526 \end{bmatrix} = 0$$

Expanding this determinant, the resulting characteristic polynomial is

$$\alpha_e(z) = z^2 + (k_1 - 1.48)z + 0.502 - 0.853k_1 + 0.361k_2 = 0$$

It is clear from the coefficients that selection of the observer gains will control the observer pole locations. If the observer is to be significantly faster than the controlled plant, a good choice for pole locations would be $z = 0.4 \pm j0.2$, so the characteristic polynomial is

$$\alpha_e(z) = z^2 - 0.8z + 0.2 = 0$$

Equating the coefficients of these two versions of the characteristic equation yields the following linear equations:

$$1.48 - k_1 = 0.8$$

$$0.502 - 0.853k_1 + 0.361k_2 = 0.2$$

Solution of these linear equations yields observer gains

$$k_1 = 0.68, \qquad k_2 = 0.77$$

So the observer is of the form

$$\begin{bmatrix} \hat{x}_1 \\ \hat{x}_2 \end{bmatrix}_{k+1} = \begin{bmatrix} 0.6277 & 0.3597 \\ 0.0899 & 08526 \end{bmatrix} \begin{bmatrix} \hat{x}_1 \\ \hat{x}_2 \end{bmatrix}_k + \begin{bmatrix} 0.0251 \\ 0.1150 \end{bmatrix} u_k + \begin{bmatrix} 0.68 \\ 0.77 \end{bmatrix} [x_1(k) - \hat{x}_1(k)]$$

with a control law of

$$u_k = -[2.248 \quad 3.685] \begin{bmatrix} \hat{x}_1 \\ \hat{x}_2 \end{bmatrix}_k$$

The complete control system is shown in Fig. 8.11, while typical initial condition responses are shown in Fig. 8.12. The estimator is started from

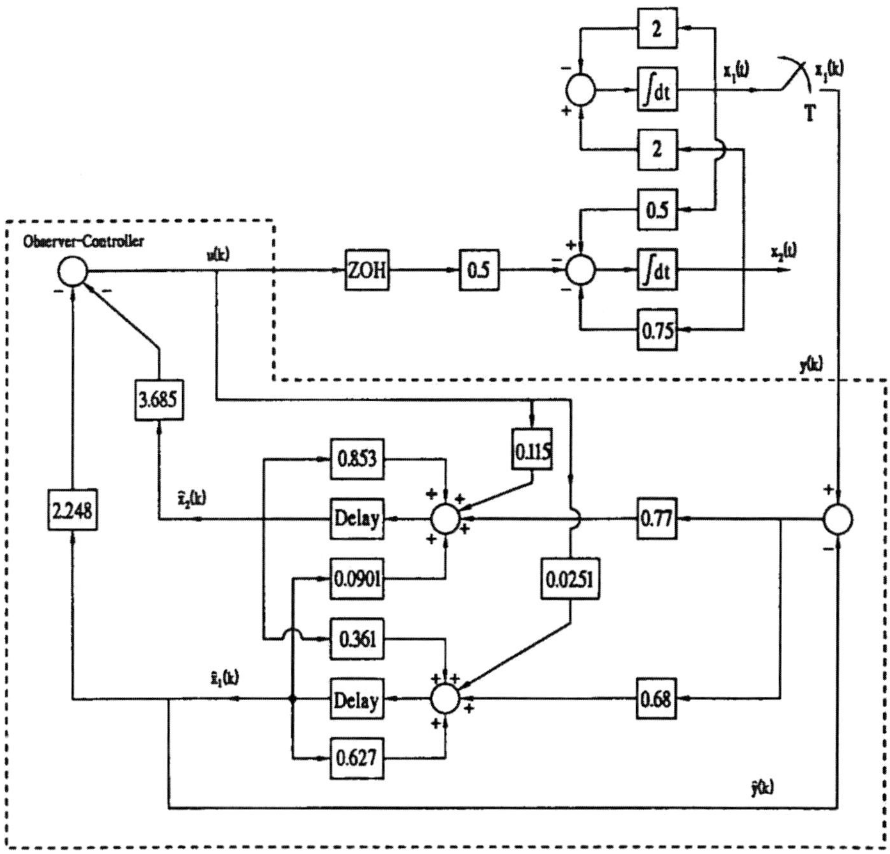


**Figure 8.11.** Complete estimated state feedback control system for thermal plant.

zero initial conditions since the initial condition of the control system is usually known to the observer. Similar response curves for the system with actual state feedback were given in Example 8.3 and are given here for comparison.

## 8.6 CURRENT ESTIMATOR OR OBSERVER

In Section 8.5 we derived an estimator that estimated the state at $(k + 1)T$ given a measurement at $kT$ and hence the term *prediction estimator*. It would seem that a better job could be done if data at the $k$th step were used to estimate the state at that same time.

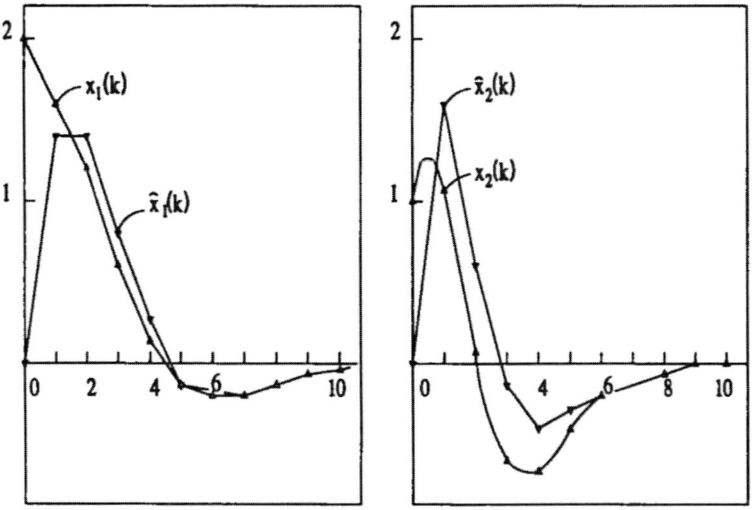


**Figure 8.12.** State and estimated state histories for thermal plant with estimated state feedback control.

Let us denote by $\bar{\mathbf{x}}_k$ the estimate of the state at time $kT$ based only on the previous state estimate $\hat{\mathbf{x}}_{k-1}$, or

$$\bar{\mathbf{x}}_k = \mathbf{A}\hat{\mathbf{x}}_{k-1} + \mathbf{B}\mathbf{u}_{k-1} \tag{8.6.1}$$

and the updated estimate of $\mathbf{x}_k$ based on measurement $\mathbf{y}_k$ will be denoted as $\hat{\mathbf{x}}_k$, which will be that given by (8.6.1) plus a term proportional to the error in the estimate of the measurement:

$$\hat{\mathbf{x}}_k = \bar{\mathbf{x}}_k + \mathbf{K}(\mathbf{y}_k - \bar{\mathbf{y}}_k) \tag{8.6.2}$$

where

$$\bar{\mathbf{y}}_k = \mathbf{C}\bar{\mathbf{x}}_k \tag{8.6.3}$$

Substitution of (8.6.1) and (8.6.3) into (8.6.2) yields an overall estimator of the form

$$\hat{\mathbf{x}}_k = \mathbf{A}\hat{\mathbf{x}}_{k-1} + \mathbf{B}\mathbf{u}_{k-1} + \mathbf{K}[\mathbf{y}_k - \mathbf{C}(\mathbf{A}\hat{\mathbf{x}}_{k-1} + \mathbf{B}\mathbf{u}_{k-1})] \tag{8.6.4}$$

and after simplification

$$\hat{\mathbf{x}}_k = (\mathbf{A} - \mathbf{KCA})\hat{\mathbf{x}}_{k-1} + (\mathbf{B} - \mathbf{KCB})\mathbf{u}_{k-1} + \mathbf{K}\mathbf{y}_k \tag{8.6.5}$$

The actual value of $\mathbf{x}_k$ is given by the state equation

$$\mathbf{x}_k = \mathbf{A}\mathbf{x}_{k-1} + \mathbf{B}\mathbf{u}_{k-1} \tag{8.6.6}$$

Subtracting (8.6.5) from (8.6.6) to compute the error $\tilde{\mathbf{x}}_k$, we get

$$\tilde{\mathbf{x}}_k = (\mathbf{A} - \mathbf{KCA})\tilde{\mathbf{x}}_{k-1} \tag{8.6.7}$$

So the estimator dynamics, as before, are controlled by the choice of the estimator gain matrix **K**. The elements of **K** should be chosen such as to make the dynamics of the estimator faster than those of the closed-loop plant.

If, as in the preceding section, we choose to feed back the estimated state with a feedback matrix **F** to provide the control effort, a proof of the separation principle can be made in a similar fashion, so that the design of the estimator is restricted to the choice of the elements of the **K** matrix.

There is a problem in the implementation of this estimator in that the computation of the estimate $\hat{\mathbf{x}}_k$ cannot be completed until the sample $\mathbf{y}_k$ is taken. After the estimate is made, the control vector must be computed by

$$\mathbf{u}_k = -\mathbf{F}\hat{\mathbf{x}}_k \tag{8.6.8}$$

Clearly, $\mathbf{u}_k$ will not be available instantaneously after the sample $\mathbf{y}_k$ is taken, and hence it might be more realistic to use the predictor estimator since after the sample is taken, a whole sampling period is available to estimate the state and compute the control vector.

## 8.7 REDUCED-ORDER ESTIMATOR OR OBSERVER

If our measurements are made directly on some, say $p$, of the state variables, we need only reconstruct the remaining $n - p$ of the state variables. Let us partition the state vector into those states that are measured directly—call this $\mathbf{x}_1$—and those that must be estimated, denoted by $\mathbf{x}_2$, or

$$\mathbf{x}_k = \begin{bmatrix} \mathbf{x}_{1k} \\ \mathbf{x}_{2k} \end{bmatrix} \tag{8.7.1}$$

With this definition we can write the state equations in partitioned form

$$\begin{bmatrix} \mathbf{x}_{1k+1} \\ \mathbf{x}_{2k+1} \end{bmatrix} = \begin{bmatrix} \mathbf{A}_{11} & \mathbf{A}_{12} \\ \mathbf{A}_{21} & \mathbf{A}_{22} \end{bmatrix} \begin{bmatrix} \mathbf{x}_{1k} \\ \mathbf{x}_{2k} \end{bmatrix} + \begin{bmatrix} \mathbf{B}_1 \\ \mathbf{B}_2 \end{bmatrix} \mathbf{u}_k \tag{8.7.2}$$

where the output equation is now

$$\mathbf{y}_k = [\mathbf{I} \mid \mathbf{0}] \begin{bmatrix} \mathbf{x}_{1k} \\ \mathbf{x}_{2k} \end{bmatrix} \tag{8.7.3}$$

where the **I** represents a $p \times p$ identity matrix. We can write the second of the equations of (8.7.2) as

$$\mathbf{x}_{2k+1} = \mathbf{A}_{22}\mathbf{x}_{2k} + \mathbf{A}_{21}\mathbf{x}_{1k} + \mathbf{B}_2\mathbf{u}_k \tag{8.7.4}$$

The last two terms of (8.7.4) can be thought of as input functions to an ordinary state equation with $\mathbf{x}_2$ as a state vector. The first equation (8.7.2) could be written as

$$\mathbf{x}_{1k+1} - \mathbf{A}_{11}\mathbf{x}_{1k} - \mathbf{B}_1\mathbf{u}_k = \mathbf{A}_{12}\mathbf{x}_{2k} \tag{8.7.5}$$

Everything on the left side of (8.7.5) is known, and hence this has the form of an output equation ($\mathbf{y}_k = \mathbf{C}\mathbf{x}_k$), and now we may apply our prediction estimator developed in a previous section to give an estimator. The changes in matrix quantities from those of a full-order observer are given in Table 8.1. Now the predictor estimator can be written down directly as

$$\begin{aligned}\hat{\mathbf{x}}_{2k+1} = {} & \mathbf{A}_{22}\hat{\mathbf{x}}_{2k} + \mathbf{A}_{21}\mathbf{x}_{1k} + \mathbf{B}_2\mathbf{u}_k \\ & + \mathbf{K}[\mathbf{x}_{1k+1} - \mathbf{A}_{11}\mathbf{x}_{1k} - \mathbf{B}_1\mathbf{u}_k - \mathbf{A}_{12}\hat{\mathbf{x}}_k]\end{aligned} \tag{8.7.6}$$

Although we used predictor estimator theory to derive this estimator, the structure of the problem makes the resulting estimator a current estimator since $\hat{\mathbf{x}}_{2k+1}$ is constructed based on a measurement of $\mathbf{x}_{1k+1}$. Now let us define, as before, the error vector

$$\tilde{\mathbf{x}}_{2k} = \mathbf{x}_{2k} - \hat{\mathbf{x}}_{2k} \tag{8.7.7}$$

If we now subtract (8.7.6) from (8.7.4) and substitute from (8.7.5) inside the brackets, we get

$$\tilde{\mathbf{x}}_{2k+1} = [\mathbf{A}_{22} - \mathbf{K}\mathbf{A}_{12}]\tilde{\mathbf{x}}_{2k} \tag{8.7.8}$$

so the $n - p$ poles of the estimator are controlled by the selection of the elements of $\mathbf{K}$. We can also prove the separation principle for this estimator,

**Table 8.1.** Matrix Quantities for Full- and Reduced-Order Estimators

| Full-order estimator | Reduced-order estimator |
|---|---|
| $\mathbf{x}_k$ | $\mathbf{x}_{2k}$ |
| $\mathbf{A}$ | $\mathbf{A}_{22}$ |
| $\mathbf{B}\mathbf{u}_k$ | $\mathbf{A}_{21}\mathbf{x}_{1k} + \mathbf{B}_2\mathbf{u}_k$ |
| $\mathbf{y}_k$ | $\mathbf{x}_{1k+1} - \mathbf{A}_{11}\mathbf{x}_{1k} - \mathbf{B}_1\mathbf{u}_k$ |
| $\mathbf{C}$ | $\mathbf{A}_{12}$ |

and here we find that the remainder of the system poles are controlled by the feedback matrix **F**, where the control law is of the form

$$\mathbf{u}_k = -\mathbf{F}\begin{bmatrix} \mathbf{x}_{1k} \\ \hat{\mathbf{x}}_{2k} \end{bmatrix}$$

## 8.8 ALGORITHM FOR GAIN CALCULATIONS FOR SINGLE-INPUT SYSTEMS

In Section 8.2 we considered the calculation of state feedback gains to place all of the $n$ closed-loop system poles at arbitrary locations in the $z$-plane. We found that expanding the closed-loop characteristic determinant with the symbolic feedback gains present was a tedious task subject to potential algebraic errors. Similar problems were encountered in Section 8.5 when placing observer poles. Ackermann (1972) and Bass and Gura (1965) have developed algorithms suitable for automating the process of pole placement in the state feedback context for problems with a single control effort or in the state estimation context involving a single system output.

We shall consider the Ackermann algorithm here, as it is slightly more efficient computationally. We are seeking a state feedback gain vector $\mathbf{f}^T$ in the control law

$$u_k = -\mathbf{f}^T\mathbf{x}_k \tag{8.8.1}$$

for the system

$$\mathbf{x}_{k+1} = \mathbf{A}\mathbf{x}_k + \mathbf{b}u_k \tag{8.8.2}$$

which will pace the closed-loop characteristic roots at arbitrary $z$-domain locations as long as complex roots occur in conjugate pairs. The closed-loop characteristic equation is

$$\alpha_c(z) = z^n + d_{n-1}z^{n-1} + \cdots + d_1 z + d_0 \tag{8.8.3}$$

This characteristic equation is also given by

$$\alpha_c(z) = \det[z\mathbf{I} - \mathbf{A} + \mathbf{b}\mathbf{f}^T] \tag{8.8.4}$$

The problem here is that if the $d_i$ are given, then find the vector $\mathbf{f}^T$ to cause the characteristic polynomials of (8.8.3) (known) and (8.8.4) to be the same without a manual expansion of (8.8.4).

We shall assume that there exists a transformation **T** that will transform the system into controllable canonical form with state vector $\mathbf{w}_k$ or

$$\mathbf{x}_k = \mathbf{T}\mathbf{w}_k \tag{8.8.5}$$

Substitute (8.8.5) into (8.8.2) to give the controllable canonical form, which is

$$\mathbf{w}_{k+1} = \mathbf{T}^{-1}\mathbf{A}\mathbf{T}\mathbf{w}_k + \mathbf{T}^{-1}\mathbf{b}u_k \tag{8.8.6}$$

and the control law in these new coordinates is

$$u_k = -\mathbf{f}^T\mathbf{T}\mathbf{w}_k \tag{8.8.7}$$

where we shall refer to vector $\mathbf{f}^T\mathbf{T}$ as $\mathbf{f}_W^T$. In the $\mathbf{x}_k$ coordinate system the controllability matrix is

$$\mathbf{C}_x = [\mathbf{A}^{n-1}\mathbf{b} \quad \mathbf{A}^{n-2}\mathbf{b} \quad \cdots \quad \mathbf{A}\mathbf{b} \quad \mathbf{b}] \tag{8.8.8}$$

and in the $\mathbf{w}_k$ coordinates it is

$$\mathbf{C}_w = [\mathbf{A}_1^{n-1}\mathbf{b}_1 \quad \mathbf{A}_1^{n-2}\mathbf{b}_1 \quad \cdots \quad \mathbf{A}_1\mathbf{b}_1 \quad \mathbf{b}_1] \tag{8.8.9}$$

where, for the controllable canonical form, the matrix $\mathbf{A}_1$ and vector $\mathbf{b}_1$ are

$$\mathbf{A}_1 = \mathbf{T}^{-1}\mathbf{A}\mathbf{T} = \begin{bmatrix} 0 & 1 & \cdots & 0 \\ 0 & 0 & & \vdots \\ \vdots & & & 1 \\ -a_0 & -a_1 & \cdots & -a_{n-1} \end{bmatrix} \quad \mathbf{b}_1 = \mathbf{T}^{-1}\mathbf{b} = \begin{bmatrix} 0 \\ \vdots \\ 0 \\ 1 \end{bmatrix} \tag{8.8.10}$$

If we note the definitions of $\mathbf{A}_1$ and $\mathbf{b}_1$ given above, the $\mathbf{w}_k$ controllability matrix is

$$\mathbf{C}_w = \mathbf{T}^{-1}[\mathbf{A}^{n-1}\mathbf{b} \quad \mathbf{A}^{n-2}\mathbf{b} \quad \cdots \quad \mathbf{A}\mathbf{b} \quad \mathbf{b}] \tag{8.8.11}$$

and upon recognizing the controllability matrix $\mathbf{C}_x$ the relation between the controllability matrices is

$$= \mathbf{T}^{-1}\mathbf{C}_x \tag{8.8.12}$$

Thus the desired inverse transformation is

$$\mathbf{T}^{-1} = \mathbf{C}_w\mathbf{C}_x^{-1} \tag{8.8.13}$$

but at this point we do not know $\mathbf{A}_1$ and $\mathbf{b}_1$, and hence $\mathbf{C}_w$ is unknown. The open-loop characteristic equation for the controllable canonical form is from $\mathbf{A}_1$ in (8.8.10):

$$\begin{aligned} \alpha_0(z) &= \det[z\mathbf{I} - \mathbf{A}_1] \\ &= z^n + a_{n-1}z^{n-1} + \cdots + a_1 z + a_0 = 0 \end{aligned} \tag{8.8.14}$$

The $\mathbf{b}_1\mathbf{f}_w^T$ vector is

$$\mathbf{b}_1\mathbf{f}_w^T = \begin{bmatrix} 0 & \cdots & 0 \\ \vdots & & \vdots \\ 0 & & 0 \\ f_{w1} & \cdots & f_{wn} \end{bmatrix} \tag{8.8.15}$$

and

$$\mathbf{A}_1 - \mathbf{b}_1\mathbf{f}_w^T = \begin{bmatrix} 0 & 1 & 0 & \cdots & 0 \\ 0 & 0 & 1 & & 0 \\ \vdots & & & & \vdots \\ 0 & & & & 1 \\ -(a_0 + f_{w1}) & & & \cdots & -(a_{n-1} + f_{wn}) \end{bmatrix} \tag{8.8.16}$$

The closed-loop characteristic equation for the $\mathbf{A}_1$, $\mathbf{b}_1$ system

$$\alpha_c(z) = \det[z\mathbf{I} - \mathbf{A}_1 + \mathbf{b}_1\mathbf{f}_w^T] = 0 \tag{8.8.17}$$

and expanding in a similar fashion to the open-loop case,

$$z^n + \underbrace{(a_{n-1} + f_{wn})}_{d_{n-1}} z^{n-1} + \cdots + \underbrace{(a_1 + f_{w2})}_{d_1} z + \underbrace{(a_0 + f_{w1})}_{d_0} = 0 \tag{8.8.18}$$

The coefficients $d_i$ are then defined as

$$d_i = a_i + f_{wi+1} \qquad i = 1, \ldots, n \tag{8.8.19}$$

The Cayley–Hamilton theorem (see Appendix B) states that a matrix satisfies its own characteristic equation, so

$$\mathbf{A}_1^n + a_{n-1}\mathbf{A}_1^{n-1} + \cdots + a_1\mathbf{A}_1 + a_0\mathbf{I} = 0 \tag{8.8.20}$$

and for the closed-loop

$$\alpha_c(\mathbf{A}_1) = \mathbf{A}_1^n + d_{n-1}\mathbf{A}_1^{n-1} + \cdots + d_1\mathbf{A}_1 + d_0\mathbf{I} \tag{8.8.21}$$

but this equation is not zero because $\mathbf{A}_1$ is not the closed-loop system matrix, while the $d_i$ are the closed-loop coefficients. From the open-loop equation (8.8.20),

$$\mathbf{A}_1^n = -a_{n-1}\mathbf{A}_1^{n-1} - \cdots - a_1\mathbf{A}_1 - a_0\mathbf{I} \tag{8.8.22}$$

Substitute this into $\alpha_c(\mathbf{A}_1)$ to get

$$\alpha_c(\mathbf{A}_1) = (d_{n-1} - a_{n-1})\mathbf{A}_1^{n-1} + \cdots + (d_1 - a_1)\mathbf{A}_1 + (d_0 - a_0)\mathbf{I} \tag{8.8.23}$$

but from (8.8.19) this becomes

$$\alpha_c(\mathbf{A}_1) = f_{wn}\mathbf{A}_1^{n-1} + \cdots + f_{w2}\mathbf{A}_1 + f_{w1}\mathbf{I} \tag{8.8.24}$$

Now let us investigate the premultiplication of $\mathbf{A}_1$ raised to the various powers by the first unit vector $\mathbf{e}_1^T$. From the form of $\mathbf{A}_1$ in expression (8.8.10),

$$\mathbf{e}_1^T\mathbf{A}_1 = \mathbf{e}_2^T \tag{8.8.25}$$

so

$$\mathbf{e}_1^T\mathbf{A}_1^2 = (\mathbf{e}_1^T\mathbf{A}_1)\mathbf{A}_1 = \mathbf{e}_2^T\mathbf{A}_1 = \mathbf{e}_3^T \tag{8.8.26}$$

and finally,

$$\mathbf{e}_1^T\mathbf{A}_1^{n-1} = \mathbf{e}_n^T \tag{8.8.27}$$

Now take $\mathbf{e}_1^T\alpha_c(\mathbf{A}_1)$ using the form from (8.8.24), or

$$\begin{aligned} \mathbf{e}_1^T\alpha_c(\mathbf{A}_1) &= f_{wn}\mathbf{e}_1^T\mathbf{A}_1^{n-1} + \cdots + f_{w2}\mathbf{e}_1^T\mathbf{A}_1 + f_{w1}\mathbf{e}_1^T\mathbf{I} \\ &= f_{wn}\mathbf{e}_n^T + \cdots + f_{w2}\mathbf{e}_2^T + f_{w1}\mathbf{e}_1^T \\ \mathbf{e}_1^T\alpha_c(\mathbf{A}_1) &= \mathbf{f}_w^T \end{aligned} \tag{8.8.28}$$

but

$$\mathbf{f}^T = \mathbf{f}_w^T\mathbf{T}^{-1} \tag{8.8.29}$$

$$= \mathbf{e}_1^T\alpha_c(\mathbf{A}_1)\mathbf{T}^{-1} \tag{8.8.30}$$

For the polynomial matrix

$$\alpha_c(\mathbf{A}_1) = \alpha_c(\mathbf{T}^{-1}\mathbf{A}\mathbf{T}) = \mathbf{T}^{-1}\alpha_c(\mathbf{A})\mathbf{T} \tag{8.8.31}$$

Then substitution of (8.8.31) into (8.8.30) yields

$$\mathbf{f}^T = \mathbf{e}_1^T\mathbf{T}^{-1}\alpha_c(\mathbf{A})\mathbf{T}\mathbf{T}^{-1} \tag{8.8.32}$$

$$= \mathbf{e}_1^T\mathbf{T}^{-1}\alpha_c(\mathbf{A}) \tag{8.8.33}$$

but $\mathbf{T}^{-1}$ is given by (8.8.13), so

$$\mathbf{f}^T = \mathbf{e}_1^T\mathbf{C}_w\mathbf{C}_x^{-1}\alpha_c(\mathbf{A}) \tag{8.8.34}$$

Because of the form of $\mathbf{A}_1$ and $\mathbf{b}_1$, the top row of $\mathbf{C}_w$ is $\mathbf{e}_1^T$; thus

$$\mathbf{e}_1^T\mathbf{C}_w = \mathbf{e}_1^T \tag{8.8.35}$$

so the feedback gain vector is

$$\mathbf{f}^T = \mathbf{e}_1^T\mathbf{C}_x^{-1}\alpha_c(\mathbf{A}) \tag{8.8.36}$$

If we now substitute the controllability matrix and the polynomial for the $\alpha_c(\mathbf{A})$, we get

$$\begin{aligned}\mathbf{f}^T = \mathbf{e}_1^T[\mathbf{A}^{n-1}\mathbf{b} \quad \mathbf{A}^{n-2}\mathbf{b} \quad \cdots \quad \mathbf{A}\mathbf{b} \quad \mathbf{b}]^{-1} \\ \times [\mathbf{A}^n + d_{n-1}\mathbf{A}^{n-1} + \cdots + d_0\mathbf{I}]\end{aligned} \tag{8.8.37}$$

where the $d_i$ are the coefficients of the desired closed-loop characteristic polynomial.

It is now unnecessary to know the transformation matrix **T**; only the **A** and **b** matrices must be known along with the closed-loop characteristic equation. The assurance that the inverse of the controllability matrix exists lies in the fact that the system must be controllable to attempt state feedback as a pole-placement technique.

In a similar fashion we can show that for a single-output system the prediction observer examined in Section 8.5 gain vector can be calculated by

$$\mathbf{k}^T = \mathbf{e}_1^T[(\mathbf{A}^T)^{n-1} \quad \mathbf{c}(\mathbf{A}^T)^{n-2}\mathbf{c} \quad \cdots \quad \mathbf{A}^T\mathbf{c} \quad \mathbf{c}]^{-1}\alpha_e(\mathbf{A}^T) \tag{8.8.38}$$

where $\alpha_e(\mathbf{A}^T)$ is the desired estimator characteristic polynomial.

Let us now consider the case of the prediction estimator for a single-output system where the output is a scalar linear combination of the states, or

$$y_k = \mathbf{c}^T\mathbf{x}_k \tag{8.8.39}$$

In this case the desired estimator characteristic equation is given by (8.5.9) as

$$\alpha_e(z) = \det[z\mathbf{I} - \mathbf{A} + \mathbf{k}\mathbf{c}^T] \tag{8.8.40}$$

If we transpose the matrix, it does not affect the value of the determinant, so $\alpha_e(z)$ can be written as

$$\alpha_e(z) = \det[z\mathbf{I} - \mathbf{A}^T + \mathbf{c}\mathbf{k}^T] \tag{8.8.41}$$

From the form of this equation and that of (8.8.4) we see that we may also use the algorithm to place the poles of a prediction observer, so the algorithm for this case is

$$\mathbf{k}^T = \mathbf{e}_1^T[(\mathbf{A}^T)^{n-1}\mathbf{c} \quad (\mathbf{A}^T)^{n-2}\mathbf{c} \cdots \mathbf{c}]^{-1}\alpha_e(\mathbf{A}^T) \tag{8.8.42}$$

where $\alpha_e(z)$ is the desired estimator characteristic equation. We see the matrix to be inverted is he observability matrix, and hence for us to be able to estimate the state, the system must be observable as stated in Section 8.5.

In the case of the current observer the characteristic equation was found to be

$$\det[z\mathbf{I} - \mathbf{A} + \mathbf{k}\mathbf{c}^T\mathbf{A}] \tag{8.8.43}$$

and we see that if we transpose this relation, we get

$$\alpha_e(z) = \det[z\mathbf{I} - \mathbf{A}^T + \mathbf{A}^T\mathbf{c}\mathbf{k}^T] \tag{8.8.44}$$

which indicates that (8.8.43) can be used to calculate current observer gains if vector **c** is replaced by vector $\mathbf{A}^T\mathbf{c}$.

It should be noted that either or both the state feedback poles and observer poles can be made to be of the finite settling type by locating them at the origin of the $z$-plane by letting all the $d_i = 0$ in $\alpha_c(z)$ or $\alpha_e(z)$, respectively.

**Example 8.5.** For the inertial plant employ Ackermann's method to place the poles at arbitrary locations characterized by a closed-loop characteristic equation

$$\alpha_c(z) = z^2 + d_1 z + d_0$$

where the $d_i$ are real. The system matrices are for arbitrary $T$

$$\mathbf{A} = \begin{bmatrix} 1 & T \\ 0 & 1 \end{bmatrix} \qquad \mathbf{b} = \begin{bmatrix} T^2/2 \\ T \end{bmatrix}$$

Then $\alpha_c(\mathbf{A})$ is

$$\alpha_c(\mathbf{A}) = \mathbf{A}^2 + d_1\mathbf{A} + d_0\mathbf{I} = \begin{bmatrix} 1 + d_1 + d_0 & (2 + d_1)T \\ 0 & 1 + d_1 + d_0 \end{bmatrix}$$

The controllability matrix is

$$[\mathbf{Ab} \quad \mathbf{b}] = \begin{bmatrix} \frac{3T^2}{2} & \frac{T^2}{2} \\ T & T \end{bmatrix}$$

The inverse of the controllability matrix is

$$[\mathbf{Ab} \quad \mathbf{b}]^{-1} = \frac{1}{T^3}\begin{bmatrix} T & -\frac{T^2}{2} \\ -T & \frac{3T^2}{2} \end{bmatrix} = \begin{bmatrix} \frac{1}{T^2} & -\frac{1}{2T} \\ -\frac{1}{T^2} & \frac{3}{2T} \end{bmatrix}$$

The feedback gain vector is then

$$\mathbf{f}^T = \mathbf{e}_1^T \begin{bmatrix} \dfrac{1}{T^2} & -\dfrac{1}{2T} \\ -\dfrac{1}{T^2} & -\dfrac{3}{2T} \end{bmatrix} \begin{bmatrix} 1 + d_1 + d_0 & (2 + d_1)T \\ 0 & 1 + d_1 + d_0 \end{bmatrix}$$

$$= \begin{bmatrix} \dfrac{1 + d_1 + d_0}{T^2} & \dfrac{3 + d_1 - d_0}{2T} \end{bmatrix}$$

It is interesting to note that the gains are varying inversely with the sampling interval.

## 8.9 REGULATION WITH NONZERO REFERENCE INPUTS

In this section we seek to control the system such that the output vector $\mathbf{y}_k$ tracks a set of reference variables $\mathbf{r}_k$ of the same dimension as $\mathbf{y}_k$. We shall also assume that the number of control efforts in vector $\mathbf{u}_k$ is the same as the number of outputs $\mathbf{y}_k$. The system is governed by

$$\mathbf{x}_{k+1} = \mathbf{A}\mathbf{x}_k + \mathbf{B}\mathbf{u}_k \tag{8.9.1}$$

with an output relation

$$\mathbf{y}_k = \mathbf{C}\mathbf{x}_k \tag{8.9.2}$$

We shall seek to accomplish this task by modifying the control law to be of the form

$$\mathbf{u}_k = -\mathbf{F}\mathbf{x}_k + \mathbf{G}\mathbf{r}_k \tag{8.9.3}$$

where $\mathbf{G}$ is a square matrix yet to be calculated. This situation is illustrated in Fig. 8.13. If the control effort of (8.9.3) is substituted into relation (8.9.1), the result is

$$\mathbf{x}_{k+1} = (\mathbf{A} - \mathbf{B}\mathbf{F})\mathbf{x}_k + \mathbf{B}\mathbf{G}\mathbf{r}_k \tag{8.9.4}$$

Let us now take the $z$-transform of (8.9.4) to give

$$\mathbf{X}(z) = (z\mathbf{I} - \mathbf{A} + \mathbf{B}\mathbf{F})^{-1}\mathbf{B}\mathbf{G}\mathbf{R}(z) \tag{8.9.5}$$

Then the transform of the output vector $\mathbf{y}_k$ is, from relation (8.9.2),

$$\mathbf{Y}(z) = \mathbf{C}(z\mathbf{I} - \mathbf{A} + \mathbf{B}\mathbf{F})^{-1}\mathbf{B}\mathbf{G}\mathbf{R}(z) \tag{8.9.6}$$

For lack of a better strategy, let us choose $\mathbf{G}$ such that the dc gain between $\mathbf{R}(z)$ and $\mathbf{Y}(z)$ is the identity matrix, or

$$\mathbf{C}(z\mathbf{I} - \mathbf{A} + \mathbf{B}\mathbf{F})^{-1}\mathbf{B}\mathbf{G}\big|_{z=1} = \mathbf{I} \tag{8.9.7}$$

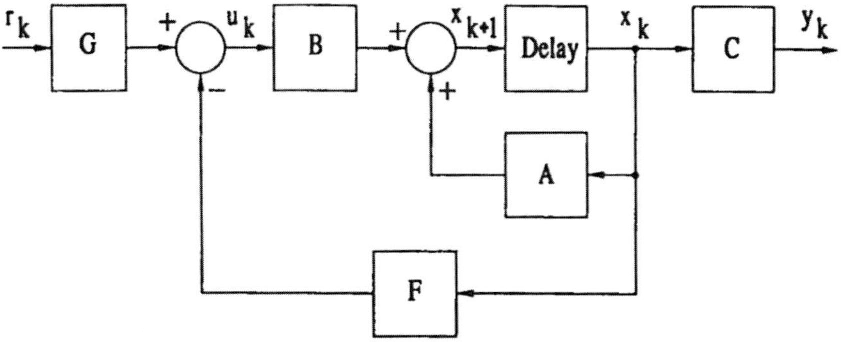


**Figure 8.13.** Complete state feedback system with reference input.

The gain matrix **G** is thus

$$\mathbf{G} = [\mathbf{C}(\mathbf{I} - \mathbf{A} + \mathbf{BF})^{-1}\mathbf{B}]^{-1} \tag{8.9.8}$$

The closed-loop poles are still placed by the selection of the elements of **F** as in the simple state feedback regulator.

**Example 8.6.** Let us design a state feedback control system to control the temperature $x_1(k)$ about some nonzero fixed point $\bar{y}$ and have the system poles located at $z = 0.5 \pm j0.2$. This is similar to the problem given in Example 8.2.

The system matrices are

$$\mathbf{A} = \begin{bmatrix} 0.6277 & 0.3597 \\ 0.0899 & 0.8526 \end{bmatrix} \qquad \mathbf{b} = \begin{bmatrix} 0.0251 \\ 0.1150 \end{bmatrix}$$

and

$$\mathbf{c}^T = [1 \quad 0]$$

The pole locations will be given by the roots of $\det(z\mathbf{I} - \mathbf{A} + \mathbf{bf}^T) = 0$, where the elements of $\mathbf{f}^T$ will be selected to place them as specified above. This was already done in Example 8.2, and the state feedback gain matrix was found to be

$$\mathbf{f}^T = [2.248 \quad 3.684]$$

Now all we need to do is to evaluate the reference input gain $G$ as given by expression (8.9.8). The calculations of relation (8.9.8) are most easily carried out by using the digital computer, and for the numerical values just given the gain is

$$G = [\mathbf{c}^T(\mathbf{I} - \mathbf{A} + \mathbf{b}\mathbf{f}^T)^{-1}\mathbf{b}]^{-1} = 6.408$$

The complete control system to control temperature sequence $x_1(k)$ about some set point $\bar{y}$ is shown in Fig. 8.14. The unit-step response for both state variables is illustrated in Fig. 8.15.

## 8.10 REFERENCE INPUTS FOR SYSTEMS WITH PREDICTION OBSERVERS

In Sections 8.5 through 8.7 we discussed the use of observers to reconstruct the state for purposes of feeding those estimates back to control system dynamics. This entire discussion was dedicated to the regulator problem wherein no reference input is present and all deviations from the steady operation conditions were caused by initial values of the state variables. In Section 8.9 reference inputs were considered in the context of state feedback, and it was found that a reference input needed to be added to the control law.

In this section we consider a similar problem with a prediction-state estimator in the feedback loop. We shall consider, for simplicity, the case where there is a single output $y(k)$ and a single input $u(k)$.

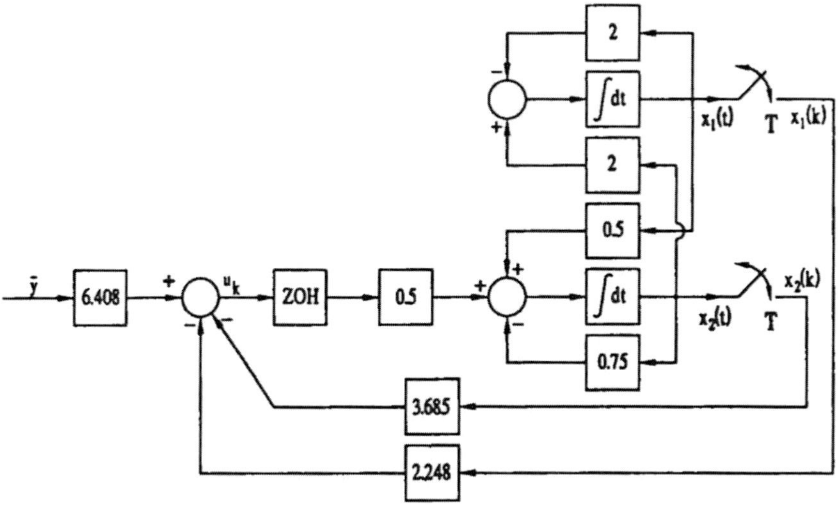


**Figure 8.14.** Complete state feedback control system to control state $x_1(k)$ about set point $y$.

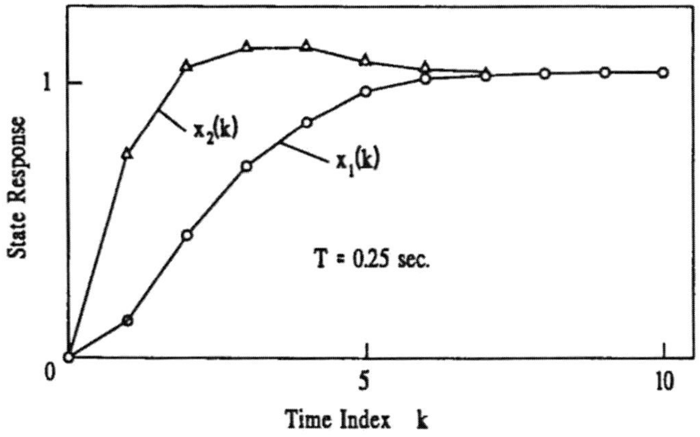


**Figure 8.15.** Step response of the thermal control system.

We are then considering the system

$$\mathbf{x}_{k+1} = \mathbf{A}\mathbf{x}_k + \mathbf{b}u_k \tag{8.10.1}$$

with scalar output

$$y_k = \mathbf{c}^T\mathbf{x}_k \tag{8.10.2}$$

We shall modify the observer of expression (8.5.18) by inclusion of a term including the reference input $r_k$, or

$$\hat{\mathbf{x}}_{k+1} = [\mathbf{A} - \mathbf{b}\mathbf{f}^T - \mathbf{k}\mathbf{c}^T]\hat{\mathbf{x}}_k + \mathbf{k}y_k + \mathbf{m}r_k \tag{8.10.3}$$

where $\mathbf{m}$ is a vector yet to be determined. We shall also modify the control law by inclusion of a term with the control effort

$$u_k = -\mathbf{f}^T\hat{\mathbf{x}}_k + Gr_k \tag{8.10.4}$$

If we subtract (8.10.3) from (8.10.1) and note (8.10.4), the resultant error equation is

$$\tilde{\mathbf{x}}_{k+1} = (\mathbf{A} - \mathbf{k}\mathbf{c}^T)\tilde{\mathbf{x}}_k + \mathbf{b}Gr_k - \mathbf{m}r_k \tag{8.10.5}$$

To make the error independent of the reference input, we can choose

$$\mathbf{m} = \mathbf{b}G \tag{8.10.6}$$

and now all that is left to be determined is the gain $G$. This strategy is illustrated in Fig. 8.16. The strategy is now to make the gain from $r_k$ to $z_k$ to be the reciprocal of the dc gain from $z_k$ to $y_k$. This condition then gives a dc gain from $r_k$ to $y_k$ of unity.

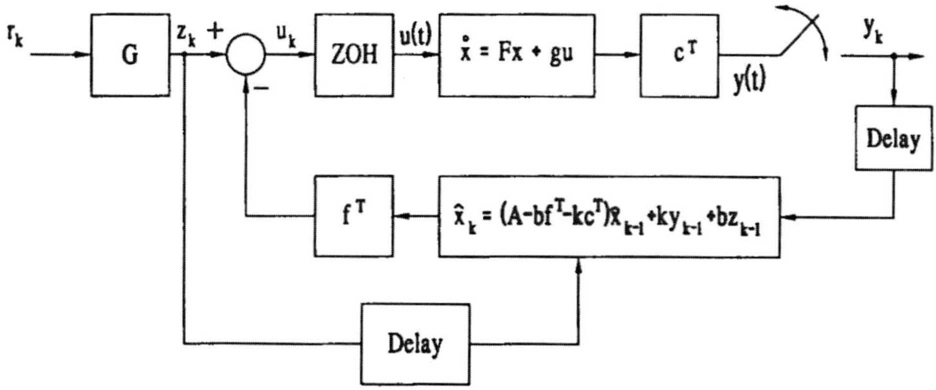


**Figure 8.16.** Estimated state feedback control with a reference input.

Algebra of considerable complication yields the following expression for the reference input gain $G$:

$$G = \frac{1 + \mathbf{c}^T(\mathbf{I} - \mathbf{A})^{-1}\mathbf{b}\mathbf{f}^T(\mathbf{I} - \mathbf{A} + \mathbf{b}\mathbf{f}^T + \mathbf{k}\mathbf{c}^T)^{-1}\mathbf{k}}{\mathbf{c}^T(\mathbf{I} - \mathbf{A})^{-1}\mathbf{b}(1 - \mathbf{f}^T(\mathbf{I} - \mathbf{A} + \mathbf{b}\mathbf{f}^T + \mathbf{k}\mathbf{c}^T)^{-1}\mathbf{b})} \tag{8.10.7}$$

For the case where the number of outputs, the number of control efforts and the number of reference inputs is the same, a result similar to (8.10.7) may be derived; however, it will involve inverting yet another matrix of dimension equal to the number of control efforts.

**Example 8.7.** Find the reference input gain for the thermal system with discrete-time system matrices.

$$\mathbf{A} = \begin{bmatrix} 0.6277 & 0.3597 \\ 0.0899 & 0.8525 \end{bmatrix} \qquad \mathbf{b} = \begin{bmatrix} 0.0251 \\ 0.1150 \end{bmatrix} \qquad \mathbf{c}^T = [1 \quad 0]$$

To place the closed-loop state feedback poles at $z = 0.5 \pm j0.2$, the required feedback gains were found in Example 8.2 to be

$$\mathbf{f}^T = [2.248 \quad 3.685]$$

In Example 8.4 the observer poles were placed at $z = 0.4 \pm j0.2$, yielding prediction observer gains of

$$\mathbf{k} = \begin{bmatrix} 0.68 \\ 0.77 \end{bmatrix}$$

With these gains all calculated, the reference input gain can be calculated with relation (8.10.7) to yield $G = 6.408$. A step response has been calculated and found to be indistinguishable from the data presented in Fig. 8.15. This indicates that the observer poles were chosen sufficiently fast that the observer tracks the state rapidly enough, particularly when they both start from zero initial conditions, as in the case of the step response.

## 8.11 SUMMARY

In this chapter we have examined the modern control theoretical approach to the design of control systems, based largely on pole-placement techniques. We first examined the design process employing state feedback but discussed the possibility that all the states might not be physical variables in the problem and the need for an economically prohibitive number of measurement channels for a large-scale system. We examined briefly the case for using output-variable feedback, but the main emphasis is that devoted to the application of estimated state feedback for control purposes. It was shown that estimator design and feedback control design operations could be done separately and the problem reduced to two pole-placement problems similar to those encountered in state feedback. The necessity of a closed-loop estimation procedure was emphasized and system models as estimators were rejected because of divergence of the estimated state, thus given, from the actual system state.

An algorithm is developed for the placement of poles for state feedback and prediction and current observers. In the final section of the chapter we examined the use of state feedback to control the output variables about a set of nonzero fixed points with specified system dynamics.

## PROBLEMS

8.1. The plant to be examined here is the same as that examined in Problems 3.6, 3.7, and 6.6, which is the fluid-level control system with discrete-time system matrices

$$\mathbf{A} = \begin{bmatrix} 0.8336 & 0.1424 \\ 0.1424 & 0.6200 \end{bmatrix} \qquad \mathbf{b} = \begin{bmatrix} 0.09117 \\ 0.00798 \end{bmatrix}$$

Calculate the state feedback gain vector $\mathbf{f}^T = [f_1 \quad f_2]$ which will locate the system poles at $z_{1,2} = 0.5 \pm j0.3$.

8.2. A digital positioning system has the state-variable form

$$\frac{d}{dt}\begin{bmatrix} x_1 \\ x_2 \end{bmatrix} = \begin{bmatrix} -1 & 0 \\ 1 & 0 \end{bmatrix}\begin{bmatrix} x_1 \\ x_2 \end{bmatrix} + \begin{bmatrix} 1 \\ 0 \end{bmatrix} u(t)$$

or, in discrete-time for sampling interval $T = 1$ sec,

$$\begin{bmatrix} x_1 \\ x_2 \end{bmatrix}_{k+1} = \begin{bmatrix} 0.368 & 0 \\ 0.632 & 1 \end{bmatrix}\begin{bmatrix} x_1 \\ x_2 \end{bmatrix}_k + \begin{bmatrix} 0.632 \\ 0.368 \end{bmatrix} u_k$$

$$y_k = [0 \quad 1]\begin{bmatrix} x_1 \\ x_2 \end{bmatrix}_k$$

The open-loop poles of the discrete-time system are located at $z = 1$ and $z = 0.368$. It is desired to design a prediction state estimator which is slightly faster than the open-loop system. Find the estimator gain $k_1$ and $k_2$ to accomplish the task.

8.3. A plant to be digitally controlled along with the A/D and D/A converters is modeled by the vector difference equation

$$\begin{bmatrix} x_1 \\ x_2 \end{bmatrix}_{k+1} = \begin{bmatrix} 1.1 & 0.1 \\ -0.6 & 0.6 \end{bmatrix}\begin{bmatrix} x_1 \\ x_2 \end{bmatrix}_k + \begin{bmatrix} 1 \\ 0 \end{bmatrix} u(k)$$

We choose to measure only state $2(\mathbf{c}^T = [0 \quad 1])$. Based on the open-loop dynamics, find the gains for a full-state observer.

8.4. For the plant of Problem 8.1, it is desired to construct a full-state prediction observer with estimator poles at $z = 0.4 \pm j0.3$. Find the estimator gain vector $\mathbf{k}$ that will accomplish this. Assume that $x_2(k)$ is measured.

8.5. Find the transfer function of the compensator that represents the control law of $\mathbf{u}_k = -\mathbf{F}\hat{\mathbf{x}}_k$ and the predictor estimator found in Problem 8.4. Use the feedback gains of Problem 8.1 in the estimated state feedback gain matrix.

8.6. The plant of the printhead positioning system was considered in Problem 6.12 and the discrete-time state-variable model was found to have $\mathbf{A}$ and $\mathbf{b}$ matrices

$$\mathbf{A} = \begin{bmatrix} 0.569 & 0.0381 \\ -15.26 & 0.4164 \end{bmatrix} \qquad \mathbf{b} = \begin{bmatrix} 0.00216 \\ 0.0763 \end{bmatrix}$$

(a) Find the state-feedback gain matrix $\mathbf{f}^T$ to place the poles at $z = 0.5 \pm j0.2$.

(b) If only the displacement of the printhead is measured (the velocity remains unmeasured), design a full-order prediction observer that has poles at $z = 0.4 \pm j0.2$.

8.7. Consider the plant of Problem 8.6 and design a "current" estimator with estimator poles located at $z = 0.4 \pm j0.2$.

8.8. For the plant of Problem 8.6, design a partial-state estimator to estimate the velocity based on measured displacement. (Note that a differentiator would be the ideal answer except for the problem of high-frequency noise.) Locate the single pole of the estimator at $z = 0.4$.

## REFERENCES

Ackermann, J., 1972. Der Entwurf linearer Regelungssystems im Zustandsraum, *Regelungstechnik und Prozess-Datenverarbeitung*, *7*: 297–300.

Anderson, B. D. O., and J. B. Moore, 1990. *Optimal Control: Linear Quadratic Methods*, Prentice Hall, Englewood Cliffs, NJ.

Astrom, K. J., and B. Whittenmark, 1984. *Computer Controlled Systems*, Prentice Hall, Englewood Cliffs, NJ.

Bass, R. W., and I. Gura, 1965. High order system design via state space considerations, *Proc. Joint Automatic Control Conference*, 311–318.

Brogan, W. L., 1985. *Modern Control Theory*, 2nd Ed., Prentice Hall, Englewood Cliffs, NJ.

Bryson, A. E., Jr., and D. G. Luenberger, 1970. Synthesis of regulator logic using state-variable concepts, *Proc. IEEE*, *58*(11): 1803–1811.

Davison, E. J., 1968. On pole assignment in multivariable linear systems, *IEEE Trans. on Automatic Control*, *AC-13*(6): 748–749.

Davison, E. J., 1970. On pole assignment in linear systems with incomplete state feedback, *IEEE Trans. on Automatic Control*, *AC-15*(3): 348–351.

Franklin, G. F., J. D. Powell, and M. L. Workman, 1990. *Digital Control of Dynamic Systems*, 2nd Ed., Addison-Wesley, Reading, MA.

Friedland, B., 1986. *Control System Design: An Introduction to State Space Methods*, McGraw-Hill, New York.

Heymann, M., 1968. Comments on pole assignments in multi-input controllable linear systems, *IEEE Trans. on Automatic Control*, *AC-13*(6): 748–749.

Kailath, T., 1980. *Linear Systems*, Prentice Hall, Englewood Cliffs, NJ.

Kwakernaak, H., and R. Sivan, 1972. *Linear Optimal Control*, Wiley, New York.

Luenberger, D. C., 1964. Observing the state of a linear system, *IEEE Trans. on Military Electronics*, *MIL-8*: 74–80.

Van Landingham, H. F., 1985. *Introduction to Digital Control Systems*, Macmillan, New York.

Wonham, W. M., 1967. On pole assignment in multi-input controllable linear systems, *IEEE Trans. on Automatic Control*, *AC-12*(6): 660–665.

# 9
# Linear Discrete-Time Optimal Control

## 9.1 INTRODUCTION

In Chapter 3 we discussed conventional system performance specifications in both the time and frequency domains. Typical specifications were rise time, percent overshoot, settling time, bandwidth, and resonant peak gain. Often, some of those specifications could be met by locating a dominant pair of second-order poles. This control system synthesis technique is approximate only for higher-order systems but is commonly used with a great deal of success. In Chapter 8 we considered the problem of control system design from the state-variable point of view, but the emphasis was still on pole placement, employing state-variable or estimated state-variable feedback.

In this chapter we consider the problem of designing an optimal control system. Instead of placing the poles of the system to achieve a conventional performance specification, we seek a control strategy that minimizes a scalar objective function or performance index. We denote this performance index by the symbol $J$.

When a dynamic system operates in the presence of unknown disturbances, the objective of a control system should be to minimize the deviations from equilibrium; however, the control effort required to drive the system is costly because both hardware and energy are required. As a result, any performance index to be minimized should involve penalties for both excessive deviation from equilibrium and excessive control efforts. The optimal control strategy will thus be a trade-off between excessive deviations of the system and excessive control efforts. The minimization of the performance index must be compatible with the constraints imposed by the dynamics of the plant to be controlled.

It turns out that after minimization is accomplished, the optimal control effort is proportional to the state of the system and the proportionality is time varying. For time-invariant systems and when sufficiently long intervals of time are considered, the optimal feedback solution becomes time invariant.

The problem addressed in this chapter is that of finding a control strategy history, $\mathbf{u}_0, \mathbf{u}_1, \ldots, \mathbf{u}_{N-1}$ such as to minimize the performance index

$$J = \frac{1}{2}\mathbf{x}_N^T\mathbf{H}\mathbf{x}_N + \frac{1}{2}\sum_{k=0}^{N-1} \mathbf{x}_k^T\mathbf{Q}\mathbf{x}_k + \mathbf{u}_k^T\mathbf{R}\mathbf{u}_k \tag{9.1.1}$$

subject to the constraints imposed by the plant equation

$$\mathbf{x}_{k+1} = \mathbf{A}\mathbf{x}_k + \mathbf{B}\mathbf{u}_k \tag{9.1.2}$$

It should be noted that to ensure that $J$ is a positive, nonvanishing, concave, scalar function, $\mathbf{Q}$ and $\mathbf{H}$ must be positive semidefinite matrices and $\mathbf{R}$ must be a positive definite matrix. The initial term in $J$ penalizes terminal deviations from the origin of the state space, the first term in the sum is a penalty for deviations from the origin at the various times of interest, and the final term penalizes excessive control efforts at all times of interest.

A square matrix $\mathbf{M}$ is said to be *positive definite* if for all nonnull vectors $\mathbf{v}$ the triple scalar product $\mathbf{v}^T\mathbf{M}\mathbf{v}$ is positive. Similarly, a matrix $\mathbf{M}$ is said to be *positive semidefinite* if for all nonnull vectors $\mathbf{v}$ the triple scalar product $\mathbf{v}^T\mathbf{M}\mathbf{v}$ is positive or zero.

The *Sylvester criterion* is a test for the definiteness of a matrix $\mathbf{M}$. It states that for a matrix to be positive definite, the minor determinants of all orders involving the major diagonal must all be positive. For a matrix $\mathbf{M}$ defined by

$$\mathbf{M} = \begin{bmatrix} m_{11} & m_{12} & \cdots & m_{1n} \\ m_{21} & m_{22} & \cdots & \\ \vdots & & & \\ m_{n1} & & & m_{nn} \end{bmatrix} \tag{9.1.3}$$

Then to be positive definite, the following must hold:

$$
\begin{aligned}
\mathbf{m}_{11} &> 0 \\
\begin{vmatrix} \mathbf{m}_{11} & \mathbf{m}_{12} \\ \mathbf{m}_{21} & \mathbf{m}_{22} \end{vmatrix} &> 0 \\
&\ \vdots \\
|\mathbf{M}| &> 0
\end{aligned}
\tag{9.1.4}
$$

The condition for **M** to be positive semidefinite is that these minor determinants be positive or zero.

The selection of the elements of the weighting matrices **H**, **Q**, and **R** is not clearly spelled out at this writing, and usually a trial-and-error selection process is recommended to give a meaningful optimization problem. The excellent text of Bryson and Ho (1975) suggests a technique for choosing the elements of **H**, **Q**, and **R**, and this is presented here as a technique for a first guess: (1) Make, for lack of better knowledge, the off-diagonal elements of **H**, **Q**, and **R** zero. (2) Pick the diagonal elements as follows:

$$
\begin{aligned}
H_{ii} &= \frac{1}{x_i^2(N)_{\max}} \\
Q_{ii} &= \frac{1}{x_i^2(k)_{\max}} \\
R_{ii} &= \frac{1}{u_i^2(k)_{\max}}
\end{aligned}
\tag{9.1.5}
$$

## 9.2 DISCRETE LINEAR REGULATOR PROBLEM

Consider the discrete-time dynamic system described by the vector-matrix difference equation

$$\mathbf{x}_{k+1} = \mathbf{A}\mathbf{x}_k + \mathbf{B}\mathbf{u}_k \qquad k = 0, 1, 2, \ldots \tag{9.2.1}$$

with some arbitrary initial condition $\mathbf{x}_0$. If the sampling interval is $T$ seconds, we shall consider a period of time $N$ sampling intervals long, commonly called the *planning horizon*. The goal is to find a sequence of control vectors, $\mathbf{u}_0, \mathbf{u}_1, \ldots, \mathbf{u}_{N-1}$, to minimize the following performance index:

$$J = \frac{1}{2}\mathbf{x}_N^T\mathbf{H}\mathbf{x}_N + \sum_{k=0}^{N-1} \frac{1}{2}\mathbf{x}_k^T\mathbf{Q}\mathbf{x}_k + \frac{1}{2}\mathbf{u}_k^T\mathbf{R}\mathbf{u}_k \tag{9.2.2}$$

where **Q** and **H** are positive semidefinite matrices, **R** is a positive definite matrix, and all are real and symmetric. Let us augment $J$ by adjoining the state equation with a set of arbitrary vector Lagrangian multipliers $\boldsymbol{\lambda}_1, \boldsymbol{\lambda}_2, \ldots, \boldsymbol{\lambda}_N$

$$J = \frac{1}{2}\mathbf{x}_N^T\mathbf{H}\mathbf{x}_N + \sum_{k=0}^{N-1}\left[\frac{1}{2}\mathbf{x}_k^T\mathbf{Q}\mathbf{x}_k + \frac{1}{2}\mathbf{u}_k^T\mathbf{R}\mathbf{u}_k + \boldsymbol{\lambda}_{k+1}^T(\mathbf{A}\mathbf{x}_k + \mathbf{B}\mathbf{u}_k - \mathbf{x}_{k+1})\right] \tag{9.2.3}$$

This does not change the minimization problem since we have added zero to $J$. Now we want to choose $\mathbf{x}_k$ and $\mathbf{u}_k$ to minimize $J$ of Eq. (9.2.3), so let us form the differential of $J$:

$$dJ = \mathbf{x}_N^T\mathbf{H}\,d\mathbf{x}_N + \sum_{k=0}^{N-1}\left[\mathbf{x}_k^T\mathbf{Q}\,d\mathbf{x}_k + \mathbf{u}_k^T\mathbf{R}\,d\mathbf{u}_k + \boldsymbol{\lambda}_{k+1}^T(\mathbf{A}\,d\mathbf{x}_k + \mathbf{B}\,d\mathbf{u}_k - d\mathbf{x}_{k+1})\right] \tag{9.2.4}$$

Combining like terms and noting that since $\mathbf{x}_0$ is arbitrary but not variable in the optimization process, $d\mathbf{x}_0$ must be zero, or

$$dJ = (\mathbf{x}_N^T\mathbf{H} - \boldsymbol{\lambda}_N^T)d\mathbf{x}_N + \sum_{k=0}^{N-1}(\mathbf{u}_k^T\mathbf{R} + \boldsymbol{\lambda}_{k+1}^T\mathbf{B})d\mathbf{u}_k + \sum_{k=1}^{N-1}[\mathbf{x}_k^T\mathbf{Q} + \boldsymbol{\lambda}_{k+1}^T\mathbf{A} - \boldsymbol{\lambda}_k^T]d\mathbf{x}_k \tag{9.2.5}$$

For a minimum, the differential of $J$ must vanish for arbitrary small $d\mathbf{x}_k$, $d\mathbf{u}_k$, and $d\mathbf{x}_N$, and hence each of the coefficients in the parentheses must vanish, or

$$\mathbf{Q}\mathbf{x}_k - \boldsymbol{\lambda}_k + \mathbf{A}^T\boldsymbol{\lambda}_{k+1} = 0 \tag{9.2.6}$$

$$\mathbf{R}\mathbf{u}_k + \mathbf{B}^T\boldsymbol{\lambda}_{k+1} = 0 \tag{9.2.7}$$

and

$$\mathbf{H}\mathbf{x}_N - \boldsymbol{\lambda}_N = 0 \tag{9.2.8}$$

Rewriting these, we get

$$\boldsymbol{\lambda}_k = \mathbf{Q}\mathbf{x}_k + \mathbf{A}^T\boldsymbol{\lambda}_{k+1} \tag{9.2.9}$$

$$\mathbf{u}_k = -\mathbf{R}^{-1}\mathbf{B}^T\boldsymbol{\lambda}_{k+1} \tag{9.2.10}$$

and

$$\boldsymbol{\lambda}_N = \mathbf{H}\mathbf{x}_N \tag{9.2.11}$$

Substitution of (9.2.10) into (9.2.1) gives

$$\mathbf{x}_{k+1} = \mathbf{A}\mathbf{x}_k - \mathbf{B}\mathbf{R}^{-1}\mathbf{B}^T\boldsymbol{\lambda}_{k+1} \tag{9.2.12}$$

and (9.2.9) is

$$\boldsymbol{\lambda}_k = \mathbf{Q}\mathbf{x}_k + \mathbf{A}^T\boldsymbol{\lambda}_{k+1} \tag{9.2.13}$$

Equation (9.2.13) may be rewritten as

$$\boldsymbol{\lambda}_{k+1} = \mathbf{A}^{T-1}(\boldsymbol{\lambda}_k - \mathbf{Q}\mathbf{x}_k) \tag{9.2.14}$$

Substitution of (9.2.14) into (9.2.12) yields

$$\mathbf{x}_{k+1} = \mathbf{A}\mathbf{x}_k - \mathbf{B}\mathbf{R}^{-1}\mathbf{B}^T\mathbf{A}^{T-1}[\boldsymbol{\lambda}_k - \mathbf{Q}\mathbf{x}_k] \tag{9.2.15}$$

Expressions (9.2.14) and (9.2.15) are linear forward difference equations in $\mathbf{x}_k$, so that a transition matrix equation may be written between the terminal or $N$th point in time and the general or $k$th point:

$$\begin{bmatrix} \mathbf{x}_k \\ \boldsymbol{\lambda}_k \end{bmatrix} = \begin{bmatrix} \boldsymbol{\beta}_{11} & \boldsymbol{\beta}_{12} \\ \boldsymbol{\beta}_{21} & \boldsymbol{\beta}_{21} \end{bmatrix} \begin{bmatrix} \mathbf{x}_N \\ \boldsymbol{\lambda}_N \end{bmatrix} \tag{9.2.16}$$

where the submatrices $\boldsymbol{\beta}_{ij}$ are functions of $N$ and $k$. Also recall expression (9.2.11), which relates $\mathbf{x}_N$ to $\boldsymbol{\lambda}_N$:

$$\boldsymbol{\lambda}_N = \mathbf{H}\mathbf{x}_N \tag{9.2.17}$$

The second expression of (9.2.16) gives

$$\boldsymbol{\lambda}_k = \boldsymbol{\beta}_{21}\mathbf{x}_N + \boldsymbol{\beta}_{22}\boldsymbol{\lambda}_N \tag{9.2.18}$$

and with substitution of (9.2.17), we get

$$\boldsymbol{\lambda}_k = (\boldsymbol{\beta}_{21} + \boldsymbol{\beta}_{22}\mathbf{H})\mathbf{x}_N \tag{9.2.19}$$

The first expression of (9.2.16) after substitution of (9.2.17) is

$$\mathbf{x}_k = (\boldsymbol{\beta}_{11} + \boldsymbol{\beta}_{12}\mathbf{H})\mathbf{x}_N \tag{9.2.20}$$

and solving for $\mathbf{x}_N$ gives

$$\mathbf{x}_N = (\boldsymbol{\beta}_{11} + \boldsymbol{\beta}_{12}\mathbf{H})^{-1}\mathbf{x}_k \tag{9.2.21}$$

Then substituting this into (9.2.19) gives

$$\boldsymbol{\lambda}_k = (\boldsymbol{\beta}_{21} + \boldsymbol{\beta}_{22}\mathbf{H})(\boldsymbol{\beta}_{11} + \boldsymbol{\beta}_{12}\mathbf{H})^{-1}\mathbf{x}_k \tag{9.2.22}$$

Since the $\boldsymbol{\beta}_{ij}$ are functions of $k$ and $N$ and $N$ is fixed, expression (9.2.22) may be written generally as

$$\boldsymbol{\lambda}_k = \mathbf{P}_k\mathbf{x}_k \qquad k = 0, 1, 2, \ldots, N \tag{9.2.23}$$

This expression is sometimes referred to as the *Riccati transformation*. Now let us substitute (9.2.23) into (9.2.12) to yield

$$\mathbf{x}_{k+1} = \mathbf{A}\mathbf{x}_k - \mathbf{B}\mathbf{R}^{-1}\mathbf{B}^T\mathbf{P}_{k+1}\mathbf{x}_{k+1} \tag{9.2.24}$$

or

$$\mathbf{x}_{k+1} = [\mathbf{I} + \mathbf{B}\mathbf{R}^{-1}\mathbf{B}^T\mathbf{P}_{k+1}]^{-1}\mathbf{A}\mathbf{x}_k \tag{9.2.25}$$

and substitution of (9.2.23) into (9.2.13) yields

$$\mathbf{P}_k\mathbf{x}_k = \mathbf{Q}\mathbf{x}_k + \mathbf{A}^T\mathbf{P}_{k+1}\mathbf{x}_{k+1} \tag{9.2.26}$$

Now substitute (9.2.25) into (9.2.26) to give

$$\mathbf{P}_k\mathbf{x}_k = \mathbf{Q}\mathbf{x}_k + \mathbf{A}^T\mathbf{P}_{k+1}[\mathbf{I} + \mathbf{B}\mathbf{R}^{-1}\mathbf{B}^T\mathbf{P}_{k+1}]^{-1}\mathbf{A}\mathbf{x}_k \tag{9.2.27}$$

which for arbitrary $\mathbf{x}_k$ implies that

$$\mathbf{P}_k = \mathbf{Q} + \mathbf{A}^T\mathbf{P}_{k+1}[\mathbf{I} + \mathbf{B}\mathbf{R}^{-1}\mathbf{B}^T\mathbf{P}_{k+1}]^{-1}\mathbf{A} \tag{9.2.28}$$

Now moving the $\mathbf{P}_{k+1}$ outside the brackets to the inside, we get

$$\mathbf{P}_k = \mathbf{Q} + \mathbf{A}^T[\mathbf{P}_{k+1}^{-1} + \mathbf{B}\mathbf{R}^{-1}\mathbf{B}^T]^{-1}\mathbf{A} \tag{9.2.29}$$

The matrix to be inverted here is $n \times n$, and to reduce the amount of computation, simply introduce the matrix inversion lemma, which is proven in Appendix C to yield

$$\mathbf{P}_k = \mathbf{A}^T[\mathbf{P}_{k+1} - \mathbf{P}_{k+1}\mathbf{B}(\mathbf{B}^T\mathbf{P}_{k+1}\mathbf{B} + \mathbf{R})^{-1}\mathbf{B}^T\mathbf{P}_{k+1}]\mathbf{A} + \mathbf{Q} \tag{9.2.30}$$

From (9.2.10) we can write the control law as

$$\mathbf{u}_k = -\mathbf{R}^{-1}\mathbf{B}^T\boldsymbol{\lambda}_{k+1} \tag{9.2.31}$$

and substitution of (9.2.14) yields

$$\mathbf{u}_k = -\mathbf{R}^{-1}\mathbf{B}^T\mathbf{A}^{T-1}(\boldsymbol{\lambda}_k - \mathbf{Q}\mathbf{x}_k) \tag{9.2.32}$$

and further substitution of (9.2.23) yields

$$\mathbf{u}_k = -\mathbf{R}^{-1}\mathbf{B}^T\mathbf{A}^{T-1}(\mathbf{P}_k - \mathbf{Q})\mathbf{x}_k \tag{9.2.33}$$

This is a negative-feedback control law, which may be written as

$$\mathbf{u}_k = -\mathbf{F}_k\mathbf{x}_k \tag{9.2.34}$$

where from (9.2.33) the feedback gain matrix $\mathbf{F}_k$ is defined by

$$\mathbf{F}_k = \mathbf{R}^{-1}\mathbf{B}^T\mathbf{A}^{T-1}(\mathbf{P}_k - \mathbf{Q}) \tag{9.2.35}$$

and $\mathbf{A}^{T-1}$ denotes the inverse of the transpose of $\mathbf{A}$. This control system is shown in Fig. 9.1. Let us define a quantity $\mathbf{G}_{k+1}$ as the bracketed quantity of (9.2.30), to yield

$$\mathbf{G}_{k+1} = \mathbf{P}_{k+1} - \mathbf{P}_{k+1}\mathbf{B}[\mathbf{B}^T\mathbf{P}_{k+1}\mathbf{B} + \mathbf{R}]^{-1}\mathbf{B}^T\mathbf{P}_{k+1} \tag{9.2.36}$$

and then (9.2.30) can be written as

$$\mathbf{P}_k = \mathbf{A}^T\mathbf{G}_{k+1}\mathbf{A} + \mathbf{Q} \tag{9.2.37}$$

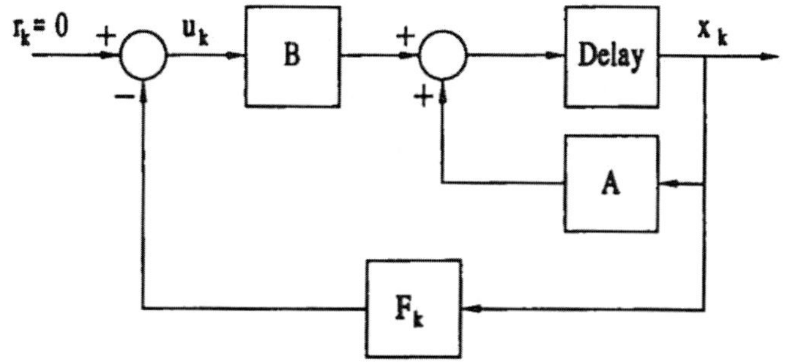


**Figure 9.1.** Discrete linear regulator.

Expression (9.2.30) is commonly called the *discrete-time matrix Riccati equation*. The condition required to start the computation (9.2.35), (9.2.36), and (9.2.37) is given by examining (9.2.11) and (9.2.23), which gives

$$\mathbf{P}_N = \mathbf{H} \tag{9.2.38}$$

Given this terminal condition, we can evaluate $\mathbf{G}_N$ from (9.2.36) and then $\mathbf{P}_{N-1}$ from (9.2.37) and hence $\mathbf{F}_{N-1}$ from (9.2.35) and then return to (9.2.36) to calculate $\mathbf{G}_{N-1}$ and continue cyclically until the gain matrix is evaluated at all points in time to yield $\mathbf{F}_0, \mathbf{F}_1, \mathbf{F}_2, \ldots, \mathbf{F}_{N-1}$. These calculations are best carried out by digital computer and a routine exists in the MATLAB™ Control Systems Toolbox (Saadat, 1993) to evaluate these expressions. In a later section of this chapter, a technique is presented that is computationally far more efficient for large-scale systems.

In all the previous derivations, the system matrices **A** and **B** and the weighting matrices **Q** and **R** have been assumed to be independent of the temporal index $k$. This, however, need not be the case, and the equations could be rederived by carrying along the subscripts in $\mathbf{A}_k$, $\mathbf{B}_k$, $\mathbf{Q}_k$, and $\mathbf{R}_k$. Computationally this is still possible, but it becomes a significant data management problem, as these matrices must be updated each step in time.

**Example 9.1.** Find the optimal feedback gain sequence $\mathbf{F}_k$ for the following problem. The plant is first order and of the form

$$x_{k+1} = 0.8x_k + u_k$$

Here $A = 0.8$ and $B = 1$ and the performance index is

$$J = \frac{1}{2}2x_4^2 + \frac{1}{2}\sum_{k=0}^{3} x_k^2 + u_k^2$$

so $H = 2$, $Q = 1$, $R = 1$, and $N = 4$. Writing expressions (9.2.35), (9.2.36), and (9.2.37) for this scalar problem gives

$$F_k = 1.25(P_k - 1)$$

$$G_{k+1} = P_{k+1} - \frac{P_{k+1}^2}{P_{k+1} + 1}$$

$$P_k = 0.64G_{k+1} + 1$$

with a final condition

$$P_4 = H = 2$$

Sequentially evaluating the foregoing equations backward in time, we get the $P_k$ and $F_k$ histories illustrated in Fig. 9.2. For non-time-varying systems and constant **Q** and **R** the sequences for $\mathbf{P}_k$ and $\mathbf{F}_k$ will become stationary if evaluated far enough backward in time.

The general formulation for computation of the optimal feedback gain can be specialized to a general first-order system. Consider the system governed by the first-order difference equation

$$x_{k+1} = ax_k + bu_k \tag{9.2.39}$$

and an index of performance

$$J = \frac{1}{2}\sum_{k=0}^{\infty} qx_k^2 + ru_k^2 \tag{9.2.40}$$

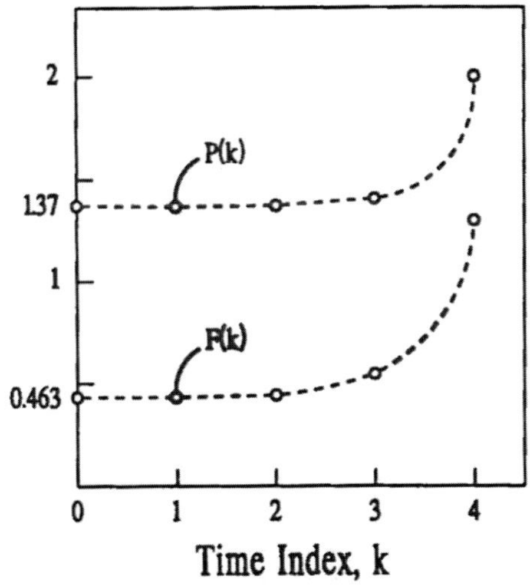


**Figure 9.2.** Feedback gain and Riccati variable histories.

The general form of the scalar Riccati equation from (9.2.30) is

$$P_k = a^2\left(P_{k+1} - \frac{P_{k+1}^2 b^2}{b^2 P_{k+1} + r}\right) + q \qquad (9.2.41)$$

with the associated scalar feedback gain given by (9.2.35),

$$F_k = \frac{b}{ar}(P_k - q) \qquad (9.2.42)$$

If we are searching for the stationary solution to the problem, it may be found by letting $P_{k+1} = P_k = P$ in relation (9.2.41), which results in a quadratic equation that can be solved for $P$.

$$P^2 - \frac{[(a^2 - 1)r + qb^2]}{b^2} P - \frac{qr}{b} = 0 \qquad (9.2.43)$$

Choosing the positive solution to this, we get

$$P = \frac{(a^2 - 1)r + qb^2}{2b^2} + \sqrt{\left[\frac{(a^2 - 1)r + qb^2}{2b^2}\right]^2 + \frac{qr}{b^2}} \qquad (9.2.44)$$

Thus the optimal feedback gain is, from (9.2.42),

$$F = \frac{a^2 - 1}{2ab} - \frac{qb}{2ra} + \sqrt{\left[\frac{(a^2 - 1)r}{2ab} + \frac{qb^2}{2ra}\right]^2 + \frac{q}{ra^2}} \qquad (9.2.45)$$

It is relatively easy to calculate the nondimensional gain $Fb/a^2$ as a function of $qb^2/r$ for various values of the nondimensional parameter $a$, several plots of which are given in Fig. 9.3.

**Example 9.2.** Let us consider the optimal control of the thermal system shown in Fig. 9.4 and used in Example 6.4. The discrete-state equations found in Example 6.4 were

$$\begin{bmatrix} x_1(k+1) \\ x_2(k+1) \end{bmatrix} = \begin{bmatrix} 0.6277 & 0.3597 \\ 0.0899 & 0.8526 \end{bmatrix} \begin{bmatrix} x_1(k) \\ x_2(k) \end{bmatrix} + \begin{bmatrix} 0.025 \\ 0.115 \end{bmatrix} u(k)$$

The performance index is

$$J = \frac{1}{2} \sum_{k=0}^{9} \mathbf{x}_k^T \mathbf{Q} \mathbf{x}_k + R u_k^2$$

where the planning horizon is 10 sampling intervals and the weighting

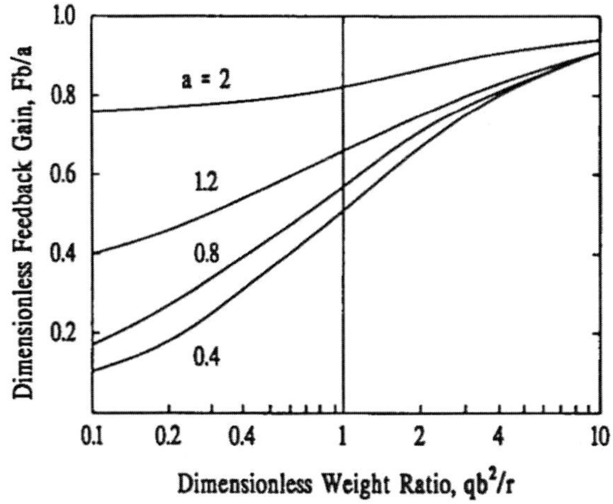


**Figure 9.3.** Dimensionless stationary optimal feedback gains for a first-order plant.

matrices are

$$\mathbf{Q} = \begin{bmatrix} 50 & 0 \\ 0 & 10 \end{bmatrix} \qquad R = 1$$

Since there is no terminal penalty, then

$$\mathbf{P}_{10} = \mathbf{H} = \begin{bmatrix} 0 & 0 \\ 0 & 0 \end{bmatrix}$$

Evaluating expressions (9.2.35), (9.2.36), and (9.2.37) yields the feedback gain histories shown in Fig. 9.5. It is interesting to note that in both of these examples the $\mathbf{P}_k$ and $\mathbf{F}_k$ sequences become constant valued as the

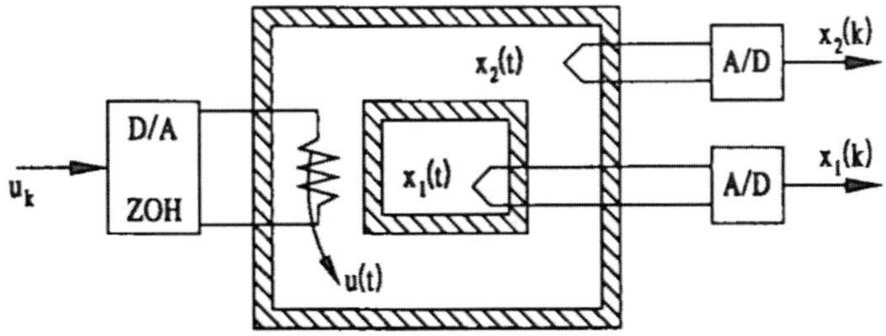


**Figure 9.4.** Digitally controlled thermal system.

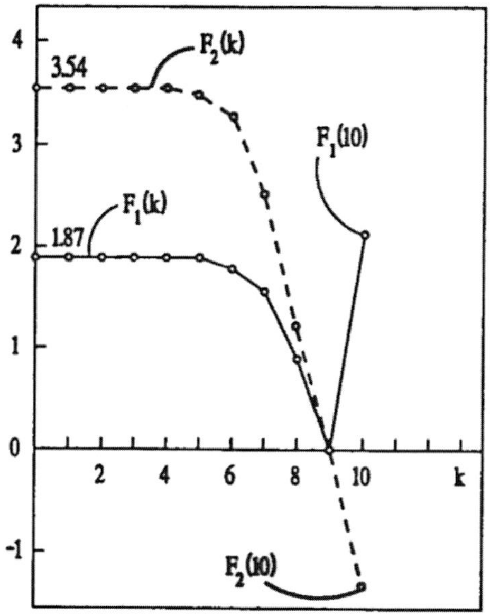


**Figure 9.5.** Feedback gain histories for thermal problem with state feedback.

process is carried out backward a sufficient number of steps. In the practical design problem, the stationary value of $\mathbf{F}_k$ is often used in the implementation of real digital controllers. The stationary feedback gain matrix is

$$\mathbf{f}^T = [1.872 \quad 3.543]$$

If we implement the control system using these gains, the resultant control system is as shown in Fig. 9.6. Typical responses for initial conditions $x_1(0) = 2$ and $x_2(0) = 1$ are shown in Fig. 9.7.

## 9.3 COST OF CONTROL

In Section 9.2 the augmented performance index was given in (9.2.2) to be

$$J = \frac{1}{2}\mathbf{x}_N^T\mathbf{H}\mathbf{x}_N + \sum_{k=0}^{N-1}\left[\frac{1}{2}\mathbf{x}_k^T\mathbf{Q}\mathbf{x}_k + \frac{1}{2}\mathbf{u}_k^T\mathbf{R}\mathbf{u}_k + \boldsymbol{\lambda}_{k+1}^T(\mathbf{A}\mathbf{x}_k + \mathbf{B}\mathbf{u}_k - \mathbf{x}_{k+1})\right] \tag{9.3.1}$$

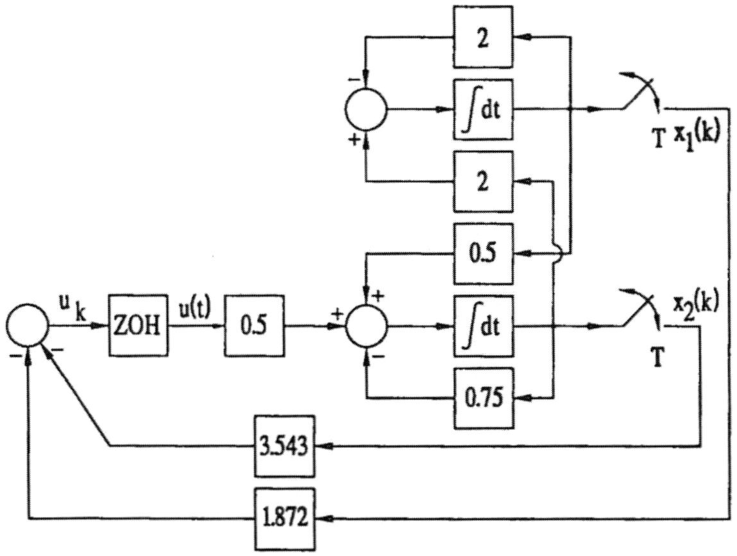


Figure 9.6. Optimal feedback control of the thermal system.

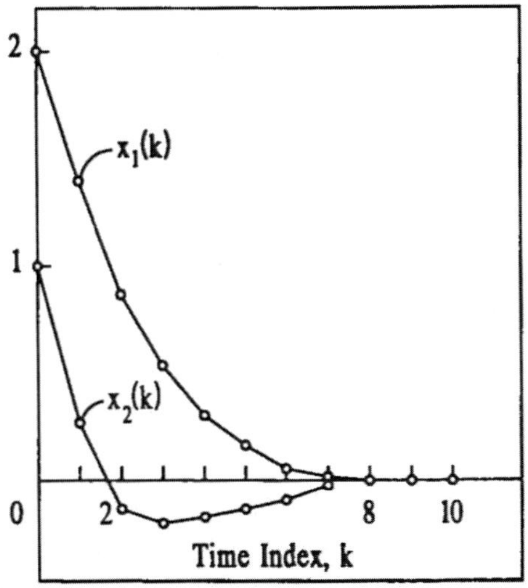


Figure 9.7. State histories for the optimally controlled thermal system.

but we know from relation (9.2.7) that

$$\boldsymbol{\lambda}_{k+1}^T\mathbf{B} = -\mathbf{u}_k^T\mathbf{R} \tag{9.3.2}$$

and from relation (9.2.6),

$$\boldsymbol{\lambda}_{k+1}^T\mathbf{A} = \boldsymbol{\lambda}_k^T - \mathbf{x}_k^T\mathbf{Q} \tag{9.3.3}$$

The augmented performance index is, after substitution of (9.3.2) and (9.3.3) for $\boldsymbol{\lambda}_{k+1}^T$ in the appropriate places,

$$J = \frac{1}{2}\mathbf{x}_N\mathbf{H}\mathbf{x}_N + \sum_{k=0}^{N-1}\left[\frac{1}{2}\mathbf{x}_k\mathbf{Q}\mathbf{x}_k + \frac{1}{2}\mathbf{u}_k^T\mathbf{R}\mathbf{u}_k \right.$$
$$\left. + (\boldsymbol{\lambda}_k^T - \mathbf{x}_k^T\mathbf{Q})\mathbf{x}_k - \mathbf{u}_k^T\mathbf{R}\mathbf{u}_k - \boldsymbol{\lambda}_{k+1}^T\mathbf{x}^{k+1}\right] \tag{9.3.4}$$

or

$$J = \frac{1}{2}\mathbf{x}_N^T\mathbf{H}\mathbf{x}_N + \sum_{k=0}^{N-1}\left[-\frac{1}{2}\mathbf{x}_k^T\mathbf{Q}\mathbf{x}_k - \frac{1}{2}\mathbf{u}_k\mathbf{R}\mathbf{u}_k\right]$$
$$+ \sum_{k=0}^{N-1}[\boldsymbol{\lambda}_k^T\mathbf{x}_k - \boldsymbol{\lambda}_{k+1}^T\mathbf{x}_{k+1}] \tag{9.3.5}$$

Add the original $J$ of expression (9.2.2) to both sides of (9.3.5) to yield

$$2J = \mathbf{x}_N^T\mathbf{H}\mathbf{x}_N + \sum_{k=0}^{N-1}\boldsymbol{\lambda}_k^T\mathbf{x}_k - \boldsymbol{\lambda}_{k+1}^T\mathbf{x}_{k+1} \tag{9.3.6}$$

Noting that only two terms are left from the summation or

$$2J = \mathbf{x}_N^T\mathbf{H}\mathbf{x}_N + \boldsymbol{\lambda}_0^T\mathbf{x}_0 - \boldsymbol{\lambda}_N^T\mathbf{x}_N \tag{9.3.7}$$

or

$$2J = (\mathbf{x}_N^T\mathbf{H} - \boldsymbol{\lambda}_N^T)\mathbf{x}_N + \boldsymbol{\lambda}_0^T\mathbf{x}_0 \tag{9.3.8}$$

But the boundary condition (9.2.11) causes the first term to vanish, or

$$2J = \boldsymbol{\lambda}_0^T\mathbf{x}_0 \tag{9.3.9}$$

but from the Riccati transformation $\boldsymbol{\lambda}_0^T = \mathbf{x}_0^T\mathbf{P}_0$, the total cost of control over the planning horizon is

$$J = \frac{1}{2}\mathbf{x}_0^T\mathbf{P}_0\mathbf{x}_0 \tag{9.3.10}$$

where $\mathbf{P}_0$ is the solution to the discrete matrix Riccati equation at $k = 0$. This expression allows the system designer to assign a single number quantitatively to a design given by a selection of $\mathbf{Q}$ and $\mathbf{R}$. This design can then be compared to that given by a different choice of the elements of **Q** and **R**.

**Example 9.3.** Consider the thermal control system considered in Example 9.2 and illustrated in Fig. 9.3. If this system is to be optimally driven back toward zero state from initial values $x_1(0) = 2$ and $x_2(0) = 1$, evaluate the total cost for a planning horizon of 10 sampling intervals. From the solution of Example 9.2 the initial **P** matrix is

$$\mathbf{P}_0 = \begin{bmatrix} 77.076 & 21.216 \\ 21.216 & 41.579 \end{bmatrix}$$

The total cost is then given by relation (9.3.24) to be

$$J = \frac{1}{2}[2 \quad 1]\begin{bmatrix} 77.076 & 21.216 \\ 21.216 & 41.479 \end{bmatrix}\begin{bmatrix} 2 \\ 1 \end{bmatrix} = 217.373$$

where the units on this cost are arbitrary.

## 9.4 RECIPROCAL EIGENVALUES AND RECIPROCAL ROOT LOCUS

Let us consider the regulator problem posed in previous sections. Let us examine the two difference equations (9.2.12) and (9.2.13), which must be solved simultaneously:

$$\mathbf{x}_{k+1} = \mathbf{A}\mathbf{x}_k - \mathbf{B}\mathbf{R}^{-1}\mathbf{B}^T\boldsymbol{\lambda}_{k+1} \tag{9.4.1}$$

and

$$\boldsymbol{\lambda}_k = \mathbf{Q}\mathbf{x}_k + \mathbf{A}^T\boldsymbol{\lambda}_{k+1} \tag{9.4.2}$$

Now take the $z$ transform of (9.4.1) and (9.4.2), ignoring the initial values, which yields the matrix form

$$\begin{bmatrix} z\mathbf{I} - \mathbf{A} & \mathbf{B}\mathbf{R}^{-1}\mathbf{B}^T \\ -\mathbf{Q} & z^{-1}\mathbf{I} - \mathbf{A}^T \end{bmatrix}\begin{bmatrix} \mathbf{x}(z) \\ z\boldsymbol{\lambda}(z) \end{bmatrix} = \begin{bmatrix} 0 \\ 0 \end{bmatrix} \tag{9.4.3}$$

Transpose this relation while noting that from their symmetry $\mathbf{Q}^T = \mathbf{Q}$ and $(\mathbf{R}^{-1})^T = \mathbf{R}^{-1}$, which gives

$$[\mathbf{x}^T(z) \quad z\boldsymbol{\lambda}^T(z)]\begin{bmatrix} z\mathbf{I} - \mathbf{A}^T & -\mathbf{Q} \\ \mathbf{B}\mathbf{R}^{-1}\mathbf{B}^T & z^{-1}\mathbf{I} - \mathbf{A} \end{bmatrix} = [0 \quad 0] \tag{9.4.4}$$

Reversing the elements of the row vector yields

$$[z\boldsymbol{\lambda}^T(z) \quad \mathbf{x}^T(z)]\begin{bmatrix} z^{-1}\mathbf{I} - \mathbf{A} & \mathbf{B}\mathbf{R}^{-1}\mathbf{B}^T \\ -\mathbf{Q} & z\mathbf{I} - \mathbf{A}^T \end{bmatrix} = [0 \quad 0] \tag{9.4.5}$$

We note that the matrices of (9.4.3) and (9.4.5) must be singular for nonnull solution vectors to exist; thus the determinant of each of the matrices must be zero, yielding a characteristic equation. Note also that the matrix in (9.4.5) is the same as that of (9.4.3) with $z$ replaced by $z^{-1}$, and hence if $z_i$ is an eigenvalue of the system (9.4.1) and (9.4.2), its reciprocal $z_i^{-1}$ is also an eigenvalue.

The *reciprocal root locus* is a technique that is useful in selecting the performance index parameters based on conventional $z$-plane pole placement. In other words, we may use the reciprocal root locus to guide the selection of a quadratic performance index such that dominant poles are placed in desired locations. Recall now that from the optimality condition, the optimal control is given by (9.2.7) as

$$\mathbf{R}\mathbf{u}_k = -\mathbf{B}^T\boldsymbol{\lambda}_{k+1} \tag{9.4.6}$$

while the plant equation is given by (9.2.1) as

$$\mathbf{x}_{k+1} = \mathbf{A}\mathbf{x}_k + \mathbf{B}\mathbf{u}_k \tag{9.4.7}$$

and the adjoint equations is given by (9.2.9) as

$$\boldsymbol{\lambda}_k = \mathbf{Q}\mathbf{x}_k + \mathbf{A}^T\boldsymbol{\lambda}_{k+1} \tag{9.4.8}$$

We may take the $z$ transform of these three equations to yield

$$\mathbf{R}\mathbf{u}(z) = -\mathbf{B}^T z\boldsymbol{\lambda}(z) \tag{9.4.9}$$

$$\mathbf{x}(z) = (z\mathbf{I} - \mathbf{A})^{-1}\mathbf{B}\mathbf{u}(z) \tag{9.4.10}$$

$$\boldsymbol{\lambda}(z) = (z^{-1}\mathbf{I} - \mathbf{A}^T)^{-1}z^{-1}\mathbf{Q}\mathbf{x}(z) \tag{9.4.11}$$

Premultiply (9.4.11) by $\mathbf{B}^T z$ and substitute (9.4.9) for the left-hand side to yield

$$-\mathbf{R}\mathbf{u}(z) = \mathbf{B}^T(z^{-1}\mathbf{I} - \mathbf{A}^T)^{-1}\mathbf{Q}\mathbf{x}(z) \tag{9.4.12}$$

Now let us substitute (9.4.10) for $\mathbf{x}(z)$ to give

$$[\mathbf{R} + \mathbf{B}^T(z^{-1}\mathbf{I} - \mathbf{A}^T)^{-1}\mathbf{Q}(z\mathbf{I} - \mathbf{A})^{-1}\mathbf{B}]\mathbf{u}(z) = \mathbf{0} \tag{9.4.13}$$

For the nontrivial control problem, $\mathbf{u}(z)$ should not vanish, so the matrix in the brackets must be singular or have a zero determinant:

$$\det[\mathbf{R} + \mathbf{B}^T(z^{-1}\mathbf{I} - \mathbf{A}^T)^{-1}\mathbf{Q}(z\mathbf{I} - \mathbf{A})^{-1}\mathbf{B}] = 0 \tag{9.4.14}$$

This is often called the *reciprocal root characteristic equation* (RRCE), and those roots that lie inside the unit circle are the closed-loop optimal system poles.

Now let us simplify this result to the optimal control of a single-input/

single-output system where the scalar output sequence is given by

$$y_k = \mathbf{c}^T\mathbf{x}_k \tag{9.4.15}$$

with a performance index that penalizes only the output variable and not the states

$$J = \frac{1}{2}\sum_{k=0}^{N-1} qy_k^2 + ru_k^2 \tag{9.4.16}$$

If we substitute (9.4.15) for the $y_k$, we get

$$J = \frac{1}{2}\sum_{k=0}^{N-1} q\mathbf{x}_k^T\mathbf{c}\mathbf{c}^T\mathbf{x}_k + ru_k^2 \tag{9.4.17}$$

It is clear from inspection that the matrix $\mathbf{Q}$ is the same as $q\mathbf{c}\mathbf{c}^T$, so expression (9.4.14) becomes

$$-r = q\mathbf{b}^T(z^{-1}\mathbf{I} - \mathbf{A})^{-1}\mathbf{c}\mathbf{c}^T(z\mathbf{I} - \mathbf{A})^{-1}\mathbf{b} \tag{9.4.18}$$

Now if we note from Chapter 6 that the input–output transfer function is

$$G(z) = \mathbf{c}^T(z\mathbf{I} - \mathbf{A})^{-1}\mathbf{b} \tag{9.4.19}$$

then (9.4.18) can be rewritten as

$$\frac{-r}{q} = G(z^{-1})G(z) \tag{9.4.20}$$

or the more familiar form of

$$1 + \frac{q}{r}G(z^{-1})G(z) = 0 \tag{9.4.21}$$

which is the exact Evan's form for construction of a root locus as a function of the parameter $q/r$, where $r$ and $q$ are the weighting constants in the performance index (9.4.16). The steps for plotting a reciprocal root locus are as follows:

1. Plot the poles and zeros of the $G(z^{-1})G(z)$ function having carefully rationalized the two functions into strict functions of the variable $z$.
2. Sketch the locus as one would in conventional control system design, and choose a point where one would desire the dominant open-loop pole on the locus. Here care must be taken to determine whether one wants the negative or positive gain root locus because in the process rationalizing $G(z^{-1})$ additional minus signs may be generated.

3. Calculate the ratio of $q/r$ by taking the products of the distance from the design point to all zeros of $G(z^{-1})G(z)$ divided by the products of all the distances from the design point to the poles of $G(z^{-1})G(z)$.

**Example 9.4.** Consider the inertial plant described in Example 6.1. In Chapter 3 this system was found to have a $z$-domain transfer function of

$$G(z) = \frac{T}{2}\frac{z^{-1}(1 + z^{-1})}{(1 - z^{-1})^2} = \frac{T^2}{2}\frac{z + 1}{(z - 1)^2}$$

and

$$G(z^{-1}) = \frac{T^2}{2}\frac{z^{-1} + 1}{(z^{-1} - 1)^2} = \frac{T^2}{2}\frac{z(z + 1)}{(z - 1)^2}$$

Now as we plot the poles and zeros of these two functions we get a pole of order 4 at $z = 1$, two zeros at $z = -1$, and a zero at the origin. Since after reduction to forms in the variable $z$ neither of the functions $G(z)$ nor $G(z^{-1})$ possesses an explicit negative sign, we want the positive gain root locus. From any point on the locus a value of $q/r$ could be obtained. The reciprocal root locus for this problem is shown in Fig. 9.8 with an expanded view of the interior of the unit circle shown in Fig. 9.9.

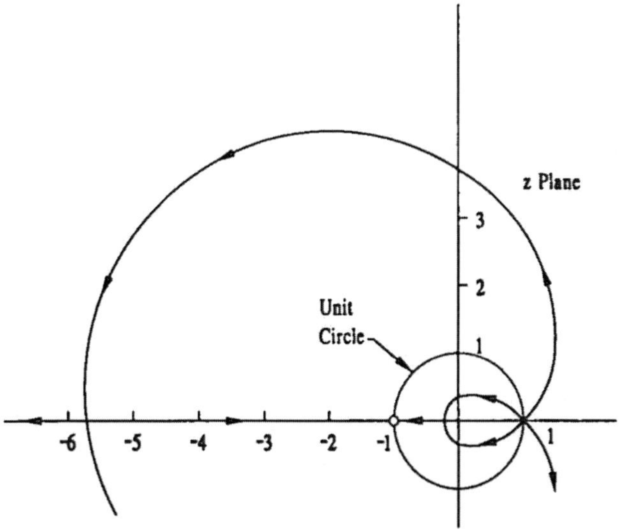


**Figure 9.8.** Reciprocal root locus for inertial plant. (Arrows indicate the direction of increasing $q/r$.)

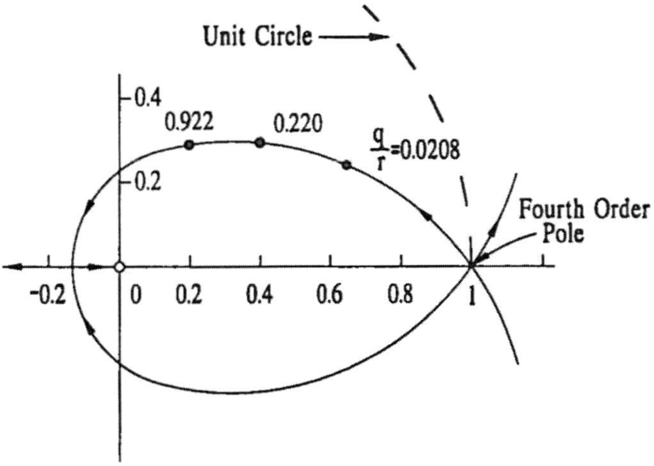


**Figure 9.9.** Close-up of reciprocal root locus for the optimal control of the inertial system.

## 9.5 STEADY-STATE REGULATOR PROBLEM BY EIGENVECTOR DECOMPOSITION

Here we shall present a computational technique that will give the steady-state feedback gain matrix without solving the Riccati equation backward in time as developed by Vaughn (1970). We have shown in Section 9.4 that the characteristic roots of the adjoint equation, plant equation, and control law must occur in reciprocal pairs; thus if $z_i$ is a characteristic root, so must be $z_i^{-1}$.

Consider now the regulator problem posed in Section 9.4. We would like to obtain the stationary feedback gain matrix $\overline{\mathbf{F}}$ when the planning horizon is long or when $N$ becomes very large.

From Section 9.4, Eq. (9.4.1) could be rewritten as

$$\mathbf{x}_k = \mathbf{A}^{-1}\mathbf{x}_{k+1} + \mathbf{A}^{-1}\mathbf{B}\mathbf{R}^{-1}\mathbf{B}^T\boldsymbol{\lambda}_{k+1} \tag{9.5.1}$$

and with substitution of (9.5.1) into (9.4.2), we get

$$\boldsymbol{\lambda}_k = \mathbf{Q}\mathbf{A}^{-1}\mathbf{x}_{k+1} + \mathbf{Q}\mathbf{A}^{-1}\mathbf{B}\mathbf{R}^{-1}\mathbf{B}^T\boldsymbol{\lambda}_{k+1} + \mathbf{A}^T\boldsymbol{\lambda}_{k+1} \tag{9.5.2}$$

Expressions (9.5.1) and (9.5.2) are a set of linear backward different equations, and in matrix form they are

$$\begin{bmatrix} \mathbf{x}_k \\ \boldsymbol{\lambda}_k \end{bmatrix} = \begin{bmatrix} \mathbf{A}^{-1} & \mathbf{A}^{-1}\mathbf{B}\mathbf{R}^{-1}\mathbf{B}^T \\ \mathbf{Q}\mathbf{A}^{-1} & \mathbf{A}^T + \mathbf{Q}\mathbf{A}^{-1}\mathbf{B}\mathbf{R}^{-1}\mathbf{B}^T \end{bmatrix} \begin{bmatrix} \mathbf{x}_{k+1} \\ \boldsymbol{\lambda}_{k+1} \end{bmatrix} \tag{9.5.3}$$

We can define the large $2n \times 2n$ backward transition matrix as $\mathbf{\Phi}$. Since the system (9.5.3) represents the same system as (9.4.1) and (9.4.2), it must have the same eigenvalues $z_i$ and $z_i^{-1}$ in pairs and associated eigenvectors. Define the similarity transformation matrix $\mathbf{W}$ by

$$\mathbf{W} = \left[\begin{array}{c|c} \mathbf{X}_u & \mathbf{X}_s \\ \hline \mathbf{\Lambda}_u & \mathbf{\Lambda}_s \end{array}\right] \tag{9.5.4}$$

where the vectors of

$$\left[\begin{array}{c} \mathbf{X}_u \\ \hline \mathbf{\Lambda}_u \end{array}\right]$$

are the eigenvectors associated with the unstable eigenvalues and

$$\left[\begin{array}{c} \mathbf{X}_s \\ \hline \mathbf{\Lambda}_s \end{array}\right]$$

is a matrix of eigenvectors associated with the stable eigenvalues of $\mathbf{\Phi}$. Also denote the diagonal matrix of eigenvalues to be $\mathbf{\Phi}^*$:

$$\mathbf{\Phi}^* = \begin{bmatrix} \mathrm{E} & 0 \\ 0 & \mathrm{E}^{-1} \end{bmatrix} \tag{9.5.5}$$

where E is a diagonal matrix of the unstable eigenvalues and $\mathrm{E}^{-1}$ is a diagonal matrix of the stable eigenvalues. We know from the eigenvalue problem of linear algebra that $\mathbf{\Phi}$ can be diagonalized into $\mathbf{\Phi}^*$ as follows:

$$\mathbf{\Phi}^* = \mathbf{W}^{-1}\mathbf{\Phi}\mathbf{W} \tag{9.5.6}$$

Also, $\mathbf{W}^{-1}$ can be used to transform the coordinates $\mathbf{x}$ and $\boldsymbol{\lambda}$ to normal coordinates $\mathbf{x}^*$ and $\boldsymbol{\lambda}^*$:

$$\left[\begin{array}{c} \mathbf{x}_k^* \\ \hline \boldsymbol{\lambda}_k^* \end{array}\right] = \mathbf{W}^{-1}\left[\begin{array}{c} \mathbf{x}_k \\ \hline \boldsymbol{\lambda}_k \end{array}\right] \tag{9.5.7}$$

and the inverse relation

$$\left[\begin{array}{c} \mathbf{x}_k \\ \hline \boldsymbol{\lambda}_k \end{array}\right] = \mathbf{W}\left[\begin{array}{c} \mathbf{x}_k^* \\ \hline \boldsymbol{\lambda}_k^* \end{array}\right] = \left[\begin{array}{c|c} \mathbf{X}_u & \mathbf{X}_s \\ \hline \mathbf{\Lambda}_u & \mathbf{\Lambda}_s \end{array}\right]\left[\begin{array}{c} \mathbf{x}_k^* \\ \hline \boldsymbol{\lambda}_k^* \end{array}\right] \tag{9.5.8}$$

In the normal coordinates the set of difference equations can be written

$$\left[\begin{array}{c} \mathbf{x}_k^* \\ \hline \boldsymbol{\lambda}_k^* \end{array}\right] = \left[\begin{array}{c|c} \mathbf{E} & \mathbf{0} \\ \hline \mathbf{0} & \mathbf{E}^{-1} \end{array}\right]\left[\begin{array}{c} \mathbf{x}_{k+1}^* \\ \hline \boldsymbol{\lambda}_{k+1}^* \end{array}\right] \tag{9.5.9}$$

and from the $k$th point to the final point can be written as

$$\begin{bmatrix} \mathbf{x}_k^* \\ \hline \boldsymbol{\lambda}_k^* \end{bmatrix} = \left[\begin{array}{c|c} \mathbf{E}^l & \mathbf{0} \\ \hline \mathbf{0} & \mathbf{E}^{-l} \end{array}\right] \begin{bmatrix} \mathbf{x}_N^* \\ \hline \boldsymbol{\lambda}_N^* \end{bmatrix} \tag{9.5.10}$$

where $k + l = N$. Since we want the steady-state solution, we will be interested in the case where $N$ becomes large. In this case $\mathbf{E}^l$ diverges and causes difficulties, so let us rewrite (9.5.10) in a different form and change the index to $l$ only so that $l = 0$ implies that $k = N$ and large values of $l$ imply small values of $k$. Thus

$$\begin{bmatrix} \mathbf{x}_l^* \\ \hline \boldsymbol{\lambda}_l^* \end{bmatrix} = \left[\begin{array}{c|c} \mathbf{E}^l & \mathbf{0} \\ \hline \mathbf{0} & \mathbf{E}^{-l} \end{array}\right] \begin{bmatrix} \mathbf{x}_0^* \\ \hline \boldsymbol{\lambda}_0^* \end{bmatrix} \tag{9.5.11}$$

The first equation of (9.5.11) can be rewritten as

$$\mathbf{x}_0^* = \mathbf{E}^{-l}\mathbf{x}_l^* \tag{9.5.12}$$

Thus the two relations of (9.5.11) may now be written as

$$\begin{bmatrix} \mathbf{x}_0^* \\ \hline \boldsymbol{\lambda}_l^* \end{bmatrix} = \left[\begin{array}{c|c} \mathbf{E}^{-l} & \mathbf{0} \\ \hline \mathbf{0} & \mathbf{E}^{-l} \end{array}\right] \begin{bmatrix} \mathbf{x}_l^* \\ \hline \boldsymbol{\lambda}_0^* \end{bmatrix} \tag{9.5.13}$$

and now we have only negative powers of $\mathbf{E}$.

From Section 9.2 recall that the boundary condition is $\boldsymbol{\lambda}_N = \mathbf{H}\mathbf{x}_N$ and that in the steady-state problem, $\mathbf{H} = \mathbf{Q}$. The boundary condition using subscript $l$ instead of $k$ gives us

$$\boldsymbol{\lambda}_0 = \mathbf{H}\mathbf{x}_0 \tag{9.5.14}$$

From the transformation between the coordinates (9.5.8), we can write

$$\mathbf{x}_0 = \mathbf{X}_u\mathbf{x}_0^* + \mathbf{X}_s\boldsymbol{\lambda}_0^* \tag{9.5.15}$$

and

$$\boldsymbol{\lambda}_0 = \boldsymbol{\Lambda}_u\mathbf{x}_0^* + \boldsymbol{\Lambda}_s\boldsymbol{\lambda}_0^* = \mathbf{Q}[\mathbf{X}_u\mathbf{x}_0^* + \mathbf{X}_s\boldsymbol{\lambda}_0^*] \tag{9.5.16}$$

From the two right-hand expressions of (9.5.16) we can write

$$\boldsymbol{\lambda}_0^* = -[\boldsymbol{\Lambda}_s - \mathbf{Q}\mathbf{X}_s]^{-1}[\boldsymbol{\Lambda}_u - \mathbf{Q}\mathbf{X}_u]\mathbf{x}_0^* \tag{9.5.17}$$

or

$$\boldsymbol{\lambda}_0^* = \mathbf{T}\mathbf{x}_0^* \tag{9.5.18}$$

The relation between $\boldsymbol{\lambda}_l^*$ and $\mathbf{x}_l^*$ can be written as

$$\boldsymbol{\lambda}_l^* = \mathbf{G}_l\mathbf{x}_l^* \tag{9.5.19}$$

where $\mathbf{G}_l$ is obtained as follows:

$$\boldsymbol{\lambda}_0^* = \mathbf{T}\mathbf{x}_0^* \tag{9.5.20}$$

Premultiplication by $\mathbf{E}^{-l}$ gives

$$\mathbf{E}^{-l}\boldsymbol{\lambda}_0^* = \mathbf{E}^{-l}\mathbf{T}\mathbf{x}_0^* \tag{9.5.21}$$

and, from (9.5.13),

$$\mathbf{x}_0^* = \mathbf{E}^{-l}\mathbf{x}_l^* \tag{9.5.22}$$

and

$$\mathbf{E}^{-l}\boldsymbol{\lambda}_0^* = \boldsymbol{\lambda}_l^* \tag{9.5.23}$$

so

$$\boldsymbol{\lambda}_l^* = \mathbf{E}^{-l}\mathbf{T}\mathbf{E}^{-l}\mathbf{x}_l^* = \mathbf{G}_l\mathbf{x}_l^* \tag{9.5.24}$$

and the matrix $\mathbf{G}_l$ has been defined. We know that $\boldsymbol{\lambda}_l$ and $\mathbf{x}_l$ are related by

$$\boldsymbol{\lambda}_l = \mathbf{P}_l\mathbf{x}_l \tag{9.5.25}$$

From (9.5.8) we see that

$$\mathbf{x}_l = \mathbf{X}_u\mathbf{x}_l^* + \mathbf{X}_s\boldsymbol{\lambda}_l^* \tag{9.5.26}$$

and

$$\boldsymbol{\lambda}_l = \boldsymbol{\Lambda}_u\mathbf{x}_l^* + \boldsymbol{\Lambda}_s\boldsymbol{\lambda}_l^* \tag{9.5.27}$$

Substitution of (9.5.24) into (9.5.27) gives

$$\boldsymbol{\lambda}_l = (\boldsymbol{\Lambda}_u + \boldsymbol{\Lambda}_s\mathbf{G}_l)\mathbf{x}_l^* \tag{9.5.28}$$

and substitution of (9.5.24) into (9.5.26) gives

$$\mathbf{x}_l = (\mathbf{X}_u + \mathbf{X}_s\mathbf{G}_l)\mathbf{x}_l^* \tag{9.5.29}$$

Solving (9.5.29) for $\mathbf{x}_l^*$ yields

$$\mathbf{x}_l^* = (\mathbf{X}_u + \mathbf{X}_s\mathbf{G}_l)^{-1}\mathbf{x}_l \tag{9.5.30}$$

Now substituting (9.5.30) into (9.5.28) yields

$$\boldsymbol{\lambda}_l = (\boldsymbol{\Lambda}_u + \boldsymbol{\Lambda}_s\mathbf{G}_l)(\mathbf{X}_u + \mathbf{X}_s\mathbf{G}_l)^{-1}\mathbf{x}_l \tag{9.5.31}$$

Comparison of (9.5.31) and (9.5.25) reveals

$$\mathbf{P}_l = (\boldsymbol{\Lambda}_u + \boldsymbol{\Lambda}_s\mathbf{G}_l)(\mathbf{X}_u + \mathbf{X}_s\mathbf{G}_l)^{-1} \tag{9.5.32}$$

Now when $l$ approaches infinity, expression (9.5.24) indicates that $\mathbf{G}_l$ vanishes, so the steady-state value of **P** is

$$\bar{\mathbf{P}} = \mathbf{\Lambda}_u \mathbf{X}_u^{-1} \tag{9.5.33}$$

The steady-state regulator gain matrix from expression (9.2.35) is given to be

$$\bar{\mathbf{F}} = \mathbf{R}^{-1}\mathbf{B}^T\mathbf{A}^{T-1}(\bar{\mathbf{P}} - \mathbf{Q}) \tag{9.5.34}$$

so upon noting (9.5.33), it becomes

$$\bar{\mathbf{F}} = \mathbf{R}^{-1}\mathbf{B}^T(\mathbf{A}^T)^{-1}[\mathbf{\Lambda}_u\mathbf{X}_u^{-1} - \mathbf{Q}] \tag{9.5.35}$$

Computational tools are readily available for efficiently computing the eigenvalues and eigenvectors of quite general matrices. A decade ago these computations required large software packages running on mainframe and minicomputers. Currently, easy-to-use software packages that run on personal computers will extract eigenvalues and eigenvectors.

Also, these feedback gain calculations may now be accomplished most easily by application of MATLAB (Saadat, 1993), wherein some simple programming is required. On the other hand, the accompanying package, the Control System Toolbox, has routines available to compute these gains with no programming.

**Example 9.5.** Consider the plant that was considered in Example 9.1, which was governed by the difference equation

$$x_{k+1} = 0.8x_k + u_k$$

with a performance index over the semiinfinite planning horizon

$$J = \frac{1}{2}\sum_{k=0}^{\infty} x_k^2 + u_k^2$$

In this problem $A = 0.8$, $B = 1$, $Q = 1$, and $R = 1$. We are interested in employing the technique of eigenvector decomposition to solve for the stationary optimal-state feedback gain. Inserting the specified parameters into the matrix of (9.5.3) yields

$$\mathbf{\Phi} = \begin{bmatrix} 1.25 & 1.25 \\ 1.25 & 0.8 + 1.25 \end{bmatrix}$$

The eigenvalues of this matrix are the roots of

$$(z - 1.25)(z - 2.05) - (1.25)^2 = z^2 - 3.3z + 1 = 0$$

The roots are

$$z_{1,2} = 0.337 \text{ and } 2.96$$

which we note are reciprocals. If we evaluate the eigenvector associated with the unstable eigenvalue (2.96), or

$$\begin{bmatrix} X_u \\ \Lambda_u \end{bmatrix} = \begin{bmatrix} 1 \\ 1.368 \end{bmatrix}$$

The stationary value of $\overline{P}$ is given from (9.5.33) to be

$$\overline{P} = \Lambda_u X_u^{-1} = (1.368)(1)^{-1} = 1.368$$

which is the same value formed by solution of the Riccati equation backward in time in Example 9.1. From expression (9.5.34) the stationary feedback gain is

$$\overline{F} = (1)^{-1}(1)(1.25)(1.368 - 1) = 0.46$$

which is the same gain as that given in Example 9.1.

## 9.6 OPTIMAL CONTROL ABOUT NONZERO SET POINTS

In a large number of applications it is desirable to control deviations of the system output about some fixed value. This is especially the case in the chemical-processing industry, where pressures, temperatures, and chemical concentrations are controlled about fixed values. The same is true for a flight control system for straight and level flight of an aircraft.

Let us consider the system that we studied in Section 9.2. It is governed by the vector difference equation

$$\mathbf{x}_{k+1} = \mathbf{A}\mathbf{x}_k + \mathbf{B}\mathbf{u}_k \tag{9.6.1}$$

with the outputs given by

$$\mathbf{y}_k = \mathbf{C}\mathbf{x}_k \tag{9.6.2}$$

We are interested in controlling the output vector $\mathbf{y}_k$ about some fixed vector $\overline{\mathbf{y}}$. We shall also assume that equilibrium conditions $\overline{\mathbf{x}}$ and $\overline{\mathbf{u}}$ exist which satisfy steady-state versions of (9.6.1) and (9.6.2), which are

$$\overline{\mathbf{x}} = \mathbf{A}\overline{\mathbf{x}} + \mathbf{B}\overline{\mathbf{u}} \tag{9.6.3}$$

and

$$\overline{\mathbf{y}} = \mathbf{C}\overline{\mathbf{x}} \tag{9.6.4}$$

Let us denote new variables which are the perturbations from the equilibrium conditions with primes; these variables are defined as

$$\begin{aligned} \mathbf{x}'_k &= \mathbf{x}_k - \bar{\mathbf{x}} \\ \mathbf{u}'_k &= \mathbf{u}_k - \bar{\mathbf{u}} \\ \mathbf{y}'_k &= \mathbf{y}_k - \bar{\mathbf{y}} \end{aligned} \tag{9.6.5}$$

Substitution of relations (9.6.5) into (9.6.1) and (9.6.2) yields

$$\mathbf{x}'_{k+1} + \bar{\mathbf{x}} = \mathbf{A}(\mathbf{x}'_k + \bar{\mathbf{x}}) + \mathbf{B}(\mathbf{u}'_k + \bar{\mathbf{u}}) \tag{9.6.6}$$

and

$$\mathbf{y}'_k + \bar{\mathbf{y}} = \mathbf{C}(\mathbf{x}'_k + \bar{\mathbf{x}}) \tag{9.6.7}$$

It is easy to see that we may subtract the equilibrium relations (9.6.3) and (9.6.4) from (9.6.1) and (9.6.2) to yield a new difference equation,

$$\mathbf{x}'_{k+1} = \mathbf{A}\mathbf{x}'_k + \mathbf{B}\mathbf{u}'_k \tag{9.6.8}$$

and an output relation

$$\mathbf{y}'_k = \mathbf{C}\mathbf{x}'_k \tag{9.6.9}$$

A performance index may be imposed which penalizes perturbations in the states and control efforts from the equilibrium values, or

$$J = \frac{1}{2}\mathbf{x}'^T_N\mathbf{H}\mathbf{x}'_N + \frac{1}{2}\sum_{k=0}^{N-1} \mathbf{x}'^T_k\mathbf{Q}\mathbf{x}'_k + \mathbf{u}'^T_k\mathbf{R}\mathbf{u}'_k \tag{9.6.10}$$

The minimization of (9.6.10) subject to the constraint of (9.6.8) is the same as the linear regulator problem and yields a linear feedback law in the perturbation quantities, which is

$$\mathbf{u}'_k = -\mathbf{F}_k\mathbf{x}'_k \tag{9.6.11}$$

This is difficult to implement because the system variables are really $\mathbf{x}_k$ and $\mathbf{u}_k$. From relations (9.6.5) and (9.6.11) we see that the actual control effort can be written as

$$\mathbf{u}_k = \bar{\mathbf{u}} - \mathbf{F}_k(\mathbf{x}_k - \bar{\mathbf{x}}) \tag{9.6.12}$$

and

$$\mathbf{u}_k = (\bar{\mathbf{u}} + \mathbf{F}_k\bar{\mathbf{x}}) - \mathbf{F}_k\mathbf{x}_k \tag{9.6.13}$$

We can define a reference quantity $\mathbf{q}_k$ as

$$\mathbf{q}_k = \bar{\mathbf{u}} + \mathbf{F}_k\bar{\mathbf{x}} \tag{9.6.14}$$

to the control law is

$$\mathbf{u}_k = \mathbf{q}_k - \mathbf{F}_k\mathbf{x}_k \tag{9.6.15}$$

which is shown implemented in Fig. 9.10. It remains to relate the reference input $\mathbf{q}_k$ to the set point $\bar{\mathbf{y}}$. Substitute (9.6.14) into equilibrium relation (9.6.3) to yield

$$\bar{\mathbf{x}} = \mathbf{A}\bar{\mathbf{x}} + \mathbf{B}(\mathbf{q}_k - \mathbf{F}_k\bar{\mathbf{x}}) \tag{9.6.16}$$

or, consolidating the $\bar{\mathbf{x}}$ terms,

$$(\mathbf{I} - \mathbf{A} + \mathbf{B}\mathbf{F}_k)\bar{\mathbf{x}} = \mathbf{B}\mathbf{q}_k \tag{9.6.17}$$

Solving for $\bar{\mathbf{x}}$ gives

$$\bar{\mathbf{x}} = (\mathbf{I} - \mathbf{A} + \mathbf{B}\mathbf{F}_k)^{-1}\mathbf{B}\mathbf{q}_k \tag{9.6.18}$$

and we may calculate the set point as

$$\bar{\mathbf{y}} = \mathbf{C}(\mathbf{I} - \mathbf{A} + \mathbf{B}\mathbf{F}_k)^{-1}\mathbf{B}\mathbf{q}_k \tag{9.6.19}$$

It is clear from relation (9.6.14) that the dimension of $\mathbf{q}_k$ is the same as that of $\mathbf{u}_k$. To have a unique solution $\mathbf{q}_k$ to the set of linear equations (9.6.19) there must be at least the same number of variables in $\mathbf{u}_k$ as in $\bar{\mathbf{y}}$. Hopefully, if we are trying to control $m$ outputs, we will have at least $m$ manipulable variables to do the task. If, in fact, the dimensions of $\mathbf{u}_k$ and $\bar{\mathbf{y}}$ are the same, we can write the solution of (9.6.19) as

$$\mathbf{q}_k = [\mathbf{C}(\mathbf{I} - \mathbf{A} + \mathbf{B}\mathbf{F}_k)^{-1}\mathbf{B}]^{-1}\bar{\mathbf{y}} \tag{9.6.20}$$

This completes the diagram of Fig. 9.10 as shown in Fig. 9.11.

If one chooses to implement the time-invariant control algorithm, the input block and feedback block will be time invariant, as will be reference vector $\mathbf{q}_k$.

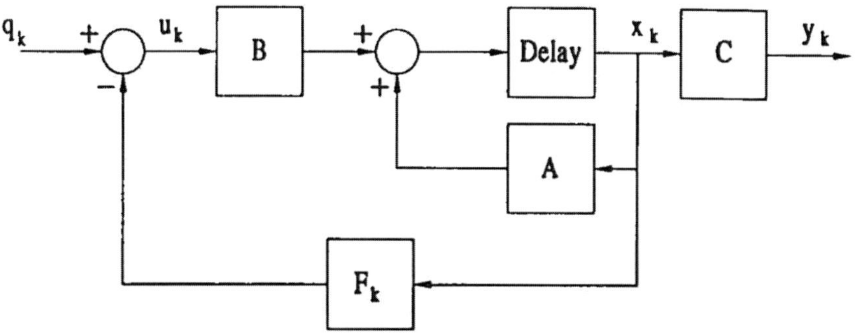


**Figure 9.10.** Feedback control system.

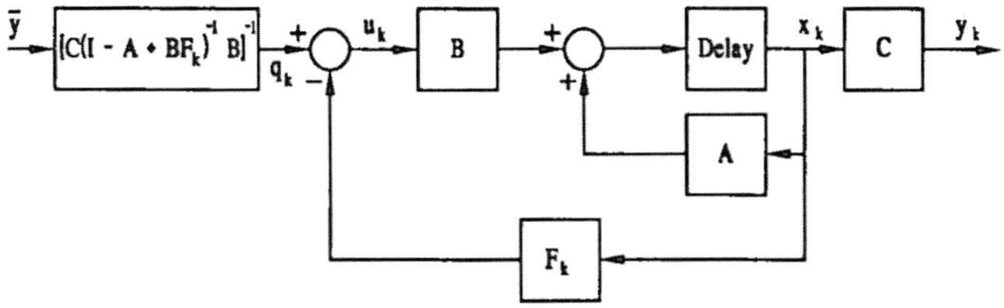


**Figure 9.11.** Complete nonzero set point control system.

**Example 9.6.** We should like to control temperature $x_1$ optimally about some arbitrary fixed point. The output matrix is in this case

$$\mathbf{c}^T = [1 \quad 0]$$

If the performance index in the perturbation quantities is the same as that given in Example 9.2, the stationary optimal feedback gain matrix is

$$\mathbf{f}^T = [1.872 \quad 3.543]$$

We now have sufficient data to calculate the gain of the reference input block from expression (9.6.20) since we know all the matrices and the stationary values of the gains. Using the digital computer to perform the necessary calculations, the input block gain is found to be 5.92. The complete control system structure is shown in Fig. 9.12.

## 9.7 SUBOPTIMAL CONTROL EMPLOYING ESTIMATED STATE FEEDBACK

In Chapter 8 we explored the possibility of using a state estimator or observer to estimate the state from incomplete state information. The control effort was then formed from a linear combination of the estimated states. The technique for picking feedback and observer gains was based on pole placement, with the poles of the observer being located such that they are faster than the poles of the feedback-controlled plant.

The topic addressed in this section is that of using the theory developed in the earlier parts of this chapter to calculate the constant-valued feedback gains, then to feed back the estimated states rather than the actual states in the case where measurements of the complete state vector is not feasible or the state variables have been chosen to be nonphysical. The physical

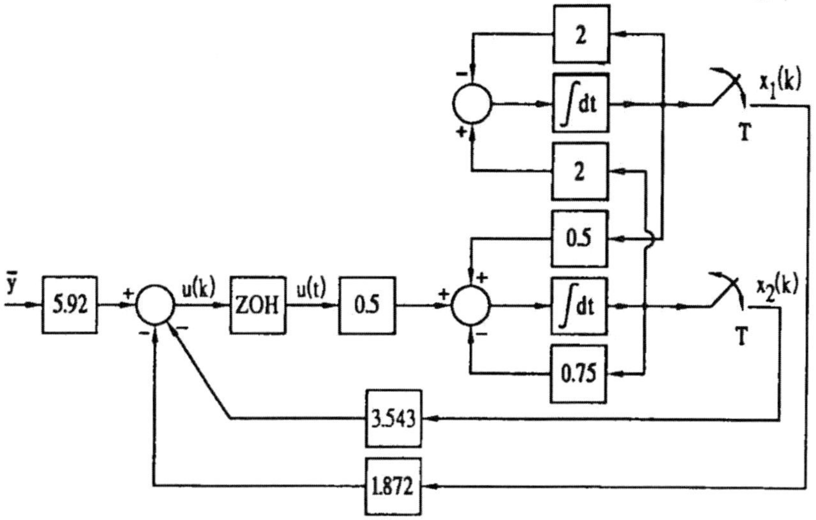


**Figure 9.12.** Control system for controlling $x_1(k)$ about some arbitrary fixed point.

situation is given in block diagram form in Fig. 9.13. In this case there will be some increased cost associated with the estimation.

If constant feedback gains are employed, the separation principle of Chapter 8 holds and half of the $2n$ system poles are at locations dictated by the characteristic equation

$$\det[z\mathbf{I} - \mathbf{A} + \mathbf{BF}] = 0 \qquad (9.7.1)$$

where **F** is the result of the regulator problem. Once these pole locations have been determined, it remains to place the remaining $n$ poles given by the determinant

$$\det[z\mathbf{I} - \mathbf{A} + \mathbf{KC}] = 0 \qquad (9.7.2)$$

where **C** is the output matrix and **K** is the matrix of observer gains. These should be placed such that these poles have faster responses. A rule of thumb for this is that they should be twice as fast. The question to be answered here is: How much does the estimator in the control loop degrade the performance index? It is clear at the outset that the degradation of the performance will be a function of the observer gains. Also, it is no longer clear that those feedback gains which were optimal for the regulator (state feedback) are those which minimize the same performance index with the estimator in the loop. This total minimization problem is a difficult optimization problem for the deterministic control problem; however, it is the

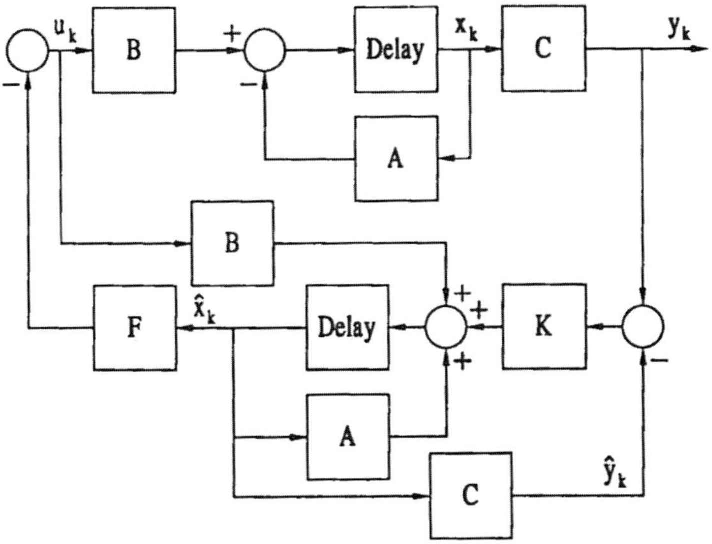


**Figure 9.13.** Estimated state feedback control.

essence of the stochastic control problem of Chapter 12. An example problem for which we have already obtained the feedback solution is logical to investigate.

**Example 9.7.** Consider the problem of Examples 9.2 and 9.3. This is the thermal problem illustrated in Fig. 9.4, except that here we shall measure only temperature $x_1(k)$. Evaluate the increase in cost for using a prediction observer of the form

$$\hat{\mathbf{x}}_{l+1} = \mathbf{A}\hat{\mathbf{x}}_l + \mathbf{B}\mathbf{u}_l + \mathbf{k}[\mathbf{x}_1(l) - \hat{\mathbf{x}}_1(l)]$$

where **A** and **B** are the same as in the plant to be controlled and the **k** vector is the observer gain vector. The feedback gain matrix was given in Example 9.2 to be

$$\mathbf{f}^T = [1.872 \quad 3.543]$$

The locations of the poles due to feedback control are given by Eq. (9.7.1) to be the roots of

$$z^2 - 1.0263z + 0.293 = 0$$

or

$$z_{1,2} = 0.513 \pm j0.173$$

These poles are illustrated in Fig. 9.14. We would like the observer poles to be faster, and the characteristic equation for the observer from (9.7.2) is

$$z^2 - z(1.4903 - k_1) + 0.5027 - 0.8526k_1 + 0.3597k_2 = 0 \qquad \text{(a)}$$

Let us locate the observer poles at

$$z_{3,4} = 0.4 \pm j0.2$$

which gives a characteristic equation of

$$z^2 - 0.8z + 0.2 = 0 \qquad \text{(b)}$$

These poles are also illustrated in Fig. 9.14.

Equating equations (a) and (b) yields an observer gain vector of

$$\mathbf{k} = \begin{bmatrix} 0.6803 \\ 0.7709 \end{bmatrix}$$

Using this observer and the estimated state feedback, the responses and estimates for initial conditions

$$\mathbf{x}(0) = \begin{bmatrix} 2 \\ 1 \end{bmatrix}$$

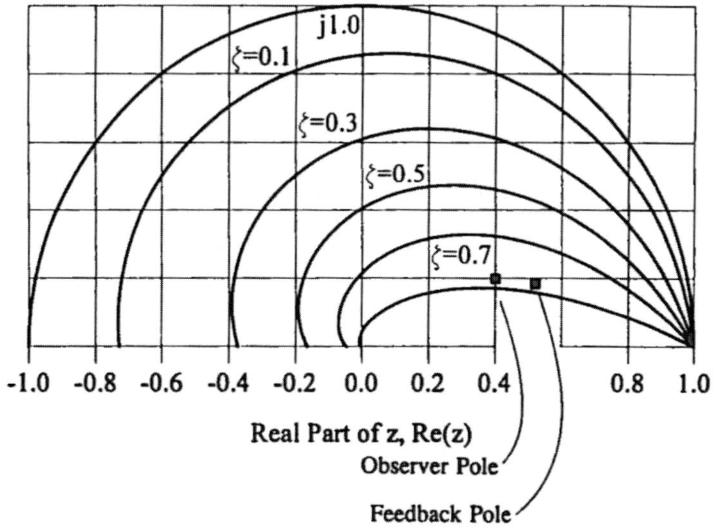


**Figure 9.14.** $z$-Plane pole locations for feedback loop and observer.

are illustrated in Fig. 9.15. Note that typically the observer must be started with an unknown initial condition, so in this case, both observer states were started at zero. The performance index of Example 9.2 was evaluated by simulation to be 270.62, where that for actual state feedback was 217.65, for an increase of about 24%. Another choice of observer gains and feedback gains might give a smaller performance index for these particular initial conditions.

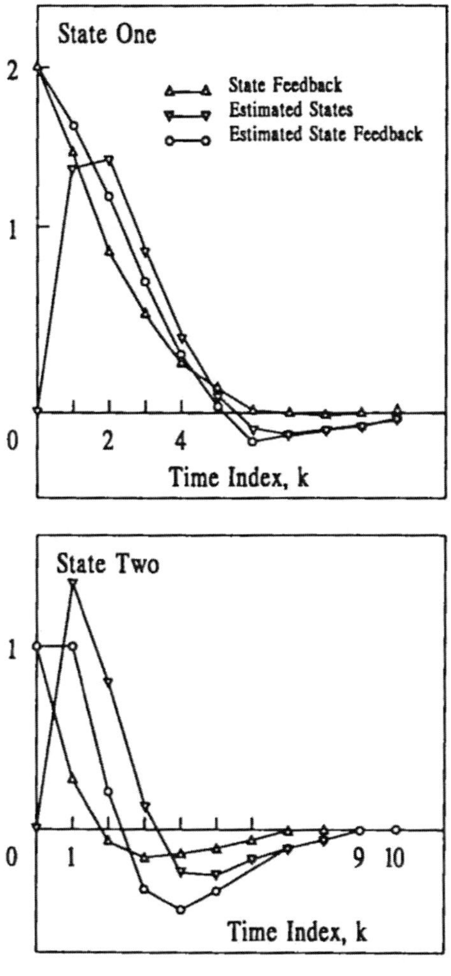


**Figure 9.15.** Responses of the thermal plant using state and estimated state feedback.

## 9.8 SUMMARY

In this chapter we have considered the so-called modern approach to control system design. Here we developed the control strategy that minimizes an objective function which is quadratic in the system state variables and the control efforts. This control strategy was one that constructed the control efforts as a linear combination of the state variables. The cost of control over any desired time interval was also evaluated.

The eigenvalues of the adjoint system and plant are shown to occur in reciprocal pairs, and the reciprocal root locus is given as a tool to guide the selection of performance indices for single-input/single-output systems in terms of desired control system pole locations. An efficient algorithm for calculation of feedback gains is given, optimal control about some fixed output values is discussed, and the zero fixed-point problem (regulator) is extended to cover this case. In the final section the deterioration of the system performance by using estimated state feedback is examined.

## PROBLEMS

9.1. A first-order plant driven by a zero-order hold and followed by an output sampler has $z$-domain transfer function

$$G(z) = \frac{X(z)}{U(z)} = \frac{0.2z^{-1}}{1 - 0.8z^{-1}}$$

(a) Find the governing difference equation.
(b) Given the performance index

$$J = \frac{1}{2}\sum_{k=0}^{39} 10x_k^2 + u_k^2$$

find the steady-state solution to the discrete-time Riccati equation. Note that $p_{k+1} = p_k = p$, and we are seeking the positive solution.
(c) Find the associated optimal feedback coefficient $f$.

9.2. A first-order plant is described by a transfer function

$$G(z) = \frac{0.2z^{-1}}{1 - 0.6z^{-1}}$$

(a) Find the equivalent difference equation.

(b) Given the performance index

$$J = \frac{1}{2}\sum_{k=0}^{10} 4x_k^2 + u_k^2$$

9.3. The two-tank system is governed by the matrix-vector difference equation

$$\begin{bmatrix} x_1 \\ x_2 \end{bmatrix}_{k+1} = \begin{bmatrix} 0.8336 & 0.1424 \\ 0.1424 & 0.6200 \end{bmatrix} \begin{bmatrix} x_1 \\ x_2 \end{bmatrix}_k + \begin{bmatrix} 0.09117 \\ 0.00798 \end{bmatrix} u(k)$$

For the following performance index matrices and a planning horizon of 2 s ($N = 20$), find the state feedback gain histories:

$$\mathbf{H} = \begin{bmatrix} 0 & 0 \\ 0 & 0 \end{bmatrix} \qquad \mathbf{Q} = \begin{bmatrix} 10 & 0 \\ 0 & 15 \end{bmatrix} \qquad \mathbf{R} = 1$$

9.4. Plot the reciprocal root locus for the system of Problem 9.1. (Note here that you must straighten out the signs of the reciprocal root characteristic equation in order to find the correct angle criterion.)

9.5. For the thermal system of Example 9.2, the transfer function is

$$G(x) = \frac{X_1(z)}{U(z)} = \frac{0.025(z + 0.816)}{(z - 0.952)/(z - 0.528)}$$

Plot the reciprocal root locus for this system if the output is $y(k) = x_1(k)$ and the performance index is

$$J = \frac{1}{2}\sum_{k=0}^{\infty} qx_1^2(k) + ru^2(k)$$

9.6. In Section 6.3 the inertial plant was found to be governed by the state equations

$$\begin{bmatrix} x_1(k+1) \\ x_2(k+1) \end{bmatrix} = \begin{bmatrix} 1 & T \\ 0 & 1 \end{bmatrix} \begin{bmatrix} x_1(k) \\ x_2(k) \end{bmatrix} + \begin{bmatrix} \frac{T^2}{2} \\ T \end{bmatrix} u(k)$$

(a) Let us sample at 5 Hz or $T = 0.2$ s. Evaluate the numerical values of the **A** and **b** matrices.

(b) For a quadratic performance index with infinite planning horizon and **Q** and $R$ specified as

$$\mathbf{Q} = \begin{bmatrix} 100 & 0 \\ 0 & 50 \end{bmatrix} \qquad R = 1$$

find the state feedback gain matrix $\mathbf{f}^T$ that minimizes the quadratic index. For the stationary gains also find the pole locations.

9.7. From Problem 8.6, the discrete-time state-variable model of the lightly sprung damped mass (print head model) driven by a zero-order hold for $T = 0.05$ s is

$$\begin{bmatrix} x_1(k+1) \\ x_2(k+1) \end{bmatrix} = \begin{bmatrix} 0.569 & 0.0381 \\ -15.26 & 0.4164 \end{bmatrix} \begin{bmatrix} x_1(k) \\ x_2(k) \end{bmatrix} + \begin{bmatrix} 0.00216 \\ 0.0763 \end{bmatrix} u(k)$$

For a performance index of

$$J = \frac{1}{2} \sum_{k=0}^{19} \mathbf{x}^T(k)\mathbf{Q}\mathbf{x}(k) + ru^2(k)$$

and penalty matrices

$$\mathbf{Q} = \begin{bmatrix} 1 & 0 \\ 0 & 300 \end{bmatrix} \qquad r = 1$$

find the feedback gain matrix for each $k$ on the planning horizon. For the stationary values of the feedback gains find the closed-loop pole locations from

$$\det[z\mathbf{I} - \mathbf{A} + \mathbf{b}\mathbf{f}^T] = 0$$

9.8. Find the input variable gain to be inserted for optimal control of the system of Problem 9.7 about some arbitrary fixed position $\bar{y}$.

## REFERENCES


Anderson, B. D. O., and J. B. Moore, 1990. *Optimal Control: Linear Quadratic Methods*, Prentice Hall, Englewood Cliffs, NJ.

Borrie, J. A., 1986. *Modern Control Systems: A Manual of Design Methods*, Prentice Hall, Englewood Cliffs, NJ.

Bryson, A. E., Jr., 1979. Some connections between modern and classical control concepts, *Trans. ASME: Journal of Dynamic Systems, Measurement and Control, 101*(2): 91–98.

Bryson, A. E., Jr., and Y.-C. Ho, 1975. *Applied Optimal Control*, Halsted, New York.

Franklin, G. F., J. D. Powell, and M. L. Workman, 1990. *Digital Control of Dynamic Systems*, 2nd Ed., Addison-Wesley, Reading, MA.

Gourlay, A. R., and G. A. Watson, 1973. *Computational Methods for Matrix Eigenproblems*, Wiley, New York.

Kirk, D. E., 1970. *Optimal Control Theory*, Prentice Hall, Englewood Cliffs, NJ.

Kwakernaak, H., and R. Sivan, 1972. *Linear Optimal Control Systems*, Wiley, New York.

Lewis, F. L., 1986. *Optimal Control*, Wiley, New York.

Phillips, C. L., and H. T. Nagle, 1990. *Digital Control System Analysis and Design*, 2nd Ed., Prentice Hall, Englewood Cliffs, NJ.

Saadat, H., 1993. *Computational Aids in Control Systems using MATLAB*, McGraw-Hill, New York.

Sage, A. P., and C. C. White, III, 1977. *Optimum Systems Control*, 2nd Ed., Prentice Hall, Englewood Cliffs, NJ.

Van Landingham, H. F., 1985. *Introduction to Digital Control Systems*, Macmillan, New York.

Vaughn, D. R., 1970. A nonrecursive algebraic solution for the discrete Riccati equation, *IEEE Trans. on Automatic Control*, *AC-15*(5): 598–599.

# 10
# Discrete-Time Stochastic Systems

## 10.1 INTRODUCTION

With the exception of the random errors discussed in Chapter 7, we have in this book considered only systems in which the discrete-time sequences involved were purely deterministic. In many real control systems the plant to be controlled is subjected to random disturbances. Such a situation is the effect of gusts and random winds on the flight of an aircraft or launched missile. In this chapter we examine the techniques for predicting the average properties of the randomly excited responses of such a system. In many control applications measurements of these random responses are often contaminated by additive random measurement noise created in the measuring or telemetry system. In this case the state of the system must be inferred by statistical estimation techniques using known statistical information about the disturbances and measurement noise. This estimation task is the topic of Chapter 11. In Chapter 12 the optimal control of these randomly disturbed systems based on noisy measurements is considered. This is the combined control and estimation problem, and fortunately, these operations may be conducted separately. To discuss average properties of these random sequences we first need to review some concepts from the theory of probability.

## 10.2 PROBABILITY AND RANDOM VARIABLES

When we deal with problems that have numerical answers and there is some uncertainty involved, we need to employ probabilistic methods to obtain as complete a description of the problem as possible. By use of probabilistic methods we may discuss the outcome of the uncertain experiment only in terms of average results or the probability that the result will lie in a certain range of values. If a large number of identical experiments are conducted, the arithmetic mean of the results is the statistical or ensemble average. We refer to the concept of the ensemble later when we discuss responses of discrete-time systems to random forcing functions.

We define any numerical random experimental result as a "random variable" and in this context we shall be concerned with continuous-valued random variables rather than those which take only discrete values. In the case of the continuous random variable the probability of one particular numerical value occurring is zeo. We must, instead, talk about the probability that the random variable lies within some interval among all possible values.

For a random variable $X$ we define the marginal probability density function $p(x)$ to be a function that describes the probability of the random variable lying in the small interval $(x, x + dx)$, or

$$\text{Prob}[x < X \leq x + dx] = p(x)\, dx \tag{10.2.1}$$

With this definition of the probability density function there are several properties that a probability density function must satisfy:

$$p(x) \geq 0 \tag{10.2.2}$$

$$\int_{-\infty}^{\infty} p(x)\, dx = 1 \tag{10.2.3}$$

$$\text{Prob}[x_1 < X \leq x_2] = \int_{x_1}^{x_2} p(x)\, dx \tag{10.2.4}$$

The first of these properties is simply a statement that all probabilities are positive numbers, while the second states that it is certain that the random variable lies somewhere on the real line. The third property is a result of the definition of the density function given in (10.2.1). We now give several examples of density functions commonly encountered in engineering work.

## Uniformly Distributed Random Variable

In the case of the uniformly distributed random variable the random variable lies in a bounded interval $(a, b)$ and within that interval the probability of the random variable lying within all subintervals of equal length is the same. This hypothesis leads to the uniform density function of

$$p(x) = \begin{cases} \dfrac{1}{b-a} & a < x \le b \\ 0 & \text{otherwise} \end{cases} \tag{10.2.5}$$

This density function is illustrated in Fig. 10.1.

We have already seen this density function used in Chapter 7, where we discussed the errors associated with finite-word-length representation of sampled data from the analog-to-digital conversion process.

### Gaussian or Normally Distributed Random Variables

In experiments with random numerical results we sometimes find that they tend to be clustered in some region of the axis and that the associated density function is symmetric. One density function that meets these requirements is the Gaussian or normal density function, which is

$$p(x) = \frac{1}{\sqrt{2\pi}\,\sigma} \exp\left[ -\frac{(x-m)^2}{2\sigma^2} \right] \tag{10.2.6}$$

where $m$ and $\sigma$ are positive parameters which we discuss later. This density function is illustrated in Fig. 10.2 and is symmetric about $m$. The probability that the random variable lies between $m - q$ and $m + q$ is given by the area under the density function between those values, or

$$\text{Prob}[m - q < X < m + q] = \int_{m-q}^{m+q} p(x)\, dx \tag{10.2.7}$$

These probabilities are tabulated in Table 10.1 for several values of $q$. This probability model is commonly used to describe problems of turbulence, random vibration, kinetic gas theory, electrical noise, and many others.

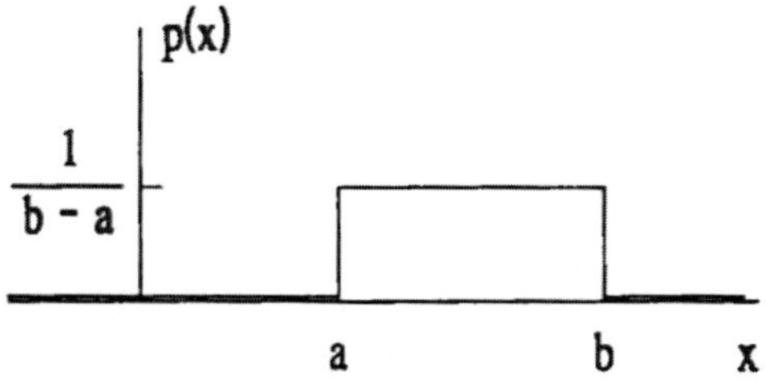


**Figure 10.1.** Uniform probability density function.

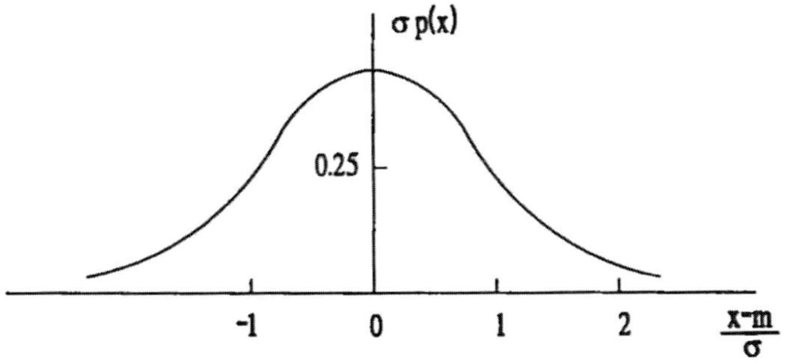


**Figure 10.2.** Gaussian density function.

### Rayleigh-Distributed Random Variable

For random variables that cannot be negative and still exhibit the property of having a cluster of values in some region of the positive axis, the Rayleigh density is an appropriate description. The expression for Rayleigh density is

$$p(x) = \frac{2x}{\alpha} \exp\left(-\frac{x^2}{\alpha}\right) \qquad x \geq 0 \tag{10.2.8}$$

where $\alpha$ is a positive parameter. The Rayleigh density (Fig. 10.3) is a density associated with the peak values of a narrow-band Gaussian process.

We have considered only a few probability density functions in this section and, of course, there are a host of others, which will be found in Papoulis (1991) and Feller (1957).

## 10.3 EXPECTATION OPERATOR AND STATISTICAL MOMENTS

Often, we are interested in a cruder description of a random variable than that given by a density function. One of these descriptions is that of the

**Table 10.1.** Probability of a Random Variable Lying in a Given Range

| $q$ | $\text{Prob}[m - q < X < m + q]$ |
|---|---|
| $\sigma$ | 0.683 |
| $2\sigma$ | 0.955 |
| $3\sigma$ | 0.997 |

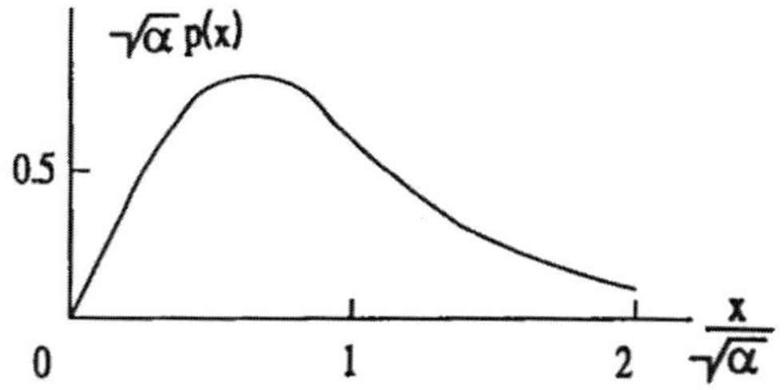


**Figure 10.3.** Rayleigh probability density function.

statistical moments or averages of the random variable. We define a function of the random variable $X$ with some expression

$$Y = f(X) \tag{10.3.1}$$

We also define the expected value of that new random variable $Y$ as

$$E[Y] = E[f(X)] = \int_{-\infty}^{\infty} f(x)p(x)\,dx \tag{10.3.2}$$

where $p(x)$ is the density function associated with the random variable $X$.

When we let the function of $X$ be $X$ itself, we refer to the resulting number as "*average value*," "*mean value*," or "*first moment*" of the random variable $X$, which will be denoted by $\bar{x}$, or

$$\bar{x} = E[X] = \int_{-\infty}^{\infty} xp(x)\,dx \tag{10.3.3}$$

Another useful moment is the *second central moment* or *variance*, which will be denoted by $\sigma^2$ and is the second moment about the mean value, or

$$\sigma^2 = E[(X - \bar{x})^2] = \int_{-\infty}^{\infty} (x - \bar{x})^2 p(x)\,dx \tag{10.3.4}$$

If we expand the quadratic function, we get

$$\sigma^2 = \int_{-\infty}^{\infty} x^2 p(x)\,dx - 2\bar{x}\int_{-\infty}^{\infty} xp(x)\,dx + \bar{x}^2\int_{-\infty}^{\infty} p(x)\,dx \tag{10.3.5}$$

Noting that the area under $p(x)$ is unity and that the second integral is the mean value $\bar{x}$, the expression (10.3.5) becomes

$$\sigma^2 = E[X^2] - \bar{x}^2 \tag{10.3.6}$$

**Example 10.1.** Evaluate the mean and variance associated with the uniform probability density function on the interval $[a,b]$. The mean value is, from relation (10.3.2),

$$E[X] = \bar{x} = \int_a^b \frac{1}{b - a} x \, dx$$

$$= \frac{1}{b - a} \frac{x^2}{2} \bigg|_a^b = \frac{b^2 - a^2}{2(b - a)}$$

$$\bar{x} = \frac{a + b}{2}$$

The mean-square value is

$$E[X^2] = \frac{1}{b - a} \int_a^b x^2 \, dx = \frac{1}{b - a} \frac{b^2 - a^3}{3}$$

$$= \frac{1}{3} [b^2 + ab + a^2]$$

The variance from expression (10.3.6) is

$$\sigma^2 = \frac{1}{3} [b^2 + ab + a^2] - \frac{1}{4} [a^2 + 2ab + b^2]$$

or

$$\sigma^2 = \frac{1}{12} [a^2 - 2ab + b^2] = \frac{1}{12} (a - b)^2$$

Since the density function was symmetric, the mean value lies at the center of the distribution, and this will be the case for all symmetric density functions.

**Example 10.2.** Evaluate the mean and variance of the Gaussian density function. From the statement above and Fig. 10.2, we see that the mean value is

$$\bar{x} = m$$

The evaluation of the second moment is not at all straightforward; this may be found in Papoulis (1991), and the result is

$$E[X^2] = \sigma^2 + m^2$$

So the variance is, in fact, $\sigma^2$.

## 10.4 DEPENDENCE, INDEPENDENCE, AND CONDITIONAL PROBABILITIES

In a large number of problems we are asked to deal with several random variables at one time. We shall define the *joint density function* $p(x,y)$ of two random variables $X$ and $Y$ as

$$p(x,y)\,dx\,dy = \text{Prob}[x < X \leq x + dx \text{ and } y < Y \leq y + dy] \tag{10.4.1}$$

Let us also define the conditional density function $p(x|y)$, where the vertical bar is read as "given that"

$$p(x|y)\,dx = \text{Prob}[x < X \leq x + dx|Y = y] \tag{10.4.2}$$

and we could in a similar fashion define $p(y|x)$, which is not the same as $p(x|y)$.

If the random variables are statistically independent, the marginal and conditional probabilities are the same, or

$$p(x|y) = p(x), \qquad p(y|x) = p(y) \tag{10.4.3}$$

Let us return to the case of dependent random variables $X$ and $Y$. The law of conditional probabilities states that

$$p(x,y) = p(x|y)p(y) \tag{10.4.4}$$

and similarly,

$$p(x,y) = p(y|x)p(x) \tag{10.4.5}$$

If $X$ and $Y$ are statistically independent, relations (10.4.3) hold and may be substituted into relations (10.4.4) and (10.4.5), so that

$$p(x,y) = p(x)p(y) \tag{10.4.6}$$

for independent random variables.

Expressions (10.4.4) and (10.4.5) may be equated to yield

$$p(y|x) = \frac{p(x|y)p(y)}{p(x)} \tag{10.4.7}$$

which is known as *Bayes' rule*. This rule is a very useful tool in statistical work and decision theory and is the foundation for the state estimation technique due to Kalman (1960). This technique is developed fully in Chapter 11.

## 10.5 JOINT GAUSSIAN RANDOM VARIABLES

We have already discussed the marginal density function for a Gaussian random variable. If two random variables $X$ and $Y$ are jointly Gaussian, their joint density function is

$$p(x,y) = \frac{1}{2\pi\sigma_x\sigma_y\sqrt{1-\rho^2}} \exp\left\{\frac{-1}{1-\rho^2}\left[\frac{(x-\bar{x})^2}{2\sigma_x^2} - \frac{\rho(x-\bar{x})(y-\bar{y})}{\sigma_x\sigma_y} + \frac{(y-\bar{y})^2}{2\sigma_y^2}\right]\right\} \tag{10.5.1}$$

where the parameters in this joint density are defined by

$$\begin{aligned} \bar{x} &= E[X] \qquad \sigma_x^2 = E[(x-\bar{x})^2] \\ \bar{y} &= E[Y] \qquad \sigma_y^2 = E[(y-\bar{y})^2] \end{aligned} \tag{10.5.2}$$

and

$$\rho = \frac{E[(x-\bar{x})(y-\bar{y})]}{\sigma_x\sigma_y} \tag{10.5.3}$$

where $\rho$ is referred to as the *correlation coefficient*. It is interesting to note that if the correlation coefficient $\rho$ is zero ($X$ and $Y$ are linearly independent), the joint density may be written as

$$p(x,y) = \frac{1}{2\pi\sigma_x\sigma_y} \exp\left[\frac{-(x-\bar{x})^2}{2\sigma_x^2} - \frac{(y-\bar{y})^2}{2\sigma_y^2}\right] \tag{10.5.4}$$

which can be factored into

$$p(x,y) = \frac{1}{\sqrt{2\pi}\,\sigma_x} \exp\left[\frac{-(x-\bar{x})^2}{2\sigma_x^2}\right] \frac{1}{\sqrt{2\pi}\,\sigma_y} \exp\left[\frac{-(y-\bar{y})^2}{2\sigma_y^2}\right] \tag{10.5.5}$$

One immediately recognizes this as the product of the respective marginal densities. This result is important because when two Gaussian random variables are linearly independent, this implies that they are also statistically independent. This is one convenient property of Gaussian random variables—that linear independence implies total statistical independence.

It is also of interest to note that the joint density function of (10.5.1) can be rewritten in matrix form as

$$p(x,y) = \frac{1}{2\pi|\mathbf{P}|^{1/2}} \exp\left\{-\frac{1}{2}\begin{bmatrix} x-\bar{x} \\ y-\bar{y} \end{bmatrix}^T \mathbf{P}^{-1} \begin{bmatrix} x-\bar{x} \\ y-\bar{y} \end{bmatrix}\right\} \tag{10.5.6}$$

where the matrix $\mathbf{P}$ is defined as

$$\mathbf{P} = E\left\{\begin{bmatrix} x-\bar{x} \\ y-\bar{y} \end{bmatrix}\begin{bmatrix} x-\bar{x} \\ y-\bar{y} \end{bmatrix}^T\right\} = \begin{bmatrix} \sigma_x^2 & \rho\sigma_x\sigma_y \\ \rho\sigma_x\sigma_y & \sigma_y^2 \end{bmatrix} \tag{10.5.7}$$

and is commonly referred to as the *covariance matrix*.

The ideas just developed could be extended to a collection of $n$ random variables. We denote the collection of random variables as a vector (ordered $n$-tuple) which we shall call $\mathbf{x}$. If these random variables are jointly Gaussian, their joint density is

$$p(\mathbf{x}) = \frac{1}{(2\pi)^{n/2}|\mathbf{P}|^{1/2}} \exp\left[ -\frac{1}{2} (\mathbf{x} - \bar{\mathbf{x}})^T \mathbf{P}^{-1} (\mathbf{x} - \bar{\mathbf{x}}) \right] \qquad (10.5.8)$$

where the mean value of this random vector is denoted by the vector $\bar{\mathbf{x}}$ and the matrix $\mathbf{P}$ is the covariance matrix defined by

$$\mathbf{P} = E[(\mathbf{x} - \bar{\mathbf{x}})(\mathbf{x} - \bar{\mathbf{x}})^T] \qquad (10.5.9)$$

If the matrix $\mathbf{P}$ is a diagonal matrix, the random variables are linearly independent, which in the case of Gaussian random variables implies total statistical independence since the joint density may be written as a product of the marginal densities, or

$$p(\mathbf{x}) = \prod_{i=1}^{n} \frac{1}{\sqrt{2\pi}\,\sigma_i} \exp\left[ -\frac{(x_i - \bar{x}_i)^2}{2\sigma_i^2} \right] \qquad (10.5.10)$$

This would be the case in the two-dimensional density of (10.5.6) if the correlation coefficient $\rho$ were zero.

## 10.6 LINEAR COMBINATIONS AND LINEAR TRANSFORMATIONS OF GAUSSIAN RANDOM VARIABLES

In physical problems known Gaussian random variables are often transformed by linear operations into new random variables. In this section we discuss the probabilistic nature of random variables constructed by such a process. The tool required for the proof of these concepts is the characteristic function (Papoulis, 1991), which is the Fourier transform of the probability density function.

Consider the case where two known Gaussian random variables $X$ and $Y$ are combined in a linear fashion to form a new random variable $Z$ given by

$$Z = a_1 X + a_2 Y \qquad (10.6.1)$$

where $a_1$ and $a_2$ are real scalar constants. It may be shown by the method of the characteristic function that if $X$ and $Y$ are Gaussian, the random variable $Z$ is also Gaussian, with mean

$$\bar{z} = a_1 \bar{x} + a_2 \bar{y} \qquad (10.6.2)$$

and the variance of the resulting random variable is

$$\sigma_z^2 = a_1^2\sigma_x^2 + a_2^2\sigma_y^2 + 2a_1a_2\sigma_x\sigma_y\rho \tag{10.6.3}$$

where $\rho$ is the correlation coefficient between $X$ and $Y$. If $X$ and $Y$ are statistically independent, $\rho = 0$ and the variance of $Z$ is

$$\sigma_z^2 = a_1^2\sigma_x^2 + a_2^2\sigma_y^2 \tag{10.6.4}$$

The relationships first considered are also applicable to sums of more than two random variables.

Another special case is the case where $Y = 1$ and is not random, or

$$Z = a_1X + a_2 \tag{10.6.5}$$

and this now represents a linear transformation of random variable $X$ into random variable $Z$. In this case it may be shown by taking the expected value of (10.6.5) that the mean value of $Z$ is

$$E[Z] = \bar{z} = a_1\bar{x} + a_2 \tag{10.6.6}$$

and the variance of the new random variable is

$$\sigma_z^2 = a_1^2\sigma_x^2 \tag{10.6.7}$$

This can easily be shown by treating $Y$ as a random variable with mean value of unity and zero variance. The resulting random variable $Z$ is also Gaussian if $X$ is Gaussian.

Let us now consider an $n$ vector of random variables $\mathbf{x}$ and an $m$ vector of random variables $\mathbf{w}$, and in this case we shall assume that $\mathbf{w}$ and $\mathbf{x}$ are statistically independent and Gaussian. The vector $\mathbf{x}$ has a mean vector $\bar{\mathbf{x}}$ and covariance $\mathbf{X}$, while $\mathbf{w}$ has mean $\bar{\mathbf{w}}$ and covariance $\mathbf{W}$. Let us examine the case where these two vectors are mapped into a new $n$ vector $\mathbf{z}$ by the linear transformation

$$\mathbf{z} = \mathbf{Ax} + \mathbf{Bw} \tag{10.6.8}$$

where $\mathbf{A}$ is an $n \times n$ matrix and $\mathbf{B}$ is an $n \times m$ matrix. The mean vector will be given by taking the expected value of (10.6.8) to yield

$$E[\mathbf{z}] = \mathbf{A}E[\mathbf{x}] + \mathbf{B}E[\mathbf{w}] \tag{10.6.9}$$

or

$$\bar{\mathbf{z}} = \mathbf{A}\bar{\mathbf{x}} + \mathbf{B}\bar{\mathbf{w}} \tag{10.6.10}$$

If we form the quantity $\mathbf{z} - \bar{\mathbf{z}}$ by subtracting (10.6.10) from (10.6.8), we get

$$\mathbf{z} = \bar{\mathbf{z}} = \mathbf{A}(\mathbf{x} - \bar{\mathbf{x}}) + \mathbf{B}(\mathbf{w} - \bar{\mathbf{w}}) \tag{10.6.11}$$

If this expression is postmultiplied by its transpose, we get

$$
\begin{aligned}
(\mathbf{z} - \bar{\mathbf{z}})(\mathbf{z} - \bar{\mathbf{z}})^T = {} & \mathbf{A}(\mathbf{x} - \bar{\mathbf{x}})(\mathbf{x} - \bar{\mathbf{x}})^T\mathbf{A}^T + \mathbf{A}(\mathbf{x} - \bar{\mathbf{x}})(\mathbf{w} - \bar{\mathbf{w}})^T\mathbf{B}^T \\
& + \mathbf{B}(\mathbf{w} - \bar{\mathbf{w}})(\mathbf{x} - \bar{\mathbf{x}})^T\mathbf{A}^T + \mathbf{B}(\mathbf{w} - \bar{\mathbf{w}})(\mathbf{w} - \bar{\mathbf{w}})^T\mathbf{B}^T
\end{aligned} \tag{10.6.12}
$$

Now let us take the expected value of this relation. Recognizing the respective covariances of **z**, **x**, and **w**, we get

$$
\mathbf{Z} = \mathbf{AXA}^T + \mathbf{BWB}^T \tag{10.6.13}
$$

where we have made use of the statistical independence of the elements **x** and **w**. The results expressed in expressions (10.6.10) and (10.6.13) will be useful later when we examine the statistical dynamics of a linear discrete-time system. Again in this case the resulting vector **z** is Gaussian with a mean vector as given by (10.6.10) and covariance matrix given by (10.6.13).

## 10.7 SCALAR DISCRETE RANDOM SEQUENCES

Consider a sequence of numbers associated with the temporal index $k$, which we shall denote as $w(k)$. The value of $w(k)$ at time $kT$ is a random variable and is defined probabilistically by its associated probability density function, which is defined by

$$
p_w(w, k) = \lim_{\Delta w \to 0} \frac{\text{Prob}[w < w(k) \le w + \Delta w]}{\Delta w} \tag{10.7.1}
$$

and its joint density function between two times

$$
\begin{aligned}
& p_w(w_1, w_2; j, k) \\
& = \lim_{\substack{\Delta w_1 \to 0 \\ \Delta w_2 \to 0}} \frac{\text{Prob}\{[w_1 < w(j) \le w_1 + \Delta w_1] \cap [w_2 < w(k) \le w_2 + \Delta w_2]\}}{\Delta w_1 \, \Delta w_2}
\end{aligned} \tag{10.7.2}
$$

This sequence may be more crudely characterized by its statistical (ensemble) averages, which are not to be confused with temporal averages. Ensemble averages are made across a large family of ensemble members for a fixed time (fixed $k$), while temporal averages are made along a single ensemble member. For temporal averages to be useful, the sequence should be stationary (statistics independent of $k$) and ergodic or the ensemble member chosen must be typical of all others. The mean value is defined as

$$
E[w(k)] = \bar{w}(k) = \int_{-\infty}^{\infty} w p_w(w, k) \, dw \tag{10.7.3}
$$

while the mean-square value is

$$E[w^2(k)] = \int_{-\infty}^{\infty} w^2 p_w(w, k)\, dw \tag{10.7.4}$$

The variance is the second central moment, or

$$\sigma_w^2(k) = E\{[w(k) - \overline{w}(k)]^2\} = \int_{-\infty}^{\infty} [w - \overline{w}(k)]^2 p_w(w, k)\, dw \tag{10.7.5}$$

Expanding this expression and recognizing the definitions of relations (10.7.1) and (10.6.2), we get

$$\sigma_w^2(k) = E[w^2(k)] - \overline{w}^2(k) \tag{10.7.6}$$

The autocorrelation sequence can be defined in terms of the joint density for the sequence, or

$$R_w(j, k) = E[w(j)w(k)] = \iint_{-\infty}^{\infty} w_1 w_2 p_w(w_1, w_2; k, j)\, dw_1\, dw_2 \tag{10.7.7}$$

If the sequence $w(k)$ is stationary, the probabilistic structure is independent of temporal translation, which implies that the first-order density is independent of $k$ and hence the mean is time-invariant, or

$$E[w(k)] = \overline{w} = \text{constant} \tag{10.7.8}$$

and the variance and mean-square value are also independent of $k$, or

$$\sigma_w^2 = E[w^2(k)] - \overline{w}^2 = \text{constant} \tag{10.7.9}$$

Stationarity also implies that the joint density for the sequence is dependent only on the difference of $j$ and $k$. Then the autocorrelation function is

$$R_w(j, k) = R_w(j - k) \tag{10.7.10}$$

If the sequence $w(k)$ is Gaussian, its first-order density function is

$$p_w(w, k) = \frac{1}{\sqrt{2\pi}\, \sigma_w(k)} \exp\left\{\frac{-[w - \overline{w}(k)]^2}{2\sigma_w^2(k)}\right\} \tag{10.7.11}$$

and the joint density for the sequence is

$$\begin{aligned} p_w(w_1, w_2; j, k) = {} & \frac{1}{2\pi\sigma_w(j)\sigma_w(k)\sqrt{1 - \rho_{jk}^2}} \\ & \times \exp\left(\frac{-1}{(1 - \rho_{jk}^2)} \left\{\frac{[w_1 - \overline{w}(j)]^2}{2\sigma_w^2(j)} \right.\right. \\ & - \frac{\rho_{jk}[w_1 - \overline{w}(j)][w_2 - \overline{w}(k)]}{\sigma_w(j)\sigma_w(k)} \\ & \left.\left. + \frac{[w_2 - \overline{w}(k)]^2}{2\sigma_w^2(k)}\right\}\right) \end{aligned} \tag{10.7.12}$$

where $\rho_{jk}$ is the correlation coefficient and is defined in terms of the autocorrelation function as

$$\rho_{jk} = \frac{R_w(j, k) - \overline{w}(j)\overline{w}(k)}{\sigma_w(j)\sigma_w(k)} \tag{10.7.13}$$

If $w(j)$ and $w(k)$ are statistically independent and Gaussian, then $\rho_{jk} = 0$ and the joint density becomes

$$p_w(w_1, w_2; j, k) = \frac{1}{2\pi\sigma_w(j)\sigma_w(k)} \exp\left\{\frac{-[w_1 - \overline{w}(j)]^2}{2\sigma_w^2(j)} - \frac{[w_2 - \overline{w}(k)]^2}{2\sigma_w^2(k)}\right\} \tag{10.7.14}$$

which as indicated in relation (10.5.10) is the product of the marginal densities of $w(j)$ and $w(k)$. The conditional density for random variable $w(j)$ given that $w(k) = w_2$ is then given by the law of conditional probability:

$$p_{w(j)|w(k)}(w_1, w_2) = \frac{p_w(w_1, w_2; j, k)}{p_w(w_2, k)} \tag{10.7.15}$$

Now if we substitute relations (10.7.11) and (10.7.12) into this relation and carry out the long division, we get

$$p_{w(j)|w(k)}(w_1, w_2) = \frac{1}{\sqrt{2\pi}\,\sigma_w(j)(1 - \rho_{jk}^2)^{1/2}} \times \exp\left(\frac{-\{w_1 - \overline{w}(j) - (\sigma_w(j)/\sigma_w(k))\rho_{jk}[w_2 - \overline{w}(k)]\}^2}{2\sigma_w^2(j)(1 - \rho_{jk}^2)}\right) \tag{10.7.16}$$

If $w(j)$ and $w(k)$ are statistically independent, the conditional density becomes the marginal density for $w(j)$, similar to that given by relation (10.7.11).

## 10.8 MARKOV AND PURELY RANDOM SEQUENCES

Let us consider a vector or scalar sequence $\mathbf{y}(0), \mathbf{y}(1), \ldots, \mathbf{y}(k)$, where the sequence is a function of the temporal index $k$. The probabilistic structure of the sequence may be completely characterized by the joint probability density function $p[\mathbf{y}(k), \mathbf{y}(k-1), \ldots, \mathbf{y}(0)]$. In general for a vector sequence this joint density contains an immense amount of information about the sequence. There are a number of random sequences that have a simpler, more convenient probabilistic structure. This structure is called the *Markovian property*. A random sequence $\mathbf{y}(k)$ is said to be

*Markovian* if the conditional density is

$$p[\mathbf{y}(k+1)|\mathbf{y}(k), \mathbf{y}(k-1), \ldots, \mathbf{y}(0)] = p[\mathbf{y}(k+1)|\mathbf{y}(k)] \tag{10.8.1}$$

where $p[\mathbf{y}(k+1)|\mathbf{y}(k)]$ is called a *transition density function*. In other words, this says that the conditional probability density function for $\mathbf{y}(k+1)$ is dependent only on knowledge of $\mathbf{y}(k)$ and not on previous values of the sequence or the probabilistic structure of these previous values. This considerably simplifies the probabilistic structure.

The joint density of a Markovian random sequence $\mathbf{y}(k)$ can be written by the law of conditional probabilities,

$$\begin{aligned} &p[\mathbf{y}(k), \mathbf{y}(k-1), \ldots, \mathbf{y}(0)] \\ &\quad = p[\mathbf{y}(k)|\mathbf{y}(k-1), \ldots, \mathbf{y}(0)] \cdots p[\mathbf{y}(k-1), \ldots, \mathbf{y}(0)] \end{aligned} \tag{10.8.2}$$

or continuing to expand the marginal density,

$$\begin{aligned} = p[\mathbf{y}(k)|\mathbf{y}(k-1), \ldots, \mathbf{y}(0)]p[\mathbf{y}(k-1)|\mathbf{y}(k-2), \ldots, \\ \mathbf{y}(0)] \cdots p[\mathbf{y}(1)|\mathbf{y}(0)]p[\mathbf{y}(0)] \end{aligned} \tag{10.8.3}$$

This is a perfectly general relation, but if the sequence is Markovian, (10.8.1) holds and relation (10.8.3) simplifies to give

$$\begin{aligned} p[\mathbf{y}(k), \mathbf{y}(k-1), \ldots, \mathbf{y}(0)] = p[\mathbf{y}(k)|\mathbf{y}(k-1)]p[\mathbf{y}(k-1)|\mathbf{y}(k-2)] \\ \cdots p[\mathbf{y}(1)|\mathbf{y}(0)]p[\mathbf{y}(0)] \end{aligned} \tag{10.8.4}$$

In other words, the joint density function may be given by the product of the transition density functions and the initial marginal density.

A sequence $\mathbf{y}(k)$ is said to be a *purely random sequence* if the conditional density is the same as the marginal density, or

$$p[\mathbf{y}(k+1)|\mathbf{y}(k)] = p[\mathbf{y}(k+1)] \tag{10.8.5}$$

This type of sequence is said to have no "memory" in that the present does not depend on the past nor does the future depend on the present. If we consider samples of the temperature of the atmosphere at some point, we know that one sample of the temperature would be related to an adjacent (in time) sample because of the finite heat capacity of the atmosphere. If, on the other hand, we consider the position of a particle suspended in a fluid and sampled these at intervals of 5 min, it is clear that the position measurements would probably be unrelated at adjacent sample times, and hence we could call this sequence a purely random sequence. It is interesting to note that if we sampled at a much higher rate (smaller sampling interval), the resulting sequence would most likely not have the purely random property.

Let us now consider a scalar purely random sequence $w(k)$ and another random variable $x(0)$ which will be used to generate a new random sequence according to the difference equation

$$x(k+1) = ax(k) + w(k) \qquad k = 0, 1, \ldots \tag{10.8.6}$$

It is clear that the sequence $x(1), x(2), \ldots$ is also random, but it is not purely random because $x(k+1)$ depends on $x(k)$ (the previous value of the sequence) and the purely random sequence $w(k)$. From the definition of the Markovian property, it is clear that the sequence $x(k)$ is a Markov sequence since $x(k+1)$ depends only on the previous value of the sequence. If the probabilistic structure of $x(0)$ and $w(k)$ are known, then since $x(1)$ is a linear combination of $x(0)$ and $w(0)$, the probabilistic structure of $x(1)$ can be found, and once this is known, then with knowledge of the probabilistic structure of $w(1)$ we can find the structure of $x(2)$, and so on.

In concept, we could predict the structure of $x(k)$ in general. If $x(0)$ and the $w(k)$ are Gaussian, then from the result of Section 10.6 the sequence $x(k)$ will also be Gaussian or known as a *Gauss–Markov sequence*.

## 10.9 VECTOR RANDOM SEQUENCE

Consider an $n$-dimensional vector random sequence that is defined by the temporal index $k$, and we denote this as

$$\mathbf{x}(k) = \begin{bmatrix} x_1(k) \\ \vdots \\ x_n(k) \end{bmatrix} \qquad k = 0, 1, 2, \ldots \tag{10.9.1}$$

where the subscript denotes the element of the vector and the temporal index is in parentheses. The elements are random variables at any fixed $k$ and thus may be characterized by the joint density function between them. If the elements are each Gaussian, they are jointly Gaussian and have a joint density

$$p_x(\mathbf{x}, k) = \frac{1}{(2\pi)^{n/2}|\mathbf{X}(k)|^{1/2}} \times \exp\left\{-\frac{1}{2}[\mathbf{x} - \bar{\mathbf{x}}(k)]^T\mathbf{X}(k)^{-1}[\mathbf{x} - \bar{\mathbf{x}}(k)]\right\} \tag{10.9.2}$$

where $\bar{\mathbf{x}}(k)$ is the mean vector, which is, in general, a function of time or $k$:

$$\bar{\mathbf{x}}(k) = E[\mathbf{x}(k)] \tag{10.9.3}$$

and $n \times n$ matrix $\mathbf{X}(k)$ is the covariance matrix defined by

$$\mathbf{X}(k) = E\{[\mathbf{x}(k) - \bar{\mathbf{x}}(k)][\mathbf{x}(k) - \bar{\mathbf{x}}(k)]^T\} \tag{10.9.4}$$

where the $ij$th element of this matrix is $E\{[x_i(k) - \bar{x}_i(k)][x_j(k) - \bar{x}_j(k)]\}$. The diagonal terms of $\mathbf{X}(k)$ are variances of the elements of $\mathbf{x}(k)$.

If the vector sequence $\mathbf{x}(k)$ is stationary, the mean and covariance matrix will be independent of the temporal index $k$. To this point nothing has been said about the correlation between elements of the sequence at different instants of time. We could thus define a correlation matrix as

$$\mathbf{X}(j, k) = E\{[\mathbf{x}(j) - \bar{\mathbf{x}}(j)][\mathbf{x}(k) - \bar{\mathbf{x}}(k)]^T\} \tag{10.9.5}$$

and if the sequence is stationary, then $\mathbf{X}(j, k)$ will be a function of temporal delay alone or

$$\mathbf{X}(j, k) = \mathbf{X}(j - k) \tag{10.9.6}$$

## 10.10 RANDOM SEQUENCES IN DISCRETE-TIME DYNAMIC SYSTEMS

Consider the following time-invariant linear discrete-time system:

$$\mathbf{x}(k + 1) = \mathbf{A}\mathbf{x}(k) + \mathbf{B}\mathbf{w}(k) \tag{10.10.1}$$

where $\mathbf{A}$ is an $n \times n$ transition matrix, $\mathbf{B}$ is an $n \times m$ matrix, $\mathbf{x}(k)$ is an $n$ vector representing $\mathbf{x}(kT)$, and $\mathbf{w}(k)$ is an $m$-dimensional vector of random input sequences. Let us also consider the case where the state $\mathbf{x}(k)$ is mapped into some output vector (usually of dimension less than $n$) of dimension $r$. The output may also be contaminated by some measurement noise sequence $\mathbf{v}(k)$, or

$$\mathbf{y}(k) = \mathbf{C}\mathbf{x}(k) + \mathbf{v}(k) \tag{10.10.2}$$

where $\mathbf{C}$ is an $r \times n$ matrix and $\mathbf{y}(k)$ and $\mathbf{v}(k)$ are $r$ vectors. Let us also assume that the input sequence $\mathbf{w}(k)$ is a purely random sequence (i.e., no correlation between time intervals), so we can write

$$E\{[\mathbf{w}(j) - \bar{\mathbf{w}}(j)][\mathbf{w}(k) - \bar{\mathbf{w}}(k)]^T\} = \begin{cases} \mathbf{0} & j \neq k \\ \mathbf{W}(k) & j = k \end{cases} \tag{10.10.3}$$

so $\mathbf{W}(k)$ is the covariance matrix of $\mathbf{w}(k)$. This process is sometimes referred to as *discrete white noise*, especially if $\mathbf{W}(k)$ is independent of $k$ or stationary.

The mean value of the response sequence is given simply by taking the expected value of (10.10.1) to yield

$$E[\mathbf{x}(k + 1)] = \mathbf{A}E[\mathbf{x}(k)] + \mathbf{B}E[\mathbf{w}(k)] \tag{10.10.4}$$

or if we denote $E[\mathbf{x}(k)] = \bar{\mathbf{x}}(k)$ and $E[\mathbf{w}(k)] = \bar{\mathbf{w}}(k)$, we get a difference equation for the mean response sequence

$$\bar{\mathbf{x}}(k+1) = \mathbf{A}\bar{\mathbf{x}}(k) + \mathbf{B}\bar{\mathbf{w}}(k) \tag{10.10.5}$$

This difference equation is subject to the initial condition $\bar{\mathbf{x}}(0) = E[\mathbf{x}(0)]$ and will simply be $\mathbf{x}(0)$ if $\mathbf{x}(0)$ is not a random variable.

Let us now examine the second-order properties of the sequence $\mathbf{x}(k)$. Transpose (10.10.1) to yield

$$\mathbf{x}^T(k+1) = \mathbf{x}^T(k)\mathbf{A}^T + \mathbf{w}^T(k)\mathbf{B}^T \tag{10.10.6}$$

Multiply (10.10.1) on the right by (10.10.6) to yield

$$\begin{aligned}\mathbf{x}(k+1)\mathbf{x}^T(k+1) &= \mathbf{A}\mathbf{x}(k)\mathbf{x}^T(k)\mathbf{A}^T + \mathbf{B}\mathbf{w}(k)\mathbf{w}^T(k)\mathbf{B}^T \\ &\quad + \mathbf{B}\mathbf{w}(k)\mathbf{x}^T(k)\mathbf{A}^T + \mathbf{A}\mathbf{x}(k)\mathbf{w}^T(k)\mathbf{B}^T\end{aligned} \tag{10.10.7}$$

We note that the expected value of the last two terms of this expression are zero and after subtracting out mean-value terms compatible with relation (10.10.5), we get

$$\mathbf{X}(k+1) = \mathbf{A}\mathbf{X}(k)\mathbf{A}^T + \mathbf{B}\mathbf{W}(k)\mathbf{B}^T \tag{10.10.8}$$

This is a set of $n^2$ linear difference equations that govern the propagation of the covariance matrix of the $\mathbf{x}(k)$ sequence given the covariance history of the forcing noise sequence $\mathbf{w}(k)$. This expression is commonly referred to as the *linear variance equation*.

Equation (10.10.8) can be solved recursively for the covariance if the covariance of the initial conditions is known [i.e., $X(0)$ is specified]:

$$\mathbf{X}(0) = E\{[\mathbf{x}(0) - \bar{\mathbf{x}}(0)][\mathbf{x}(0) - \bar{\mathbf{x}}(0)]^T\} \tag{10.10.9}$$

If the system output is as given by (10.10.2) and we also assume that $\mathbf{v}(k)$ is a purely random sequence with covariance $\mathbf{V}(k)$, the covariance of the output can be found as follows:

$$\begin{aligned}E[\mathbf{y}(k)\mathbf{y}^T(k)] &= E[\mathbf{C}\mathbf{x}(k)\mathbf{x}^T(k)\mathbf{C}^T + \mathbf{C}\mathbf{x}(k)\mathbf{v}^T(k) \\ &\quad + \mathbf{v}(k)\mathbf{v}^T(k) + \mathbf{v}(k)\mathbf{x}^T(k)C^T]\end{aligned} \tag{10.10.10}$$

and since $\mathbf{x}(k)$ and $\mathbf{v}(k)$ are independent, we get, after eliminating mean values from both sides,

$$\mathbf{Y}(k) = \mathbf{C}\mathbf{X}(k)\mathbf{C}^T + \mathbf{V}(k) \tag{10.10.11}$$

where $\mathbf{X}(k)$ is found by the solution of (10.10.8). If the elements of the $\mathbf{w}(k)$ sequence are Gaussian random variables and since by expression (10.10.1) the state is a linear combination of the past $\mathbf{w}(k)$ values, the states themselves will be Gaussian with the mean vector given by (10.10.5) and a covariance matrix given by (10.10.8).

**Example 10.3.** Consider the scalar plant of Examples 9.1 and 9.4 forced by a stationary, zero-mean random sequence $w(k)$, or

$$x(k + 1) = 0.8x(k) + w(k)$$

where the variance of the $w(k)$ sequence is

$$W = 0.5$$

If the mean of the initial condition is $\bar{x}(0)$ and $w(k)$ has zero mean, then from relation (10.10.5) the mean sequence is

$$\bar{x}(k) = (0.8)^k\bar{x}(0)$$

Now let us evaluate the variance-sequence by evaluation of the linear variance equation (Eq. 10.10.8)

$$X(k + 1) = 0.64X(k) + 0.5$$

where $X(0)$ is the variance of the initial condition, which is deterministic with value $x(0)$, so the initial condition variance is $X(0) = 0$. The mean and variance sequences are shown in Fig. 10.4. Note that the mean value of the state decays to zero because the system is stable and there is zero mean in the input. Also note that since the system is stable and the input sequence is stationary, the variance approaches a stationary value of 1.389.

**Example 10.4.** Consider the thermal control system with no heating and random fluctuations in the environmental temperature, which we shall term

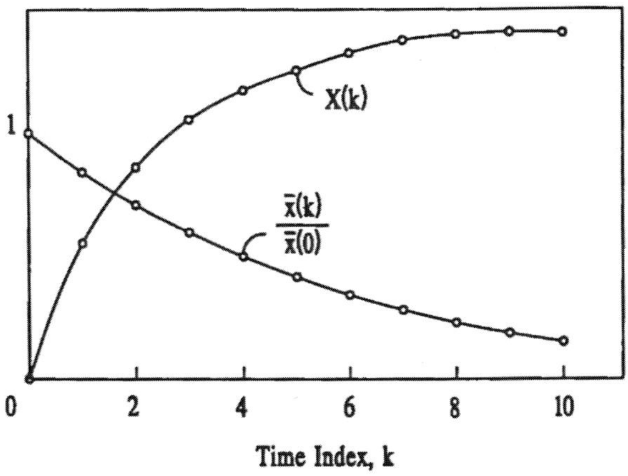


**Figure 10.4.** Mean and variance sequences for a first-order discrete-time system.

$w(k)$. The stochastic difference equation is

$$\begin{bmatrix} x_1 \\ x_2 \end{bmatrix}_{k+1} = \begin{bmatrix} 0.6277 & 0.3597 \\ 0.08993 & 0.8526 \end{bmatrix} \begin{bmatrix} x_1 \\ x_2 \end{bmatrix}_k + \begin{bmatrix} 0.0125 \\ 0.0575 \end{bmatrix} w(k)$$

Let us assume that the environmental temperature is stationary and has mean square value

$$E[w^2(k)] = W = 9(°\mathrm{C})^2$$

Let us assume that the initial temperatures are deterministic and zero, so the covariance matrix is

$$\mathbf{X}(0) = \begin{bmatrix} 0 & 0 \\ 0 & 0 \end{bmatrix}$$

and the covariance propagation equation is, from (10.10.8),

$$\begin{bmatrix} X_{11} & X_{12} \\ X_{21} & X_{22} \end{bmatrix}_{k+1} = \begin{bmatrix} 0.6277 & 0.3597 \\ 0.08993 & 0.8520 \end{bmatrix} \begin{bmatrix} X_{11} & X_{12} \\ X_{21} & X_{22} \end{bmatrix}_k \begin{bmatrix} 0.6277 & 0.08993 \\ 0.3597 & 0.8520 \end{bmatrix} + \begin{bmatrix} 0.0125 \\ 0.0575 \end{bmatrix} 9[0.0125 \quad 0.0575]$$

If we evaluate this covariance equation recursively for the covariances, we get the variances for the two chamber temperatures illustrated in Fig. 10.5.

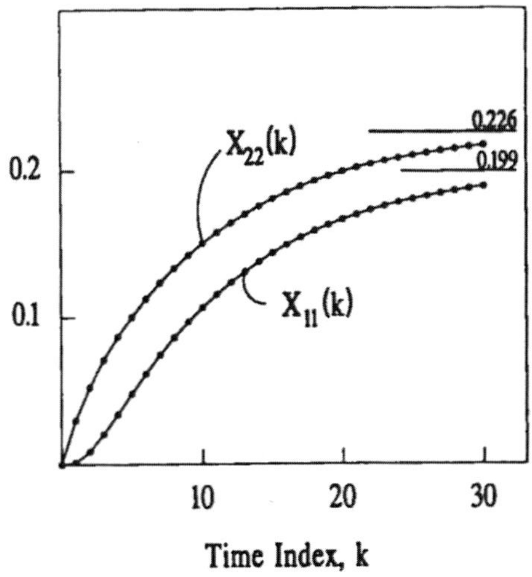


**Figure 10.5.** Variance sequences for a randomly disturbed system.

## 10.11 STATIONARY SOLUTIONS

Consider now the case where the sequence $w(k)$ is stationary, that is, the probabilistic structure is invariant under temporal translation and hence the covariance of $w(k)$ is independent of $k$, or

$$E\{[\mathbf{w}(k) - \overline{\mathbf{w}}(k)][\mathbf{w}(k) - \overline{\mathbf{w}}(k)]^T\} = \mathbf{W} \qquad \text{for all } k \tag{10.11.1}$$

Since the system of (10.10.1) is time invariant, it will have a stationary solution vector only if the system is stable or the eigenvalues of $\mathbf{A}$ lie inside the unit circle of the complex plane. The stationary solution will occur only after starting transients have decayed unless the initial condition vector is chosen from the same distribution as the stationary solution. If the $\mathbf{w}(k)$ are Gaussian random variables and since the $\mathbf{x}(k)$ are generated by linear combinations of these random variables, the covariance matrix and mean vector are sufficient to describe $\mathbf{x}(k)$ completely. If we are interested in the stationary covariance of the sequence, it is reasonable to assume in (10.10.8) that $\mathbf{X}(k + 1) = \mathbf{X}(k) = \mathbf{X}$ and $\mathbf{W}(k) = \mathbf{W}$, so (10.10.8) becomes

$$\mathbf{X} = \mathbf{AXA}^T + \mathbf{BWB}^T \tag{10.11.2}$$

This represents $n^2$ linear algebraic equations in the unknown $X_{ij}$, the elements of $\mathbf{X}$. Now let us consider the correlation between $\mathbf{x}_k$ and $\mathbf{x}_{k+m}$ $m$ steps later in time. From Eq. (6.5.5) we can form the product

$$\mathbf{x}(k + m)\mathbf{x}^T(k) = \mathbf{A}^m\mathbf{x}(k)\mathbf{x}^T(k) + \sum_{i=k}^{k+m-1} \mathbf{A}^{k+m-1-i}\mathbf{w}(i)\mathbf{x}^T(k) \tag{10.11.3}$$

The $\mathbf{w}(i)$ and $\mathbf{x}(k)$ for $i = k \ldots k + m - 1$ are uncorrelated, so the expected value is

$$E[\mathbf{x}(k + m)\mathbf{x}^T(k)] = \mathbf{A}^m\mathbf{X}(k) \tag{10.11.4}$$

where in this development we have assumed that the sequences all have zero mean.

**Example 10.5.** Consider the scalar example considered in Example 10.3. Let us seek the stationary variance sequence and the autocorrelation sequence for the system

$$x(k + 1) = 0.8x(k) + w(k)$$

where $E[w^2(k)] = 0.5$. If we examine the stationary form given by expression (10.11.2), the result is

$$X = 0.64X + 0.5$$

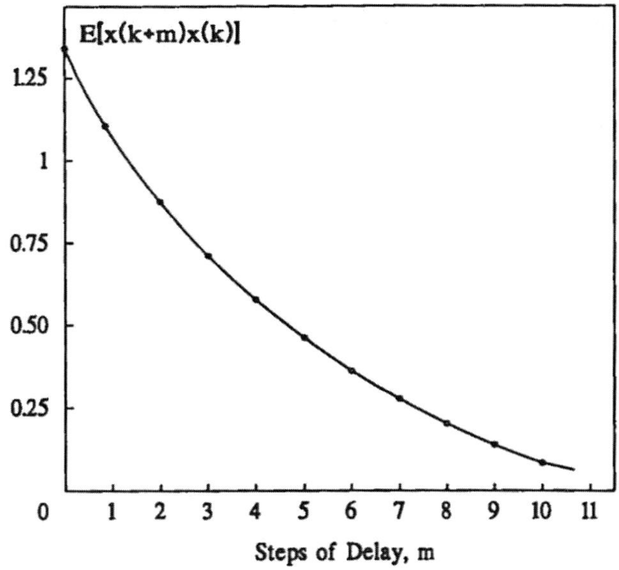


**Figure 10.6.** Stationary autocorrelation sequence for randomly excited first-order system.

Solving for the stationary solution yields

$$X = 1.389$$

which is the asymptotic solution given in Example 10.3 and illustrated in Fig. 10.4. The autocorrelation sequence of the stationary solution sequence is

$$E[x(k + m)x(k)] = (0.8)^m X = (0.8)^m(1.38)$$

which is illustrated in Fig. 10.6.

## 10.12 SUMMARY

In this chapter we have considered the problem of a linear difference system driven by an uncorrelated noise sequence. Also, the system outputs are considered to be a linear combination of the state variables and uncorrelated measurement noise. We have developed expressions for calculation of the statistical properties of the states and those of the measured system outputs for both transient problems and statistically stationary problems.

In the case where the input and measurement sequences are Gaussian, the output sequences are shown to be Gaussian and hence means and variances are sufficient to characterize these sequences.

These results will be employed in the next two chapters, where the state estimation and stochastic optimal control problems are discussed.

## PROBLEMS

10.1. Consider the following scalar stochastic discrete-time system: $x_{k+1} = 0.6x_k + w_k$, where the $w_k$ is a purely random stationary sequence with mean $\overline{w} = 1$ and variance $W = 4$. Assume also that the initial condition is random with mean value $\overline{x}_0 = 0$ and variance $X_0 = 2$.
   (a) With a hand calculator find the mean-value sequence $\overline{x}_k$ and the variance sequence $X_k$.
   (b) Continue the calculations for enough in time to show that the limiting values are $\overline{x}_\infty = 2.5$ and $X_\infty = 6.25$.
   (c) Solve the stationary mean and variance equations to show that the answers of part (b) are correct.

10.2. Investigate the mean and variance sequences for the discrete-time system $x_{k+1} = 1.2x_k + w_k$, where the sequence $w_k$ is stationary with mean $\overline{w} = 1$ and variance $W = 4$. Assume also that the initial condition is random with $\overline{x}_0 = 0$ and $X_0 = 2$.
   (a) Calculate the first 10 values of the mean sequence $\overline{x}_k$ and variance sequence $X_k$.
   (b) Can you draw any conclusion about the difference in the nature of the response of this system compared to that of Example 10.3?

10.3. Find the mean-square response sequences for the fluid-level control system for a sampling interval of $T = 0.1$ s which is governed by the vector difference equation

$$\begin{bmatrix} x_1 \\ x_2 \end{bmatrix}_{k+1} = \begin{bmatrix} 0.8336 & 0.1424 \\ 0.1424 & 0.6200 \end{bmatrix} \begin{bmatrix} x_1 \\ x_2 \end{bmatrix}_k + \begin{bmatrix} 0.0912 \\ 0.0080 \end{bmatrix} w(k)$$

where $w(k)$ is a zero-mean sequence with variance $W = 4$. The initial-state vector $\mathbf{x}_0$ will be assumed to be deterministic such that the initial-state covariance matrix $\mathbf{X}_0$ will be zero.

10.4. For the double-integrator inertial plant predict the response sequence statistics (mean and variance) for a stationary process noise sequence with mean value and variance $\overline{w} = E[w_k] = 0.5$, $E[(w_k$

$- \overline{w})^2] = 1$. The governing system equation for a sampling interval of $T = 0.2$ s is

$$\begin{bmatrix} x_1 \\ x_2 \end{bmatrix}_{k+1} = \begin{bmatrix} 1 & 0.2 \\ 0 & 1 \end{bmatrix} \begin{bmatrix} x_1 \\ x_2 \end{bmatrix}_k + \begin{bmatrix} 0.02 \\ 0.2 \end{bmatrix} w_k$$

Note the property of mean sequence and give a physical interpretation. Assume that both initial velocity and displacement are zero-mean random variables.

10.5. The second-order system comprising the dynamics of a printhead driven by a zero-order hold and followed by a sampler are described by the difference equation

$$\begin{bmatrix} x_1 \\ x_2 \end{bmatrix}_{k+1} = \begin{bmatrix} 0.569 & 0.0381 \\ -15.26 & 0.4164 \end{bmatrix} \begin{bmatrix} x_1 \\ x_2 \end{bmatrix}_k + \begin{bmatrix} 0.00216 \\ 0.0763 \end{bmatrix} w_k$$

For a zero-mean stationary random sequence $w_k$ with variance of $E[w_k^2] = 500$, find the response covariance sequences assuming that the initial covariance of the initial conditions is zero (deterministic initial states).

## REFERENCES

Brown, R. G., and P. Y. C. Hwang, 1992. *Introduction to Random Signals and Applied Kalman Filtering*, 2nd Ed., Wiley, New York.

Cooper, G. R., and C. D. McGillem, 1986. *Probabilistic Methods of Signal and System Analysis*, Holt, Rinehart and Winston, New York.

Feller, W., 1957. *An Introduction to Probability Theory and Its Applications*, 2nd Ed., Wiley, New York.

Kalman, R. E., 1960. A new approach to linear filtering and prediction problems, *Trans. ASME: Journal of Basic Engineering*, *82D*: 35–45.

Leon-Garcia, A., 1989. *Probability and Random Processes for Electrical Engineering*, Addison-Wesley, Reading, MA.

Papoulis, A., 1991. *Probability, Random Variables and Stochastic Processes*, 3rd Ed., McGraw-Hill, New York.

Peebles, P. A., Jr., 1987. *Probability, Random Variables, and Random Signal Principles*, 2nd Ed., McGraw-Hill, New York.

Williams, R. H., 1991. *Electrical Engineering Probability*, West, St. Paul, MN.

# 11
# State Estimation in the Presence of Noise

## 11.1 INTRODUCTION

The technique of estimating the parameters associated with physical phenomena based on inaccurate measurements is not new and certainly goes back in time to the effort of Gauss (1809) to estimate planetary orbit parameters based on data taken from many observatories across Europe. Gauss postulated that positive and negative errors could be penalized equally if a quadratic measure of error was employed. He also found that the best previous estimate should be updated by weighing current data according to the confidence in the accuracy of the current measurements. This is the essence of the Kalman state-estimation technique which is developed in this chapter.

In Chapters 8 and 9 the modern control theory approach was developed for the design of digital control systems. In Chapter 8 the problem was approached from the closed-loop pole-placement point of view, while in Chapter 9 the optimal state-feedback point of view was taken. In Chapter 8 state estimators were discussed in the absence of contaminating noise sources and design was accomplished by pole placement techniques. It has been shown (Kalman, 1960; Kalman and Bucy, 1961) that for optimal estimation of the state, the noisy character of signals must be considered. If the control system is to regulate properly in the presence of random disturbances and random measurement noise, it must be designed without excessive bandwidth in the closed loop, thus avoiding the propagation of noise around the loop.

The system we will be concerned with is shown in Fig. 11.1. The

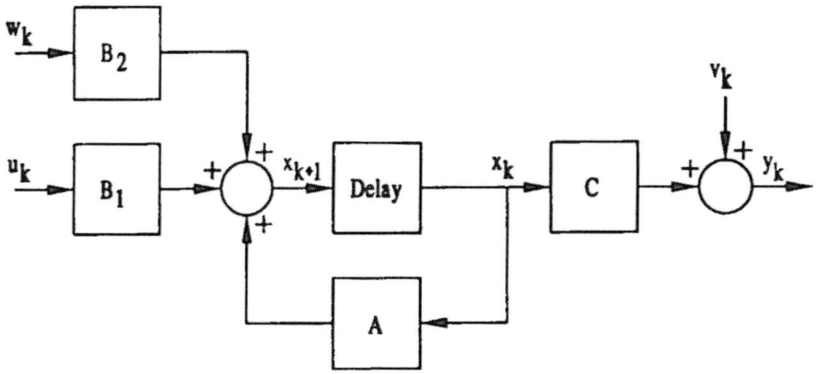


**Figure 11.1.** Linear discrete-time system with random process and measurement noises.

discrete-time dynamic system is given by

$$\mathbf{x}_{k+1} = \mathbf{A}\mathbf{x}_k + \mathbf{B}_1\mathbf{u}_k + \mathbf{B}_2\mathbf{w}_k \tag{11.1.1}$$

where $\mathbf{u}_k$ is deterministic discrete-time control effort and $\mathbf{w}_k$ is a discrete-time purely random disturbance sequence which is commonly referred to as the *process noise*. The matrices $\mathbf{A}$, $\mathbf{B}_1$, and $\mathbf{B}_2$ are constant (time-invariant), although this need not be a limitation (Bryson and Ho, 1975). The output sequence vector is a linear combination of the states contaminated by additive noise:

$$\mathbf{y}_k = \mathbf{C}\mathbf{x}_k + \mathbf{v}_k \tag{11.1.2}$$

The sequence $\mathbf{v}_k$ is also a purely random sequence and is commonly referred to as the *measurement noise* sequence. The random sequences $\mathbf{w}_k$ and $\mathbf{v}_k$ are assumed to be statistically independent, and the matrix $\mathbf{C}$ associated with the output mapping is also constant.

## 11.2 DERIVATION OF THE DISCRETE-TIME VECTOR KALMAN FILTER

Let us consider the system of expression (11.1.1), but for the purposes of this chapter, let us neglect the control effort sequence. The nonzero control effort sequence is considered in Chapter 12. The system of interest will be governed by the stochastic matrix-vector difference equation

$$\mathbf{x}_{k+1} = \mathbf{A}\mathbf{x}_k + \mathbf{B}\mathbf{w}_k \qquad k = 0, 1, 2, \ldots \tag{11.2.1}$$

with a measured output vector given a linear combination of the state contaminated with additive measurement noise:

$$\mathbf{y}_k = \mathbf{C}\mathbf{x}_k + \mathbf{v}_k \tag{11.2.2}$$

where the matrices **A**, **B**, and **C** are, respectively, $n \times n$, $n \times m$, and $r \times n$. The matrices **A**, **B**, and **C** will be assumed to be time-invariant (independent of $k$), but this need not be a limitation. The process noise $\mathbf{w}_k$ and measurement noise $\mathbf{v}_k$ are zero-mean Gaussian purely random sequences which will be assumed to be stationary, but this also need not be a limitation. The respective covariances of these sequences are

$$E[\mathbf{w}_j\mathbf{w}_k^T\} = \begin{cases} \mathbf{W} & j = k \\ \mathbf{0} & j \neq k \end{cases} \tag{11.2.3}$$

$$E[\mathbf{v}_j\mathbf{v}_k^T] = \begin{cases} \mathbf{V} & j = k \\ \mathbf{0} & j \neq k \end{cases} \tag{11.2.4}$$

where pure randomness implies the noncorrelation between intervals of time. We shall assume that the $\mathbf{w}_j$ are independent of the $\mathbf{v}_j$.

The problem to be examined here is that of obtaining the best estimate of $\mathbf{x}_k$ in terms of past and current measurements $\mathbf{y}_0, \mathbf{y}_1, \ldots, \mathbf{y}_k$. Denote this data set by

$$\mathbf{Y}_k = \{\mathbf{y}_0, \mathbf{y}_1, \ldots, \mathbf{y}_k\} \tag{11.2.5}$$

and the estimation error vector as

$$\tilde{\mathbf{x}}_k = \mathbf{x}_k - \hat{\mathbf{x}}_k \tag{11.2.6}$$

We would like to choose $\hat{\mathbf{x}}_k$ such that the conditional mean-square error will be minimized, or the minimization of the scalar objective function

$$J = \frac{1}{2} E[\tilde{\mathbf{x}}_k^T\tilde{\mathbf{x}}_k|\mathbf{Y}_k] \tag{11.2.7}$$

Define now the conditional mean as

$$\boldsymbol{\mu}_k = E[\mathbf{x}_k|\mathbf{Y}_k] \tag{11.2.8}$$

and the conditional covariance matrix as

$$\mathbf{Z}_k = E[(\mathbf{x}_k - \boldsymbol{\mu}_k)(\mathbf{x}_k - \boldsymbol{\mu}_k)^T|\mathbf{Y}_k] \tag{11.2.9}$$

Substitute (11.2.6) into (11.2.7) to yield a cost function

$$J = \frac{1}{2} E[(\mathbf{x}_k - \hat{\mathbf{x}}_k)^T(\mathbf{x}_k - \hat{\mathbf{x}}_k)|\mathbf{Y}_k] \tag{11.2.10}$$

Now let us expand this to yield

$$J = \frac{1}{2} E[\mathbf{x}_k^T\mathbf{x}_k - \hat{\mathbf{x}}_k^T\mathbf{x}_k - \mathbf{x}_k^T\hat{\mathbf{x}}_k + \hat{\mathbf{x}}_k^T\hat{\mathbf{x}}_k|\mathbf{Y}_k] \tag{11.2.11}$$

Since the expectation operator, $E[\cdot]$, is a linear operator, we may take the expected value term by term (retaining the condition) to yield

$$J = \frac{1}{2}\{E[\mathbf{x}_k^T\mathbf{x}_k|\mathbf{Y}_k] - \hat{\mathbf{x}}_k^T E[\mathbf{x}_k|\mathbf{Y}_k] - E[\mathbf{x}_k^T|\mathbf{Y}_k]\hat{\mathbf{x}}_k + \hat{\mathbf{x}}_k^T\hat{\mathbf{x}}_k\} \tag{11.2.12}$$

since $\hat{\mathbf{x}}_k$, the estimate, is a fixed vector of numbers and not random variables. Now minimize $J$ with respect to the estimate $\hat{\mathbf{x}}_k$, or

$$\frac{\partial J}{\partial \hat{\mathbf{x}}_k} = \frac{1}{2}\{- E[\mathbf{x}_k^T|\mathbf{Y}_k] - E[\mathbf{x}_k^T|\mathbf{Y}_k] + 2\hat{\mathbf{x}}_k^T\} = 0 \tag{11.2.13}$$

Recalling the definition in (11.2.8) yields

$$\hat{\mathbf{x}}_k = E[\mathbf{x}_k|\mathbf{Y}_k] = \boldsymbol{\mu}_k \tag{11.2.14}$$

Thus, given the data set $\mathbf{Y}_k$, the best least-squares estimate of $\mathbf{x}_k$ is a conditional mean. Expression (11.2.9), which defines covariance $\mathbf{Z}_k$, can be expanded to yield

$$\mathbf{Z}_k = E[\mathbf{x}_k\mathbf{x}_k^T|\mathbf{Y}_k] - \boldsymbol{\mu}_k\boldsymbol{\mu}_k^T \tag{11.2.15}$$

so the conditional expectation is

$$E[\mathbf{x}_k\mathbf{x}_k^T|\mathbf{Y}_k] = \mathbf{Z}_k + \boldsymbol{\mu}_k\boldsymbol{\mu}_k^T \tag{11.2.16}$$

We may write a relationship between conditional densities using the law of conditional probability,

$$p(\mathbf{x}_{k+1}|\mathbf{Y}_{k+1}) = p(\mathbf{x}_{k+1}|\mathbf{Y}_k, \mathbf{y}_{k+1}) \tag{11.2.17}$$

Substituting the conditional law yields

$$p(\mathbf{x}_{k+1}|\mathbf{Y}_{k+1}) = \frac{p(\mathbf{x}_{k+1}, \mathbf{Y}_k, \mathbf{y}_{k+1})}{p(\mathbf{y}_{k+1}, \mathbf{Y}_k)} \tag{11.2.18}$$

and breaking it down further gives

$$p(\mathbf{x}_{k+1}|\mathbf{Y}_{k+1}) = \frac{p(\mathbf{y}_{k+1}|\mathbf{Y}_k, \mathbf{x}_{k+1})p(\mathbf{Y}_k, \mathbf{x}_{k+1})}{p(\mathbf{y}_{k+1}|\mathbf{Y}_k)p(\mathbf{Y}_k)} \tag{11.2.19}$$

$$= \frac{p(\mathbf{y}_{k+1}|\mathbf{Y}_k, \mathbf{x}_{k+1})p(\mathbf{x}_{k+1}|\mathbf{Y}_k)p(\mathbf{Y}_k)}{p(\mathbf{y}_{k+1}|\mathbf{Y}_k)p(\mathbf{Y}_k)} \tag{11.2.20}$$

so

$$p(\mathbf{x}_{k+1}|\mathbf{Y}_{k+1}) = \frac{p(\mathbf{y}_{k+1}|\mathbf{Y}_k, \mathbf{x}_{k+1})p(\mathbf{x}_{k+1}|\mathbf{Y}_k)}{p(\mathbf{y}_{k+1}|\mathbf{Y}_k)} \tag{11.2.21}$$

This may be recognized as the statement of Bayes' theorem for probability densities. If $\mathbf{w}_k$ and $\mathbf{v}_k$ are Gaussian and since $\mathbf{x}_k$ and $\mathbf{y}_k$ are linear combinations of $\mathbf{w}_k$ and $\mathbf{v}_k$, they will also be Gaussian, so substituting the appropriate densities on both sides of (11.2.21), the result is

$$\frac{1}{(2\pi)^{n/2}|\mathbf{Z}_{k+1}|^{1/2}} \exp\left[-\frac{1}{2}(\mathbf{x}_{k+1} - \boldsymbol{\mu}_{k+1})^T\mathbf{Z}_{k+1}^{-1}(\mathbf{x}_{k+1} - \boldsymbol{\mu}_{k+1})\right]$$
$$= \frac{|\mathbf{S}_2|^{1/2}}{(2\pi)^{n/2}|\mathbf{S}_1|^{1/2}|\mathbf{S}_3|^{1/2}}$$
$$\times \exp\left\{-\frac{1}{2}[(\mathbf{x}_{k+1} - \boldsymbol{\mu}_1)^T\mathbf{S}_1^{-1}(\mathbf{x}_{k+1} - \boldsymbol{\mu}_1)\right.$$
$$- (\mathbf{y}_{k+1} - \boldsymbol{\mu}_2)^T\mathbf{S}_2^{-1}(\mathbf{y}_{k+1} - \boldsymbol{\mu}_2)$$
$$\left. + (\mathbf{y}_{k+1} - \boldsymbol{\mu}_3)^T\mathbf{S}_3^{-1}(\mathbf{y}_{k+1} - \boldsymbol{\mu}_3)]\right\} \tag{11.2.22}$$

where the respective mean vectors are defined as

$$\boldsymbol{\mu}_1 \triangleq E[\mathbf{x}_{k+1}|\mathbf{Y}_k] \tag{11.2.23}$$
$$\boldsymbol{\mu}_2 \triangleq E[\mathbf{y}_{k+1}|\mathbf{Y}_k] \tag{11.2.24}$$
$$\boldsymbol{\mu}_3 \triangleq E[\mathbf{y}_{k+1}|\mathbf{Y}_k, \mathbf{x}_{k+1}] \tag{11.2.25}$$

and the respective covariance matrices are

$$\mathbf{S}_1 = E[(\mathbf{x}_{k+1} - \boldsymbol{\mu}_1)(\mathbf{x}_{k+1} - \boldsymbol{\mu}_1)^T|\mathbf{Y}_k] \tag{11.2.26}$$
$$\mathbf{S}_2 = E[(\mathbf{y}_{k+1} - \boldsymbol{\mu}_2)(\mathbf{y}_{k+1} - \boldsymbol{\mu}_2)^T|\mathbf{Y}_k] \tag{11.2.27}$$

and

$$\mathbf{S}_3 = E[(\mathbf{y}_{k+1} - \boldsymbol{\mu}_3)(\mathbf{y}_{k+1} - \boldsymbol{\mu}_3)^T|\mathbf{Y}_k, \mathbf{x}_{k+1}] \tag{11.2.28}$$

Now we need to evaluate these means and covariances to yield

$$\boldsymbol{\mu}_1 = E[\mathbf{x}_{k+1}|\mathbf{Y}_k] = E[\mathbf{A}\mathbf{x}_k + \mathbf{B}\mathbf{w}_k|\mathbf{Y}_k] \tag{11.2.29}$$
$$= \mathbf{A}E[\mathbf{x}_k|\mathbf{Y}_k] + \mathbf{B}E[\mathbf{w}_k|\mathbf{y}_k] \tag{11.2.30}$$

and since $\mathbf{w}_k$ is independent of the $k$th and all previous outputs, the last term is zero, so the remaining term is $\mathbf{A}$ times the optimal estimate of $\mathbf{x}_k$, as defined in (11.2.14):

$$\boldsymbol{\mu}_1 = \mathbf{A}\boldsymbol{\mu}_k = \mathbf{A}\hat{\mathbf{x}}_k \tag{11.2.31}$$

We should now evaluate the covariance $\mathbf{S}_1$, so

$$\mathbf{S}_1 = E[(\mathbf{x}_{k+1} - \mathbf{A}\hat{\mathbf{x}}_k)(\mathbf{x}_{k+1} - \mathbf{A}\hat{\mathbf{x}}_k)^T|\mathbf{Y}_k] \tag{11.2.32}$$

Let us define $\mathbf{M}_{k+1} = \mathbf{S}_1$:

$$\mathbf{M}_{k+1} = E[(\mathbf{A}\mathbf{x}_k + \mathbf{B}\mathbf{w}_k - \mathbf{A}\hat{\mathbf{x}}_k)(\mathbf{A}\mathbf{x}_k + \mathbf{B}\mathbf{w}_k - \mathbf{A}\hat{\mathbf{x}}_k)^T|\mathbf{Y}_k] \tag{11.2.33}$$

$$= E\{[\mathbf{A}(\mathbf{x}_k - \hat{\mathbf{x}}_k) + \mathbf{B}\mathbf{w}_k][\mathbf{A}(\mathbf{x}_k - \hat{\mathbf{x}}_k) + \mathbf{B}\mathbf{w}_k]^T|\mathbf{Y}_k\} \tag{11.2.34}$$

But we previously defined a conditional covariance matrix in expression (11.2.15) as

$$\mathbf{Z}_k = E[(\mathbf{x}_k - \hat{\mathbf{x}}_k)(\mathbf{x}_k - \hat{\mathbf{x}}_k)^T|\mathbf{Y}_k] \tag{11.2.35}$$

so (11.2.34) may be written noting that $(\mathbf{x}_k - \hat{\mathbf{x}}_k)$ is independent of $\mathbf{w}_k$, or

$$\mathbf{M}_{k+1} = \mathbf{S}_1 = \mathbf{A}\mathbf{Z}_k\mathbf{A}^T + \mathbf{B}\mathbf{W}\mathbf{B}^T \tag{11.2.36}$$

where $\mathbf{W}$ is the covariance matrix of the $\mathbf{w}_k$ sequence. Let us continue to evaluate the parameters in expressions (11.2.23) to (11.2.28):

$$\boldsymbol{\mu}_2 = E[\mathbf{y}_{k+1}|\mathbf{Y}_k] = E[\mathbf{C}\mathbf{x}_{k+1} + \mathbf{v}_{k+1}|\mathbf{Y}_k] \tag{11.2.37}$$

$$= E[\mathbf{C}(\mathbf{A}\mathbf{x}_k + \mathbf{B}\mathbf{w}_k) + \mathbf{v}_{k+1}|\mathbf{Y}_k] \tag{11.2.38}$$

so

$$\boldsymbol{\mu}_2 = \mathbf{C}\mathbf{A}\boldsymbol{\mu}_k = \mathbf{C}\boldsymbol{\mu}_1 = \mathbf{C}\mathbf{A}\hat{\mathbf{x}}_k \tag{11.2.39}$$

Let us now evaluate the second covariance $\mathbf{S}_2$:

$$\mathbf{S}_2 = E[(\mathbf{y}_{k+1} - \boldsymbol{\mu}_2)(\mathbf{y}_{k+1} - \boldsymbol{\mu}_2)^T|\mathbf{Y}_k] \tag{11.2.40}$$

Expanding this quantity gives

$$\mathbf{S}_2 = E[(\mathbf{C}\mathbf{x}_{k+1} + \mathbf{v}_{k+1} - \mathbf{C}\boldsymbol{\mu}_1)(\mathbf{x}_{k+1}^T\mathbf{C}^T + \mathbf{v}_{k+1}^T - \boldsymbol{\mu}_1^T\mathbf{C}^T)|\mathbf{Y}_k] \tag{11.2.41}$$

or

$$\begin{aligned}\mathbf{S}_2 = {} & \mathbf{C}E[(\mathbf{x}_{k+1} - \boldsymbol{\mu}_1)(\mathbf{x}_{k+1} - \boldsymbol{\mu}_1)^T|\mathbf{Y}_k]\mathbf{C}^T \\ & + E[\mathbf{v}_{k+1}\mathbf{v}_{k+1}^T|\mathbf{Y}_k] + \mathbf{C}\overbrace{E[(\mathbf{x}_{k+1} - \boldsymbol{\mu}_1)\mathbf{v}_{k+1}^T|\mathbf{Y}_k]}^{0} \\ & + \underbrace{E[\mathbf{v}_{k+l}(\mathbf{x}_{k+1} - \boldsymbol{\mu}_1)^T|\mathbf{Y}_k]}_{0}\mathbf{C}^T\end{aligned} \tag{11.2.42}$$

Then

$$\mathbf{S}_2 = \mathbf{C}\mathbf{S}_1\mathbf{C}^T + \mathbf{V} = \mathbf{C}\mathbf{M}_{k+1}\mathbf{C}^T + \mathbf{V} \tag{11.2.43}$$

Now examine the density $p(\mathbf{y}_{k+1}|\mathbf{Y}_k, \mathbf{x}_{k+1})$. Since $\mathbf{y}_{k+1}$ depends only on $\mathbf{x}_{k+1}$, we write

$$p(\mathbf{y}_{k+1}|\mathbf{Y}_k, \mathbf{x}_{k+1}) = p(\mathbf{y}_{k+1}|\mathbf{x}_{k+1}) \tag{11.2.44}$$

Using this fact, we can now evaluate $\boldsymbol{\mu}_3$:

$$\boldsymbol{\mu}_3 = E[\mathbf{y}_{k+1}|\mathbf{x}_{k+1}] = E\{[\mathbf{C}\mathbf{x}_{k+1} + \mathbf{v}_{k+1}]|\mathbf{x}_{k+1}\} \tag{11.2.45}$$

$$= \mathbf{C}\mathbf{x}_{k+1} \tag{11.2.46}$$

and now let us evaluate $\mathbf{S}_3$:

$$\mathbf{S}_3 = E[(\mathbf{y}_{k+1} - \boldsymbol{\mu}_3)(\mathbf{y}_{k+1} - \boldsymbol{\mu}_3)^T|\mathbf{x}_{k+1}] \tag{11.2.47}$$

$$= E[\mathbf{C}\mathbf{x}_{k+1} + \mathbf{v}_{k+1} - \boldsymbol{\mu}_3)(\mathbf{C}\mathbf{x}_{k+1} + \mathbf{v}_{k+1} - \boldsymbol{\mu}_3)^T|\mathbf{x}_{k+1}] \tag{11.2.48}$$

$$\begin{aligned}
&= \mathbf{C}E[\mathbf{x}_{k+1}\mathbf{x}_{k+1}^T|\mathbf{x}_{k+1}]\mathbf{C}^T + \mathbf{C}E[\mathbf{x}_{k+1}\mathbf{v}_{k+1}^T|\mathbf{x}_{k+1}] \\
&\quad + E[\mathbf{v}_{k+1}\mathbf{x}_{k+1}^T|\mathbf{x}_{k+1}]\mathbf{C}^T - \boldsymbol{\mu}_3 E[\mathbf{x}_{k+1}^T\mathbf{C}^T|\mathbf{x}_{k+1}] \\
&\quad - E[\mathbf{C}\mathbf{x}_{k+1}|\mathbf{x}_{k+1}]\boldsymbol{\mu}_3^T + E[\mathbf{v}_{k+1}\mathbf{v}_{k+1}^T|\mathbf{x}_{k+1}] \\
&\quad - E[\mathbf{v}_{k+1}|\mathbf{x}_{k+1}]\boldsymbol{\mu}_3^T - \boldsymbol{\mu}_3 E[\mathbf{v}_{k+1}^T|\mathbf{x}_{k+1}] \\
&\quad + E[\boldsymbol{\mu}_3\boldsymbol{\mu}_3^T|\mathbf{x}_{k+1}]
\end{aligned} \tag{11.2.49}$$

Carrying out the indicated expectations, noting the definition of $\boldsymbol{\mu}_3$ yields

$$\mathbf{S}_3 = \mathbf{V} \tag{11.2.50}$$

Now substitute all these values into the right side of (11.2.22) and consider only the argument of the exponential:

$$\begin{aligned}
\text{Arg} = -\frac{1}{2}\{&(\mathbf{x}_{k+1} - \mathbf{A}\hat{\mathbf{x}}_k)^T\mathbf{M}_{k+1}^{-1}(\mathbf{x}_{k+1} - \mathbf{A}\hat{\mathbf{x}}_k) \\
&+ (\mathbf{y}_{k+1} - \mathbf{C}\mathbf{x}_{k+1})^T\mathbf{V}^{-1}(\mathbf{y}_{k+1} - \mathbf{C}\mathbf{x}_{k+1}) \\
&- (\mathbf{y}_{k+1} - \mathbf{C}\mathbf{A}\hat{\mathbf{x}}_k)^T[\mathbf{C}\mathbf{M}_{k+1}\mathbf{C}^T + \mathbf{V}]^{-1}(\mathbf{y}_{k+1} - \mathbf{C}\mathbf{A}\hat{\mathbf{x}}_k)\}
\end{aligned} \tag{11.2.51}$$

Carefully rewriting this expression gives

$$\begin{aligned}
\text{Arg} = -\frac{1}{2}\{&(\mathbf{x}_{k+1} - \mathbf{A}\hat{\mathbf{x}}_k)^T[\mathbf{M}_{k+1}^{-1} + \mathbf{C}^T\mathbf{V}^{-1}\mathbf{C}](\mathbf{x}_{k+1} - \mathbf{A}\hat{\mathbf{x}}_k) \\
&+ (\mathbf{y}_{k+1} - \mathbf{C}\mathbf{A}\hat{\mathbf{x}}_k)^T[\mathbf{V}^{-1} - (\mathbf{C}\mathbf{M}_{k+1}\mathbf{C}^T + \mathbf{V})^{-1}](\mathbf{y}_{k+1} - \mathbf{C}\mathbf{A}\hat{\mathbf{x}}_k) \\
&- (\mathbf{y}_{k+1} - \mathbf{C}\mathbf{A}\hat{\mathbf{x}}_k)^T\mathbf{V}^{-1}\mathbf{C}(\mathbf{x}_{k+1} - \mathbf{A}\hat{\mathbf{x}}_k) \\
&- (\mathbf{x}_{k+1} - \mathbf{A}\hat{\mathbf{x}}_k)^T\mathbf{C}^T\mathbf{V}^{-1}(\mathbf{y}_{k+1} - \mathbf{C}\mathbf{A}\hat{\mathbf{x}}_k)\}
\end{aligned} \tag{11.2.52}$$

Now complete the squares in the argument to yield

$$\begin{aligned}\text{Arg} = \frac{1}{2}\{[\mathbf{x}_{k+1} - \mathbf{A}\hat{\mathbf{x}}_k - \mathbf{Z}_{k+1}\mathbf{C}^T\mathbf{V}^{-1}(\mathbf{y}_{k+1} \\ - \mathbf{C}\mathbf{A}\hat{\mathbf{x}}_k)]^T[\mathbf{M}_{k+1}^{-1} + \mathbf{C}^T\mathbf{V}^{-1}\mathbf{C}] \\ \times [\mathbf{x}_{k+1} - \mathbf{A}\hat{\mathbf{x}}_k - \mathbf{Z}_{k+1}\mathbf{C}^T\mathbf{V}^{-1}(\mathbf{y}_{k+1} - \mathbf{C}\mathbf{A}\hat{\mathbf{x}}_k)]\}\end{aligned} \quad (11.2.53)$$

Now equate the means and covariances from the arguments of (11.2.53) and the left side of (11.2.22):

$$\boldsymbol{\mu}_{k+1} = \hat{\mathbf{x}}_{k+1} = \mathbf{A}\hat{\mathbf{x}}_k + \mathbf{Z}_{k+1}\mathbf{C}^T\mathbf{V}^{-1}(\mathbf{y}_{k+1} - \mathbf{C}\mathbf{A}\hat{\mathbf{x}}_k) \quad (11.2.54)$$

and the covariance

$$\mathbf{Z}_{k+1} = [\mathbf{M}_{k+1}^{-1} + \mathbf{C}^T\mathbf{V}^{-1}\mathbf{C}]^{-1} \quad (11.2.55)$$

Computation of $\mathbf{Z}_{k+1}$ involves two matrix inversions, both of size $n \times n$. The matrix inversion lemma proven in Appendix C allows for efficient computation of $\mathbf{Z}_{k+1}$ and the result is

$$\mathbf{Z}_{k+1} = \mathbf{M}_{k+1} - \mathbf{M}_{k+1}\mathbf{C}^T(\mathbf{V} + \mathbf{C}\mathbf{M}_{k+1}\mathbf{C}^T)^{-1}\mathbf{C}\mathbf{M}_{k+1} \quad (11.2.56)$$

This is the coefficient of the last term of (11.2.54), given the symbol $\mathbf{K}_{k+1}$ (K for Kalman), or

$$\mathbf{K}_{k+1} = \mathbf{Z}_{k+1}\mathbf{C}^T\mathbf{V}^{-1} = \mathbf{M}_{k+1}\mathbf{C}^T(\mathbf{V} + \mathbf{C}\mathbf{M}_{k+1}\mathbf{C}^T)^{-1} \quad (11.2.57)$$

and this matrix is commonly referred to as the *Kalman filter gain matrix*.

The value $\bar{\mathbf{x}}_0$ is the expected value of $\mathbf{x}_0$ conditional on no data for the previous data point ($k = -1$), and $\mathbf{M}_0$ is the covariance of the estimate of the initial condition based on no measurements or

$$\bar{\mathbf{x}}_0 = E[\mathbf{x}_0] \quad (11.2.58)$$

and

$$\mathbf{M}_0 = E[(\mathbf{x}_0 - \bar{\mathbf{x}}_0)(\mathbf{x}_0 - \bar{\mathbf{x}}_0)^T] \quad (11.2.59)$$

The filtering algorithm can now be started using $\mathbf{M}_0$ as an initial condition. Note that the quantity $\mathbf{A}\hat{\mathbf{x}}_k$ appears in expression (11.2.54) and that from (11.2.31) it is $\boldsymbol{\mu}_1$, which is the best least-squares estimate of the $(k + 1)$st state conditional on the data set $\mathbf{Y}_k$. We shall refer to this as $\bar{\mathbf{x}}_{k+1}$. Now let us write out the complete algorithm as

$$\bar{\mathbf{x}}_k = \mathbf{A}\hat{\mathbf{x}}_{k-1} \quad (11.2.60)$$

and the updated estimate as

$$\hat{\mathbf{x}}_k = \bar{\mathbf{x}}_k + \mathbf{K}_k(\mathbf{y}_k - \mathbf{C}\bar{\mathbf{x}}_k) \quad (11.2.61)$$

where the Kalman filter gain matrix is calculated sequentially from the following matrix difference equations:

$$\mathbf{Z}_k = \mathbf{M}_k - \mathbf{M}_k\mathbf{C}^T(\mathbf{C}\mathbf{M}_k\mathbf{C}^T + \mathbf{V})^{-1}\mathbf{C}\mathbf{M}_k \tag{11.2.62}$$

$$\mathbf{M}_k = \mathbf{A}\mathbf{Z}_{k-1}\mathbf{A}^T + \mathbf{B}\mathbf{W}\mathbf{B}^T \tag{11.2.63}$$

and

$$\mathbf{K}_k = \mathbf{Z}_k\mathbf{C}^T\mathbf{V}^{-1} \tag{11.2.64}$$

The initial values of the estimate $\bar{\mathbf{x}}_0$ and covariance $\mathbf{M}_0$ are

$$\bar{\mathbf{x}}_0 = E[\mathbf{x}_0] \tag{11.2.65}$$

and

$$\mathbf{M}_0 = E[(\mathbf{x}_0 - \bar{\mathbf{x}}_0)(\mathbf{x}_0 - \bar{\mathbf{x}}_0)^T] \tag{11.2.66}$$

This algorithm is shown schematically in Fig. 11.2.

It is interesting to look back at the definitions of $\mathbf{M}_k$ and $\mathbf{Z}_k$. From expression (11.2.32) it is clear that $\mathbf{M}_k$ is the covariance of the best least-

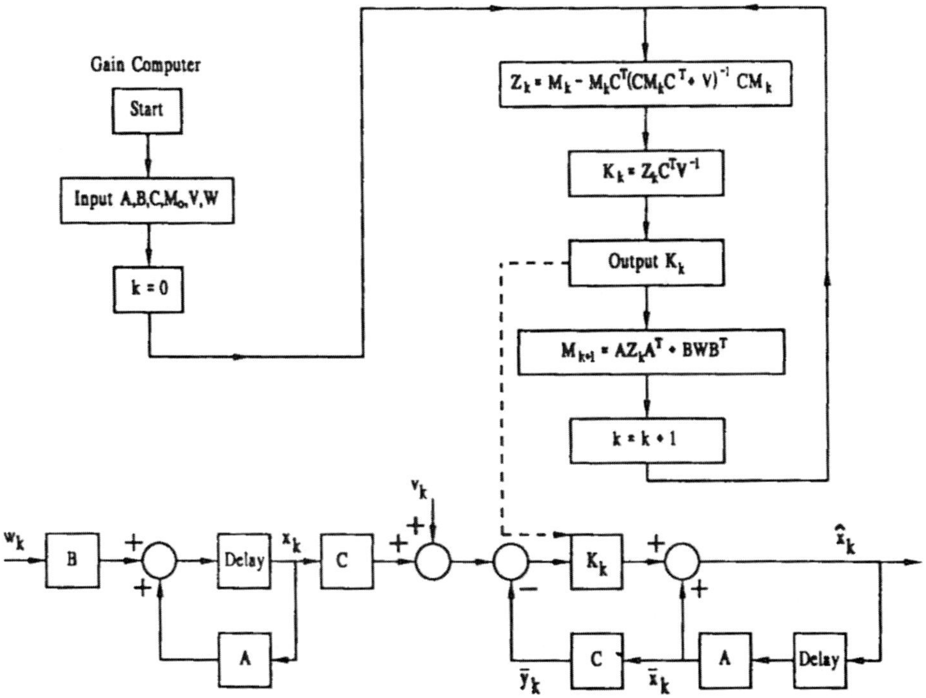


**Figure 11.2.** Vector Kalman filter.

squares estimate of $\mathbf{x}_k$ based on the measurement set $\mathbf{Y}_{k-1}$. Also from expression (11.2.15) it is apparent that $\mathbf{Z}_k$ is the covariance of the best least-squares estimate of $\mathbf{x}_k$ based on the measurement set $\mathbf{Y}_k$. Since these matrices are covariances, they must be positive semidefinite. If we examine the relationship between $\mathbf{Z}_k$ and $\mathbf{M}_k$ as given by (11.2.62), it indicates that the values in $\mathbf{Z}_k$ are smaller than the corresponding elements in $\mathbf{M}_k$, thus reflecting the improvement in the estimate of $\mathbf{x}_k$ due to the information in the measurement $\mathbf{y}_k$.

**Example 11.1.** Let us consider the scalar system with process noise sequence $w_k$ and measurement noise sequence $v_k$, where the plant is

$$x_{k+1} = ax_k + w_k$$

and the measurements are given by

$$y_k = x_k + v_k$$

where the process and measurement sequences have zero mean and variances given by

$$E[w_k^2] = W, \qquad E[v_k^2] = V$$

The Kalman filter for estimating the single-state variable is

$$\hat{x}_{k+1} = a\hat{x}_k + K_{k+1}(y_{k+1} - a\hat{x}_k)$$

where

$$K_k = \frac{Z_k}{V}$$

and the postmeasurement estimate covariance is

$$Z_k = M_k - \frac{M_k^2}{M_k + V}$$

and the premeasurement estimate covariance is

$$M_k = a^2 Z_{k-1} + W$$

The filter gain sequence is given by solving those equations simultaneously. Sequentially solving these two expressions forward in time eventually yields a set of stationary solutions that can be given by letting $Z_{k-1} = Z_k = Z$ and $M_k = M$. The stationary filter gain for various values of the parameter $a$ as a function of the ratio of mean-square process noise to mean-square measurement noise is given in Fig. 11.3.

If we fix the system parameter at $a = 0.8$ and choose the various noises to have the variances $V = 1$ and $W = 0.5$ and evaluate the transient filter

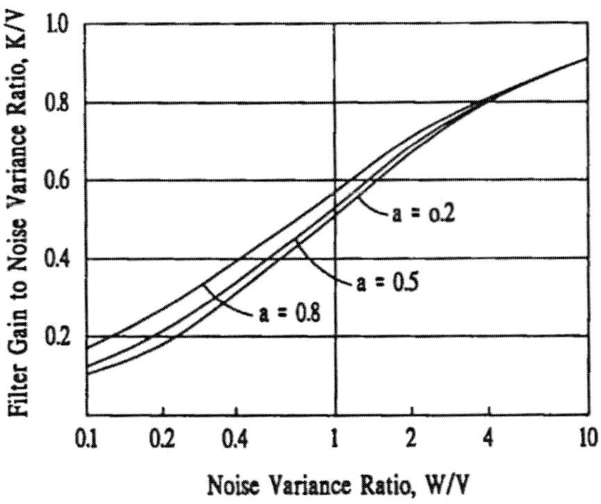


**Figure 11.3.** Filter gain/measurement noise variance ratio as a function of process/measurement noise ratio.

covariance sequences and the transient filter gain, we get the data of Fig. 11.4.

The steady-state values of the covariance and gain are $M = 0.780$ and $K = Z = 0.438$.

The performance of the filter can be evaluated heuristically by comparison of actual system states with the contaminated measurements and the state estimate given in Fig. 11.5. These data are satisfying in that the

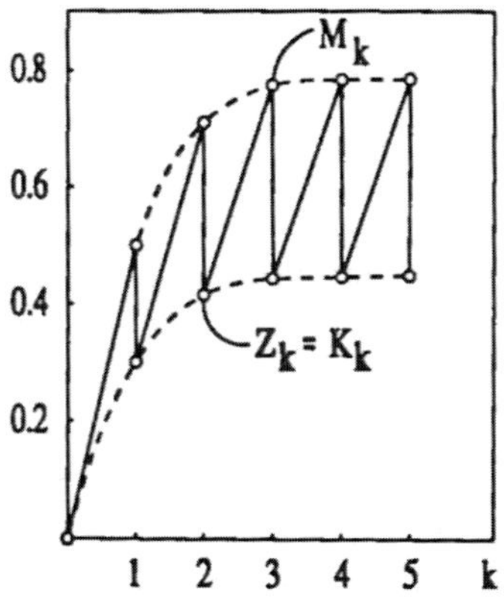


**Figure 11.4.** Transient covariance and gain sequences for the scalar Kalman filter.

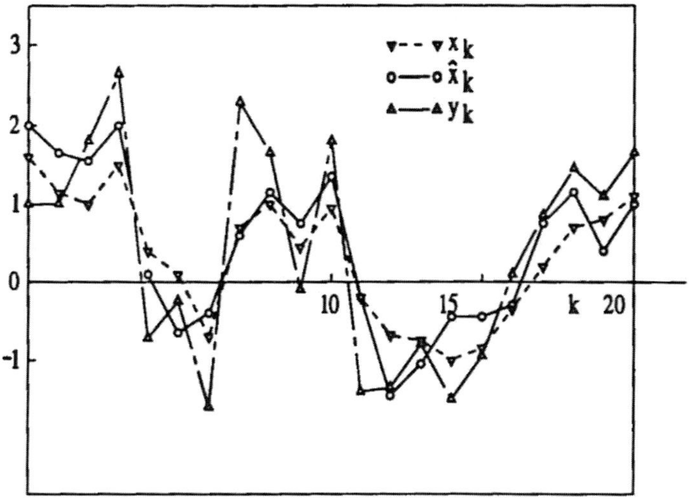


**Figure 11.5.** Various sequences in the scalar Kalman filter.

estimator essentially takes the middle ground between the measurements and the actual state.

The covariance sequences of the estimates for any Kalman filter have the character illustrated in Fig. 11.4. The $M_k$ sequence is the covariance of estimate $\bar{x}_k$ based on measurement set $Y_{k-1}$, and $Z_k$ is the covariance of estimate $\hat{x}_k$ based on the data set $Y_k$. Note that at any one time instant $k$ employing the current measurement $Y_k$ improves the estimate of $x_k$ and thus decreasing the covariance from $M_k$ to $Z_k$.

**Example 11.2.** Consider the thermal system of Examples 9.2 and 10.1, which is illustrated in Fig. 11.6. For purposes of state estimation the inner

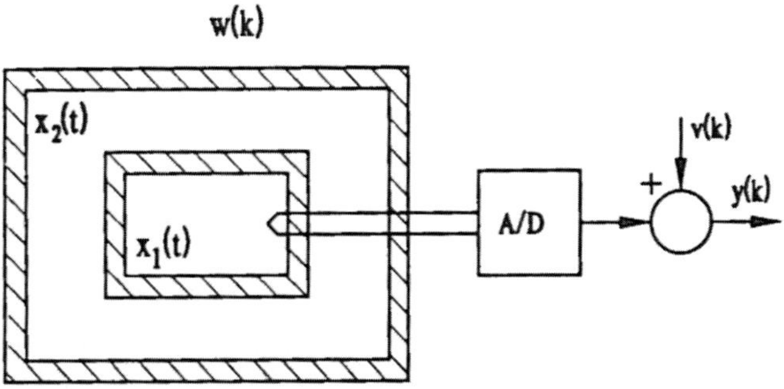


**Figure 11.6.** Thermal system for state estimation.

chamber temperature is measured and sampled. The governing difference equation was given in Example 10.4 to be

$$\begin{bmatrix} x_1 \\ x_2 \end{bmatrix}_{k+1} = \begin{bmatrix} 0.6277 & 0.3597 \\ 0.0899 & 0.8526 \end{bmatrix} \begin{bmatrix} x_1 \\ x_2 \end{bmatrix}_k + \begin{bmatrix} 0.0125 \\ 0.0575 \end{bmatrix} w(k)$$

with a measurement equation

$$y(k) = [1 \quad 0] \begin{bmatrix} x_1 \\ x_2 \end{bmatrix}_k + v(k)$$

where the $w(k)$ and $v(k)$ sequences are statistically independent stationary zero-mean sequences with variances

$$E[w^2(k)] = 9, \qquad E[v^2(k)] = 0.2$$

Let us assume that we know the initial temperatures, and hence the *a priori* covariance of the initial estimates are zero or $\mathbf{M}_0 = 0$. If we sequentially solve expressions (11.2.61), (11.2.62), and (11.2.63) forward in time using the initial condition just given, we obtain the covariance and Kalman filter gain sequences illustrated in Figs. 11.7 and 11.8. Note that as in Example 11.1, the filter gains achieve steady-state values; it has become common practice to implement a state estimator employing the stationary values of the filter gains.

Large errors occur in the initial estimates made by this estimator because the high values of gain place too much confidence in the initial few measurements. This problem is rectified in the case of this filter after about

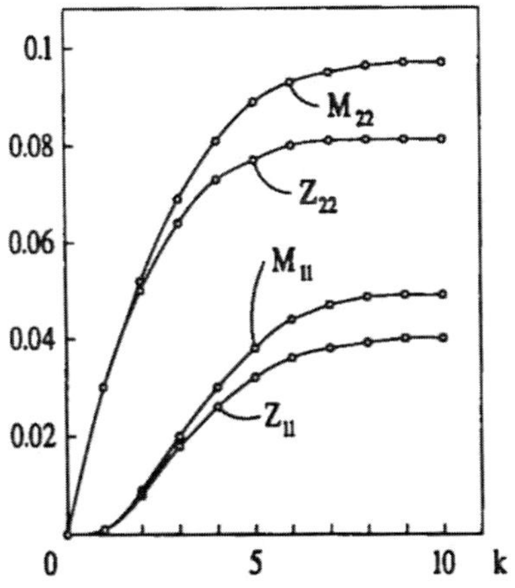


**Figure 11.7.** Second-order Kalman filter estimate variance sequences.

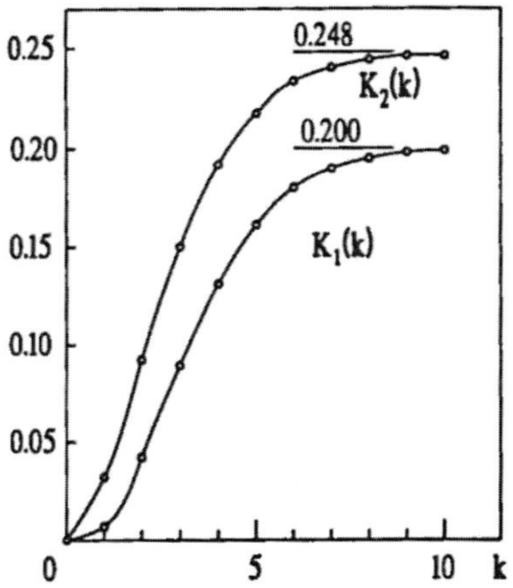


**Figure 11.8.** Second-order Kalman filter gain sequences.

10 sampling intervals. The stationary filter for this problem is

$$\begin{bmatrix} \hat{x}_1 \\ \hat{x}_2 \end{bmatrix}_{k+1} = \begin{bmatrix} 0.6277 & 0.3597 \\ 0.0899 & 0.8526 \end{bmatrix} \begin{bmatrix} \hat{x}_1 \\ \hat{x}_2 \end{bmatrix}_k + \begin{bmatrix} 0.200 \\ 0.248 \end{bmatrix} [y(k+1) - 0.6277\hat{x}_1(k) - 0.3597\hat{x}_2(k)]$$

This form can be simplified to yield the following filter:

$$\begin{bmatrix} \hat{x}_1 \\ \hat{x}_2 \end{bmatrix}_{k+1} = \begin{bmatrix} 0.05022 & 0.2878 \\ -0.0658 & 0.7634 \end{bmatrix} \begin{bmatrix} \hat{x}_1 \\ \hat{x}_2 \end{bmatrix} + \begin{bmatrix} 0.200 \\ 0.248 \end{bmatrix} y(k+1)$$

for initial conditions of

$$\mathbf{x}(0) = \begin{bmatrix} 1 \\ 0.5 \end{bmatrix}$$

This estimator has been simulated to yield the state and state estimate sequences given in Fig. 11.9.

## 11.3 STEADY-STATE KALMAN FILTER GAINS BY EIGENVECTOR DECOMPOSITION

In this section we examine a technique for computation of the steady-state Kalman filter gains which does not require the forward solution of the

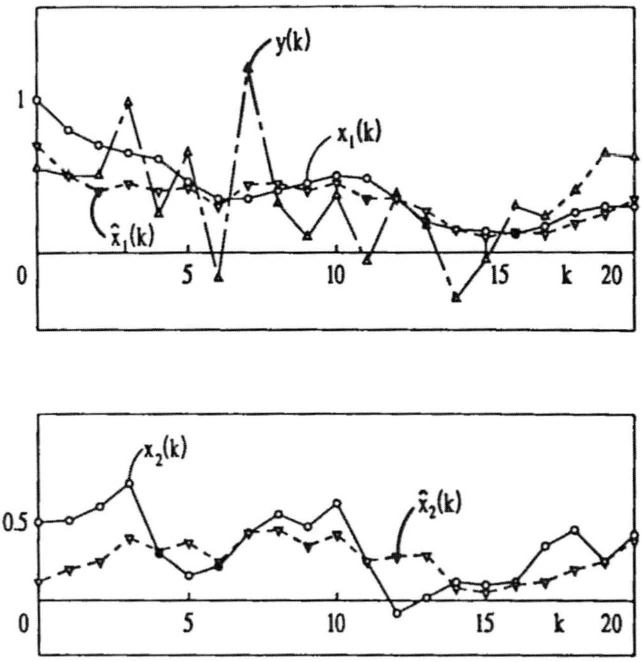


**Figure 11.9.** State and measurement sequences for a second-order thermal system.

matrix difference Eqs. (11.2.62) and (11.2.63) until stationary values of the $\mathbf{Z}_k$ and $\mathbf{M}_k$ covariance matrices are reached. This forward solution is not a difficult problem for systems of low order, but when the system order becomes significant (say, ≥4), this is an extremely time-consuming task with a large general-purpose digital computer. The technique given in this section allows us to evaluate the steady-state solutions directly in a fashion similar to that given in Section 9.5 for solution of the matrix difference equations associated with the optimal control problem. Let us recall that the equations which defined the Kalman filter gain were

$$\mathbf{K}_k = \mathbf{Z}_k\mathbf{C}^T\mathbf{V}^{-1} \tag{11.3.1}$$

where $\mathbf{Z}_k$ is the covariance of the estimate $\mathbf{x}_k$ and is given by

$$\mathbf{Z}_k = \mathbf{M}_k - \mathbf{M}_k\mathbf{C}^T(\mathbf{C}\mathbf{M}_k\mathbf{C}^T + \mathbf{V})^{-1}\mathbf{C}\mathbf{M}_k \tag{11.3.2}$$

and where the matrix $\mathbf{M}_k$ is the covariance of the estimate $\mathbf{x}_k$ and is given by

$$\mathbf{M}_k = \mathbf{A}\mathbf{Z}_{k-1}\mathbf{A}^T + \mathbf{B}\mathbf{W}\mathbf{B}^T \tag{11.3.3}$$

If the system is time-invariant and stable and $\mathbf{V}$ and $\mathbf{W}$ are not functions of time, a statistical steady state can be reached by the system and $\mathbf{M}_k$ approaches $\mathbf{M}$ and $\mathbf{Z}_k$ approaches $\mathbf{Z}$, so the equations above reduce to

$$\mathbf{K} = \mathbf{Z}\mathbf{C}^T\mathbf{V}^{-1} \tag{11.3.4}$$

$$\mathbf{Z} = \mathbf{M} - \mathbf{M}\mathbf{C}^T(\mathbf{C}\mathbf{M}\mathbf{C}^T + \mathbf{V})^{-1}\mathbf{C}\mathbf{M} \tag{11.3.5}$$

and

$$\mathbf{M} = \mathbf{A}\mathbf{Z}\mathbf{A}^T + \mathbf{B}\mathbf{W}\mathbf{B}^T \tag{11.3.6}$$

Now let us return to the discrete regulator problem in its steady state. In that problem we find that for a steady regulator, equations (9.2.36) and (9.2.37) approach a steady state, which implies that

$$\mathbf{G} = \mathbf{P} - \mathbf{P}\mathbf{B}[\mathbf{B}^T\mathbf{P}\mathbf{B} + \mathbf{R}]^{-1}\mathbf{B}^T\mathbf{P} \tag{11.3.7}$$

and

$$\mathbf{P} = \mathbf{A}^T\mathbf{G}\mathbf{A} + \mathbf{Q} \tag{11.3.8}$$

Note that expressions (11.3.5) and (11.3.6) are completely analogous to (11.3.7) and (11.3.8) but with different symbols. Table 11.1 gives the analogous matrices for the two problems. Due to this similarity, these problems have been referred to as *dual problems*. We have shown also that the solution for the steady-state value of $\mathbf{P}$ in the regulator problem could be given by the product of two matrices extracted from the unstable eigenvectors of the state and adjoint equations. Since the problems are so similar in nature, it is clear that the unstable eigenvectors of the matrix $\mathbf{\Phi}_E$ are of crucial importance, where $\mathbf{\Phi}_E$ is defined similar to the matrix associated with Eqs. (9.4.1) and (9.4.2):

$$\mathbf{\Phi}_E = \begin{bmatrix} \mathbf{A}^{T-1} & \mathbf{A}^{T-1}\mathbf{C}^T\mathbf{V}^{-1}\mathbf{C} \\ \mathbf{B}\mathbf{W}\mathbf{B}^T\mathbf{A}^{T-1} & \mathbf{A} + \mathbf{B}\mathbf{W}\mathbf{B}^T\mathbf{A}^{T-1}\mathbf{C}^T\mathbf{V}^{-1}\mathbf{C} \end{bmatrix} \tag{11.3.9}$$

**Table 11.1.** Analogous Matrix Quantities in the Linear Regulator and State Estimation Problems

| Regulator | Estimation |
|---|---|
| $\mathbf{A}$ | $\mathbf{A}^T$ |
| $\mathbf{B}$ | $\mathbf{C}^T$ |
| $\mathbf{P}$ | $\mathbf{M}$ |
| $\mathbf{Q}$ | $\mathbf{B}\mathbf{W}\mathbf{B}^T$ |
| $\mathbf{R}$ | $\mathbf{V}$ |
| $\mathbf{G}$ | $\mathbf{Z}$ |

The unstable eigenvectors of $\mathbf{\Phi}_E$ can be written as the matrix partition $[\mathbf{X}_u^T \quad \mathbf{\Lambda}_u^T]^T$. The matrix **M** is then given by

$$\mathbf{M} = \mathbf{\Lambda}_u \mathbf{X}_u^{-1} \tag{11.3.10}$$

and then the gain from Eq. (11.3.4) is

$$\mathbf{K} = \mathbf{Z}\mathbf{C}^T\mathbf{V}^{-1} \tag{11.3.11}$$

where

$$\mathbf{Z} = [\mathbf{M}^{-1} + \mathbf{C}^T\mathbf{V}^{-1}\mathbf{C}]^{-1} \tag{11.3.12}$$

Substitution of (11.3.9) into (11.3.11) and then (11.3.11) into (11.3.10) yields

$$\mathbf{K} = [\mathbf{X}_u\mathbf{\Lambda}_u^{-1} + \mathbf{C}^T\mathbf{V}^{-1}\mathbf{C}]^{-1}\mathbf{C}^T\mathbf{V}^{-1} \tag{11.3.13}$$

Since there are new computational routines built around the Q–R algorithm for the efficient computation of eigenvalues and eigenvectors of large systems, this technique is attractive for evaluation of Kalman filter gains. For scalar $w_k$, $v_k$, and $y_k$, the steady-state Kalman filter reduces to the classical causal Wiener filter (Wiener, 1949).

**Example 11.3.** Let us consider the same problem as that considered in Example 11.1, where the plant model was

$$x_{k+1} = 0.8x_k + w_k$$

with measurement

$$y_k = x_k + v_k$$

where $w_k$ and $v_k$ are zero-mean independent stationary sequences with respective variances

$$E[v_k^2] = V = 1, \qquad E[w_k^2] = W = 0.5$$

If we insert the parameters into expression (11.3.8), the estimation matrix is

$$\mathbf{\Phi}_E = \begin{bmatrix} 1.25 & 1.25 \\ 0.625 & 1.425 \end{bmatrix}$$

The eigenvalues of this matrix are the roots of the quadratic equation

$$z^2 - 2.675z + 1 = 0$$

or

$$z_{1,2} = 2.2257 \text{ and } 0.4493$$

where we note that the eigenvalues are reciprocals with one inside and one outside the unit circle. We are interested in the eigenvector associated with the unstable eigenvalue (2.2257). If we solve for the unstable eigenvector, it is

$$\begin{bmatrix} X_u \\ \Lambda_u \end{bmatrix} = \begin{bmatrix} 1 \\ 0.78 \end{bmatrix}$$

Then the stationary Kalman filter gain matrix is

$$K = [(1)(1.282) + 1]^{-1} = 0.438$$

which is the same as that given in Example 11.1.

## 11.4 SUMMARY

In this chapter we have developed the discrete-time Kalman filter as the optimal state estimator when the estimates of randomly excited states must be made from noisy measurements. This optimality is in the least-squares sense, in that it minimizes the expected errors of the states conditional on the noisy measurements taken to date. The efficient computational technique of eigenvector decomposition is given to solve the stationary Kalman filter gain problem.

## PROBLEMS

11.1. Consider the scalar system driven by zero-mean purely random noise, or $x_{k+1} = 0.6x_k + w_k$ without output measurement $y_k = x_k + v_k$, where $v_k$ is a 10 mean sequence and the two noise sequences have variances $W = 4$ and $V = 5$.
   (a) Solve for the Kalman filter gain sequences $K_k$ and the two estimate covariance sequences $M_k$ and $Z_k$.
   (b) Draw the block diagram for the resulting discrete-time estimator.

11.2. Consider the liquid-level control system, which for a sampling interval $T = 0.1$ s is governed by the matrix difference equation

$$\begin{bmatrix} x_1 \\ x_2 \end{bmatrix}_{k+1} = \begin{bmatrix} 0.8336 & 0.1424 \\ 0.1424 & 0.6200 \end{bmatrix} \begin{bmatrix} x_1 \\ x_2 \end{bmatrix}_k + \begin{bmatrix} 0.0912 \\ 0.0080 \end{bmatrix} u(k) + \begin{bmatrix} 0.0912 \\ 0.0080 \end{bmatrix} w(k)$$

with measurement

$$y(k) = [0 \quad 1]\begin{bmatrix} x_1 \\ x_2 \end{bmatrix}_k + v(k)$$

where the $v(k)$ and $w(k)$ sequences are zero-mean random sequences. The variance of the process noise $w(k)$ is $W = 4$. The variance of the measurement noise is $V = 2 \times 10^{-3}$. Find the stationary Kalman filter gain vector and the stationary estimate covariance matrices $M(\infty)$ and $Z(\infty)$.

11.3. The plant associated with the printhead positioning problem was considered in Problem 6.12 and the discrete time **A** matrix was

$$\mathbf{A} = \begin{bmatrix} 0.569 & 0.0381 \\ -15.26 & 0.4164 \end{bmatrix}$$

If the printhead undergoes a random forcing such that we may define some equivalent zero-mean process noise sequence $w_k$, we may treat the input matrix **b** as

$$\mathbf{b} = \begin{bmatrix} 0.00216 \\ 0.0763 \end{bmatrix}$$

and the equivalent random force variance is

$$E[w_k^2] = 500$$

Also assume that measurements of only $x_1(k)$ (displacement) are made but contaminated by zero-mean discrete-time measurement noise $v_k$ with noise variance $E[v_k^2] = 0.02$. Find the Kalman filter for estimation of both states $x_1(k)$ and $x_2(k)$.

## REFERENCES

Anderson, B. D. O., and J. B. Moore, 1979. *Optimal Filtering*, Prentice Hall, Englewood Cliffs, NJ.

Brown, R. G., and P. Y. C. Hwang, 1992. *Introduction to Random Signals and Applied Kalman Filtering*, 2nd Ed., Wiley, New York.

Bryson, A. E., Jr., and Y.-C. Ho, 1975. *Applied Optimal Control*, Halsted, New York.

Gauss, K. F., 1809. *Theory of Motion of the Heavenly Bodies*, Dover, New York. Reprint, 1963.

Gelb, A., Ed., 1974. *Applied Optimal Estimation*. MIT Press, Cambridge, MA.

Kalman, R. E., 1960, A new approach to linear filtering and prediction problems, *Trans. ASME: Journal of Basic Engineering, 82D:* 35–45.

Kalman, R. E., and R. C. Bucy, 1961. New results in linear filtering and prediction, *Trans. ASME: Journal of Basic Engineering, 83D:* 95–108.

Kortum, W., 1979. Computational techniques in optimal state estimation: a tutorial review, *Trans. ASME: Journal of Dynamic Systems, Measurement and Control, 107*(2): 99–107.

Lewis, F. L., 1986. *Optimal Estimation with an Introduction to Stochastic Control Theory*, Wiley, New York.

Minkler, G., and J. Minkler, 1993. *Theory and Application of Kalman Filtering*, Magellan, Palm Bay, FL.

Wiener, N. 1949. *The Interpolation and Smoothing of Stationary Time Series*, MIT Press, Cambridge, MA.

# 12
# Discrete-Time Stochastic Control Systems

## 12.1 INTRODUCTION

In Chapter 10 we discussed the response of discrete-time systems to random disturbances; in Chapter 11 we examined the problem of estimating the state of a system from limited, noise-contaminated measurements. Although estimation of the state of an open-loop system is of interest in numerous communication problems and system identification problems (Goodwin and Sin, 1984), the subject of interest in this book is system control.

If we consider the flight of an aircraft through a turbulent atmosphere, we find that the planned flight path will be modified by unknown gusts and random wind fluctuations. Ideally, we would like this aircraft to follow the preplanned flight path despite the disturbances. Since the disturbances are random, it is impossible to anticipate and plan for them; hence we compromise by generating control efforts based on the error in the flight path. This is a feedback control solution to the problem, which is not without its difficulties in that efforts to yield a precise degree of control often lead to systems that have poor dynamic character. Measurement of the aircraft's position and velocity variables are often incomplete because of the cost in building many channels of measurement hardware. The measurements that are taken by their very nature are contaminated with noise, which can come from transducers or signal-processing equipment (filters and amplifiers) or can be introduced into the signal transmission path, such as the noise present in earth-to-vehicle communications during a space mission.

The problem examined in this chapter is that outlined above in the aircraft context. Because of the random nature of the disturbances and the errors in the measurements, the problem is a statistical one to which we shall seek an optimal solution. It turns out that the solution to this problem dictates that we employ estimated state feedback similar to that discussed in Chapter 8. We also see that the tasks of design of state estimation and control may be done separately, as in the deterministic multivariable control problem.

## 12.2 OPTIMAL CONTROL WITH RANDOM DISTURBANCES AND NOISELESS MEASUREMENTS

In this section we consider the case where we wish to control a randomly disturbed system with the control strategy based on completely noiseless measurements of the system state variables. The system of interest is governed by the matrix-vector difference equation

$$\mathbf{x}_{k+1} = \mathbf{A}\mathbf{x}_k + \mathbf{B}\mathbf{u}_k + \mathbf{w}_k \tag{12.2.1}$$

where $\mathbf{u}_k$ is the yet unknown control effort and $\mathbf{w}_k$ is a vector process noise sequence. In this case we assume that the initial state has zero mean,

$$E[\mathbf{x}_0] = \mathbf{0} \tag{12.2.2}$$

and that it possesses covariance matrix

$$E[\mathbf{x}_0\mathbf{x}_0^T] = \mathbf{X}_0 \tag{12.2.3}$$

We shall also assume that the disturbance sequence $\mathbf{w}_k$ has zero mean and has autocorrelation of

$$E[\mathbf{w}_k\mathbf{w}_j^T] = \mathbf{W}\delta(k - j) \tag{12.2.4}$$

The performance index considered here is similar to that considered in Chapter 9 except that we consider the ensemble average of the quadratic index

$$J = E\left\{\frac{1}{2}\,\mathbf{x}_N^T\mathbf{H}\mathbf{x}_N + \frac{1}{2}\sum_{k=0}^{N-1} \mathbf{x}_k^T\mathbf{Q}\mathbf{x}_k + \mathbf{u}_k^T\mathbf{R}\mathbf{u}_k\right\} \tag{12.2.5}$$

We wish to choose the $\mathbf{u}_k$ sequence to minimize this function in the presence of the constraint imposed by the plant (12.2.1).

In this problem $\mathbf{w}_k$ represents the discrete-time disturbance sequence, which is uncorrelated between adjacent samples; thus it is impossible to predict $\mathbf{w}_k$ for $k > m$ with knowledge of $\mathbf{w}_k$ for $k \leq m$. Because of this we may say nothing about the future of $\mathbf{w}_k$ based on the present and the past.

Since the ensemble average of $\mathbf{w}_k$ is zero, we may in the ensemble average sense neglect the effect of $\mathbf{w}_k$, and the problem reduces to the deterministic regulator problem which was treated in Chapter 9. This may be proven by the method of dynamic programming as employed in the next section. The optimal control strategy is a state feedback strategy of the form

$$\mathbf{u}_k = -\mathbf{F}_k\mathbf{x}_k \tag{12.2.6}$$

where the feedback gain matrix $\mathbf{F}_k$ is given by the solution to

$$\mathbf{F}_k = (\mathbf{B}^T\mathbf{P}_{k+1}\mathbf{B} + \mathbf{R})^{-1}\mathbf{B}^T\mathbf{P}_{k+1}\mathbf{A} \tag{12.2.7}$$

and

$$\mathbf{P}_k = \mathbf{A}^T\mathbf{P}_{k+1}\mathbf{A} - \mathbf{F}_k^T(\mathbf{R} + \mathbf{B}^T\mathbf{P}_{k+1}\mathbf{B})\mathbf{F}_k + \mathbf{Q} \tag{12.2.8}$$

with boundary condition $\mathbf{P}_N = \mathbf{H}$. Expressions (12.2.7) and (12.2.8) can be solved backward in time using the specified boundary condition. The solution just presented makes sense from a physical point of view because we are interested in minimizing the mean-square deviations in the system state variables. In other words, it is impossible for the controller to discern whether the deviations in the system state are from deterministic initial conditions or from the random forcing sequence $\mathbf{w}_k$. In the next section we complicate matters by considering the case where the available measurements on which to base control are incomplete and contaminated with additive noise.

## 12.3 CONTROL OF RANDOMLY DISTURBED SYSTEMS WITH NOISE-CONTAMINATED MEASUREMENTS

In Section 12.2 we considered the problem of control of randomly disturbed systems when the measurements of the system state variables are complete and noiseless. The optimal solution to this problem was shown to be the same as that for the deterministic regulator considered in Chapter 9. In this section we consider the case where the measurements are not complete, and those measurements taken are contaminated with additive noise.

We consider the discrete-time plant described by the matrix-vector difference equation

$$\mathbf{x}_{k+1} = \mathbf{A}\mathbf{x}_k + \mathbf{B}\mathbf{u}_k + \mathbf{w}_k \tag{12.3.1}$$

where the vector process noise sequence $\mathbf{w}_k$ has zero mean and is uncorrelated between adjacent samples such that

$$E[\mathbf{w}_k] = 0 \tag{12.3.2}$$

and the covariance will be assumed stationary (this need not be a limitation)

$$E[\mathbf{w}_k \mathbf{w}_j^T] = \mathbf{W}\delta(k - j) \tag{12.3.3}$$

The measurements are linear combinations of the state variables contaminated with a sequence of random noise, or

$$\mathbf{y}_k = \mathbf{C}\mathbf{x}_k + \mathbf{v}_k \tag{12.3.4}$$

where the measurement noise sequence $\mathbf{v}_k$ has zero mean and is uncorrelated from sample to sample, so

$$E[\mathbf{v}_k] = 0 \tag{12.3.5}$$

and

$$E[\mathbf{v}_k \mathbf{v}_j^T] = \mathbf{V}\delta(k - j) \tag{12.3.6}$$

where stationarity has again been assumed but is not a necessary limitation on this development. We shall also assume that the initial state vector $\mathbf{x}_0$ has mean denoted by $\bar{\mathbf{x}}_0$ and initial covariance

$$E[(\mathbf{x}_0 - \mathbf{x}_0)(\mathbf{x}_0 - \bar{\mathbf{x}}_0)] = \mathbf{M}_0 \tag{12.3.7}$$

and is statistically independent of the process and measurement noise sequences

$$E[\mathbf{x}_0 \mathbf{v}_k^T] = E[\mathbf{x}_0 \mathbf{w}_k^T] = 0 \qquad \text{for all } k \tag{12.3.8}$$

We are interested in finding a control effort sequence $\mathbf{u}_0, \mathbf{u}_1, \ldots, \mathbf{u}_{N-1}$ that will minimize the ensemble average of a quadratic performance index

$$J = E\left\{\frac{1}{2}\mathbf{x}_N^T \mathbf{H} \mathbf{x}_N + \frac{1}{2}\sum_{k=0}^{N-1} \mathbf{x}_k^T \mathbf{Q} \mathbf{x}_k + \mathbf{u}_k^T \mathbf{R} \mathbf{u}_k\right\} \tag{12.3.9}$$

where, as before, $\mathbf{H}$ and $\mathbf{Q}$ are positive semidefinite symmetric matrices, and $\mathbf{R}$ is a positive definite symmetric matrix.

We derive the optimal control sequence by the method of dynamic programming, which can also employed for the deterministic problem. Define the optimal return function from time $(N - 1)T$ to time $NT$ to be

$$G_1(\mathbf{Y}_{N-1}) = \min_{\mathbf{u}_{N-1}} E\left\{\frac{1}{2}[\mathbf{x}_N^T \mathbf{H} \mathbf{x}_N + \mathbf{x}_{N-1}^T \mathbf{Q} \mathbf{x}_{N-1} + \mathbf{u}_{N-1}^T \mathbf{R} \mathbf{u}_{N-1}] | \mathbf{Y}_{N-1}\right\} \tag{12.3.10}$$

where $\mathbf{Y}_{N-1}$ denotes the set of measurements up to $t = (N - 1)T$:

$$\mathbf{Y}_{N-1} = \{\mathbf{y}_0, \mathbf{y}_1, \ldots, \mathbf{y}_{N-1}\} \tag{12.3.11}$$

Substitute the state equation for $\mathbf{x}_N$ to yield

$$G_1(\mathbf{Y}_{N-1}) = \min_{\mathbf{u}_{N-1}} E\left\{\frac{1}{2}[\mathbf{x}_{N-1}^T(\mathbf{A}^T\mathbf{H}\mathbf{A} + \mathbf{Q})\mathbf{x}_{N-1} + \mathbf{x}_{N-1}^T\mathbf{A}^T\mathbf{H}\mathbf{B}\mathbf{u}_{N-1} + \mathbf{u}_{N-1}^T\mathbf{B}^T\mathbf{H}\mathbf{A}\mathbf{x}_{N-1} + \mathbf{x}_{N-1}^T\mathbf{A}^T\mathbf{H}\mathbf{w}_{N-1} + \mathbf{w}_{N-1}^T\mathbf{H}\mathbf{A}\mathbf{x}_{N-1} + \mathbf{w}_{N-1}^T\mathbf{H}\mathbf{w}_{N-1} + \mathbf{u}_{N-1}^T(\mathbf{B}^T\mathbf{H}\mathbf{B} + \mathbf{R})\mathbf{u}_{N-1}] + \mathbf{u}_{N-1}^T\mathbf{B}^T\mathbf{H}\mathbf{w}_{N-1} + \mathbf{w}_{N-1}^T\mathbf{H}\mathbf{B}\mathbf{u}_{N-1} \middle| \mathbf{Y}_{N-1}\right\} \tag{12.3.12}$$

Since $\mathbf{w}_{N-1}$ is independent of $\mathbf{u}_{N-1}$ and $\mathbf{x}_{N-1}$ and $E[\mathbf{w}_{N-1}^T\mathbf{H}\mathbf{w}_{N-1}]$ can be evaluated in advance, we shall minimize $G_1(\mathbf{Y}_{N-1})$ with respect to $\mathbf{u}_{N-1}$, so we get

$$\frac{\partial G_1(\mathbf{Y}_{N-1})}{\partial \mathbf{u}_{N-1}} = \mathbf{0} = (\mathbf{B}^T\mathbf{H}\mathbf{B} + \mathbf{R})\mathbf{u}_{N-1} + \mathbf{B}^T\mathbf{H}\mathbf{A}E[\mathbf{x}_{N-1}|\mathbf{Y}_{N-1}] \tag{12.3.13}$$

but let us recall from Chapter 11 that the expected value given in (12.3.13) is the best least-squares estimate $\hat{\mathbf{x}}_{N-1}$ based on the past measurement data set as given by the Kalman filtering algorithm. Thus the control law is given as a feedback control law of the form

$$\mathbf{u}_{N-1} = -(\mathbf{B}^T\mathbf{H}\mathbf{B} + \mathbf{R})^{-1}\mathbf{B}^T\mathbf{H}\mathbf{A}\hat{\mathbf{x}}_{N-1} \tag{12.3.14}$$

or more concisely as

$$\mathbf{u}_{N-1} = -\mathbf{F}_{N-1}\hat{\mathbf{x}}_{N-1} \tag{12.3.15}$$

where

$$\mathbf{F}_{N-1} = (\mathbf{B}^T\mathbf{H}\mathbf{B} + \mathbf{R})^{-1}\mathbf{B}^T\mathbf{H}\mathbf{A} \tag{12.3.16}$$

Let us now substitute (12.3.15) back into expression (12.3.12), which after manipulation yields

$$G_1(\mathbf{Y}_{N-1}) = E\left[\frac{1}{2}\mathbf{x}_{N-1}^T\mathbf{P}_{N-1}\mathbf{x}_{N-1}\right] + \text{constant term} \tag{12.3.17}$$

where the matrix $\mathbf{P}_{N-1}$ is

$$\mathbf{P}_{N-1} = \mathbf{A}^T\mathbf{H}\mathbf{A} - \mathbf{F}_{N-1}^T(\mathbf{R} + \mathbf{B}^T\mathbf{H}\mathbf{B})\mathbf{F}_{N-1} + \mathbf{Q} \tag{12.3.18}$$

With some rearrangement (12.3.17) can be written as

$$G_1(\mathbf{Y}_{N-1}) = Tr\,\frac{1}{2}\{\mathbf{P}_{N-1}\mathbf{X}_{N-1} + \mathbf{H}(\mathbf{W} + \mathbf{B}\mathbf{F}_{N-1}\mathbf{Z}_{N-1}\mathbf{A}^T)\} \tag{12.3.19}$$

where $\mathbf{Z}_{N-1}$ is defined by relation (11.2.35). It can be shown that (12.3.16) and (12.3.18) are the same as Equations (9.2.35) and (9.2.29) in Chapter 9.

Now let us employ the principle of optimality for the next step backward in time, for which the return function is

$$G_2(\mathbf{Y}_{N-2}) = \min_{\mathbf{u}_{N-2}} E\left\{ G_1(\mathbf{Y}_{N-1}) + \frac{1}{2}[\mathbf{x}_{N-2}^T\mathbf{Q}\mathbf{x}_{N-2} + \mathbf{u}_{N-2}^T\mathbf{R}\mathbf{u}_{N-2}] | \mathbf{Y}_{N-2} \right\} \tag{12.3.20}$$

By the definition of the conditional mean, we see that

$$E\{G_1(\mathbf{Y}_{N-1})|\mathbf{Y}_{N-2}\} = E\left\{ E\left[ \frac{1}{2}(\mathbf{x}_{N-1}^T\mathbf{P}_{N-1}\mathbf{x}_{N-1}) \right] \times |\mathbf{y}_0, \ldots, \mathbf{y}_{N-1}|\mathbf{y}_0, \ldots, \mathbf{y}_{N-2} \right\} \tag{12.3.21}$$

Since $\mathbf{x}_{N-1}$ cannot depend on $\mathbf{y}_{N-1}$, (12.3.21) reduces to

$$E\{G_1(\mathbf{Y}_{N-1})|\mathbf{Y}_{N-2}\} = E\left\{ \frac{1}{2}\mathbf{x}_{N-1}^T\mathbf{P}_{N-1}\mathbf{x}_{N-1}|\mathbf{y}_0, \ldots, \mathbf{y}_{N-2} \right\} \tag{12.3.22}$$

and now the form of (12.3.19) reduces to a form similar to (12.3.10) with the subscript translated by one step and the cycle repeats. The variable $E[\mathbf{x}_k|\mathbf{Y}_k]$ is calculated by the usual Kalman filter, or

$$\hat{\mathbf{x}}_k = \bar{\mathbf{x}}_k + \mathbf{K}_k(\mathbf{y}_k - \mathbf{C}\bar{\mathbf{x}}_k) \tag{12.3.23}$$

and

$$\bar{\mathbf{x}}_{k+1} = \mathbf{A}\hat{\mathbf{x}}_k + \mathbf{B}\mathbf{u}_k \tag{12.3.24}$$

The Kalman filter gain matrix is given by

$$\mathbf{K}_k = \mathbf{Z}_k\mathbf{C}^T\mathbf{V}^{-1} \tag{12.3.25}$$

and the $\mathbf{Z}_k$ are the solutions to

$$\mathbf{Z}_k = \mathbf{M}_k - \mathbf{M}_k\mathbf{C}^T(\mathbf{V} + \mathbf{C}\mathbf{M}_k\mathbf{C}^T)^{-1}\mathbf{C}\mathbf{M}_k \tag{12.3.26}$$

and

$$\mathbf{M}_{k+1} = \mathbf{A}\mathbf{Z}_k\mathbf{A}^T + \mathbf{W} \tag{12.3.27}$$

with $\mathbf{M}_0$ given as the initial state covariance matrix based on no available measurements.

Since this development has been so lengthy, we now summarize the results required to accomplish the estimation and control tasks. The control law is simply the estimated state feedback

$$\mathbf{u}_k = -\mathbf{F}_k\hat{\mathbf{x}}_k \tag{12.3.28}$$

where the estimate of the state is given by the Kalman filter

$$\hat{\mathbf{x}}_k = \bar{\mathbf{x}}_k + \mathbf{K}_k(\mathbf{y}_k - \mathbf{C}\bar{\mathbf{x}}_k) \tag{12.3.29}$$

and

$$\bar{\mathbf{x}}_k = \mathbf{A}\hat{\mathbf{x}}_{k-1} + \mathbf{B}\mathbf{u}_{k-1} \tag{12.3.30}$$

The Kalman filter gain matrix $\mathbf{K}_k$ is

$$\mathbf{K}_k = \mathbf{Z}_k\mathbf{C}^T\mathbf{V}^{-1} \tag{12.3.31}$$

and the matrix $\mathbf{Z}_k$ is given by

$$\mathbf{Z}_k = \mathbf{M}_k - \mathbf{M}_k\mathbf{C}^T(\mathbf{V} + \mathbf{C}\mathbf{M}_k\mathbf{C}^T)^{-1}\mathbf{C}\mathbf{M}_k \tag{12.3.32}$$

and

$$\mathbf{M}_{k+1} = \mathbf{A}\mathbf{Z}_k\mathbf{A}^T + \mathbf{W} \tag{12.3.33}$$

with $\mathbf{M}_0$ specified. The state-feedback control gain matrix $\mathbf{F}_k$ is given by

$$\mathbf{F}_k = (\mathbf{B}^T\mathbf{P}_{k+1}\mathbf{B} + \mathbf{R})^{-1}\mathbf{B}^T\mathbf{P}_{k+1}\mathbf{A} \tag{12.3.34}$$

and

$$\mathbf{P}_k = \mathbf{A}^T\mathbf{P}_{k+1}\mathbf{A} - \mathbf{F}_k^T(\mathbf{R} + \mathbf{B}^T\mathbf{P}_{k+1}\mathbf{B})\mathbf{F}_k + \mathbf{Q} \tag{12.3.35}$$

with $\mathbf{P}_N = \mathbf{H}$, the penalty matrix for the final state. The complete stochastic control system is shown in Fig. 12.1. We have just derived the optimal control and estimation algorithms for control of a randomly disturbed system based on noise-contaminated measurements.

In the next section we consider the evaluation of average performance parameters, so we may make some statement about who well the fluctuations in the output are controlled.

**Example 12.1.** Consider the first-order system of Examples 9.1 and 11.1 governed by the scalar difference equation

$$x_{k+1} = 0.8x_k + u_k + w_k$$

with outputs

$$y_k = x_k + v_k$$

where the sequences $w_k$ and $v_k$ are zero-mean stationary sequences with respective variances

$$E[v_k^2] = 1 \quad \text{and} \quad E[w_k^2] = 0.5$$

Consider the performance index of

$$J = E\left\{\frac{1}{2}2x_4^2 + \frac{1}{2}\sum_{k=0}^{3} x_k^2 + u_k^2\right\}$$

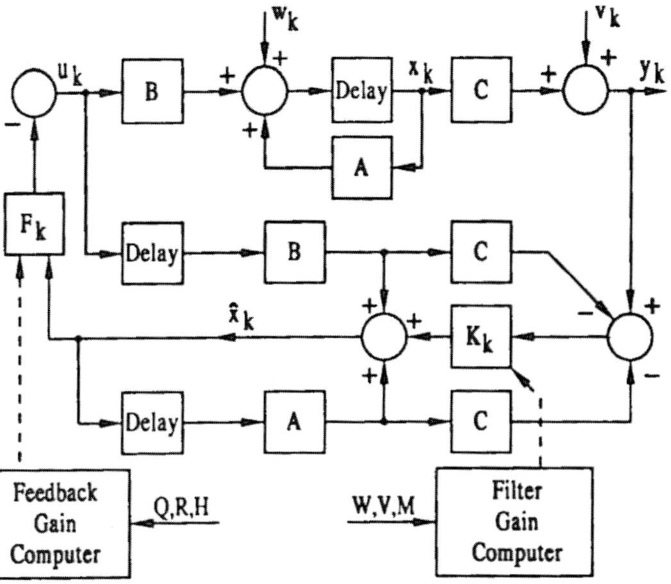


**Figure 12.1.** Complete stochastic regulator.

In Example 9.1 the stationary optimal feedback gain was found to be

$$F = 0.463$$

For the estimator let us consider the stationary solution where $\mathbf{M}_{k+1} = \mathbf{M}_k = \mathbf{M}$ and $\mathbf{Z}_k = \mathbf{Z}$ and $\mathbf{K}_k = \mathbf{K}$. The equations reduce to

$$M = 0.64Z + 0.5$$
$$Z = M - M^2(1 + M)^{-1}$$

Solving simultaneously for the positive roots for $M$ and $Z$ (since they are covariances), we get

$$Z = 0.438 \quad \text{and} \quad M = 0.780$$

while the stationary Kalman filter gain is

$$K = 0.438$$

The Kalman filter algorithm for the state estimation is

$$\hat{x}_{k+1} = 0.8\hat{x}_k + u_k + 0.438[y_{k+1} - 0.8\hat{x}_k - u_k]$$

and the stationary control algorithm

$$u_k = -0.463\hat{x}_k$$

The complete control system is shown in Fig. 12.2. Responses for a simulation of this control system are shown in Fig. 12.3 for an initial state of $x_0 = 2$. Note that the estimated state generally chooses middle ground between the actual state and the erratic noise-contaminated measurement sequence $y_k$.

The average performance of this system will be evaluated in the following section along with the ability of the feedback control to limit random fluctuations in the state variable sequence $x_k$.

**Example 12.2.** Consider the thermal system that we have used as an example in most of the chapters of this book. This system is illustrated in Fig. 12.4. This system is governed by the matrix-vector stochastic difference equation

$$\begin{bmatrix} x_1 \\ x_2 \end{bmatrix}_{k+1} = \begin{bmatrix} 0.6277 & 0.3597 \\ 0.0899 & 0.8526 \end{bmatrix} \begin{bmatrix} x_1 \\ x_2 \end{bmatrix}_k + \begin{bmatrix} 0.0251 \\ 0.1150 \end{bmatrix} u_k + \begin{bmatrix} 0.0125 \\ 0.0575 \end{bmatrix} w_k$$

with output

$$y_k = \begin{bmatrix} 1 & 0 \end{bmatrix} \begin{bmatrix} x_1 \\ x_2 \end{bmatrix}_k + v_k$$

The purely random sequences $w_k$ and $v_k$ have zero mean and respective variances

$$E[w_k^2] = 9 \quad \text{and} \quad E[v_k^2] = 0.2$$

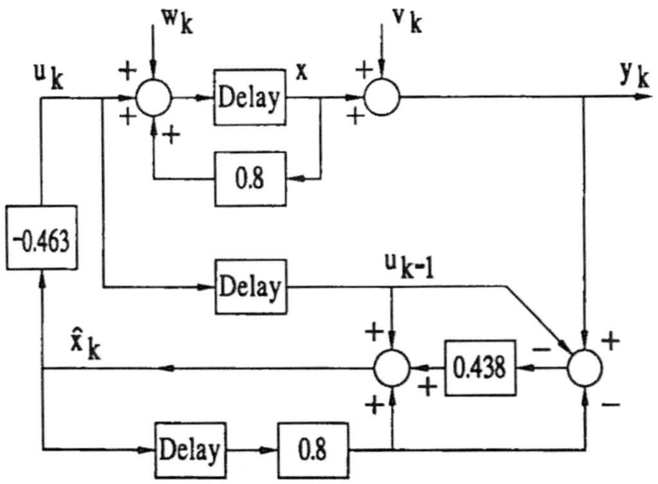


**Figure 12.2.** Complete feedback control system for stochastic first-order plant.

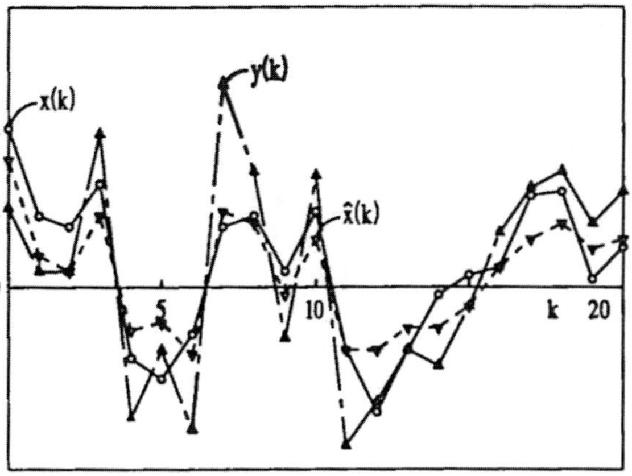


**Figure 12.3.** State, measurement, and state estimate histories for scalar Kalman filter.

We should like to control this system such as to minimize the quadratic index

$$J = E\left[\frac{1}{2}\sum_{k=1}^{9} \mathbf{x}_k^T \mathbf{Q} \mathbf{x}_k + R u_k^2\right]$$

where

$$\mathbf{Q} = \begin{bmatrix} 50 & 0 \\ 0 & 10 \end{bmatrix}, \qquad R = 1$$

From these data we see that

$$V = 0.2 \quad \text{and} \quad \mathbf{W} = \begin{bmatrix} 0.00140 & 0.00647 \\ 0.00647 & 0.0298 \end{bmatrix}$$

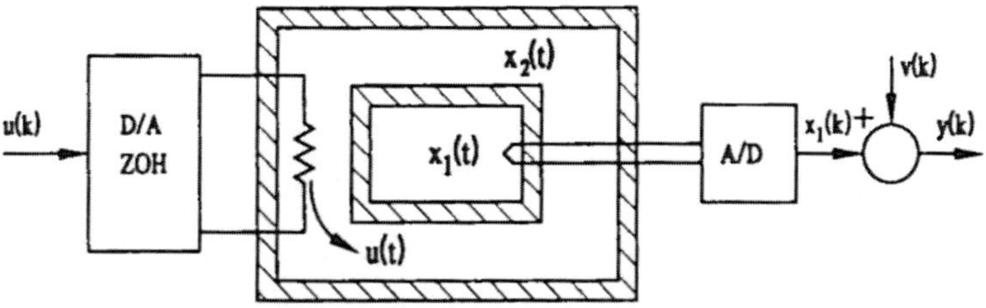


**Figure 12.4.** Stochastic thermal control system.

In Example 9.2 the optimal steady-state feedback gains for this problem were found to be

$$\mathbf{f}^T = [1.872 \quad 3.543]$$

while in Example 11.2 the optimal steady-state Kalman filter gains are

$$\mathbf{k} = \begin{bmatrix} 0.200 \\ 0.248 \end{bmatrix}$$

The Kalman filtering algorithm can be written according to expressions (12.3.29) and (12.3.30):

$$\begin{bmatrix} \hat{x}_1 \\ \hat{x}_2 \end{bmatrix}_{k+1} = \begin{bmatrix} 0.6277 & 0.3597 \\ 0.08993 & 0.8526 \end{bmatrix} \begin{bmatrix} \hat{x}_1 \\ \hat{x}_2 \end{bmatrix}_k + \begin{bmatrix} 0.025 \\ 0.115 \end{bmatrix} u(k)$$
$$+ \begin{bmatrix} 0.200 \\ 0.248 \end{bmatrix} [y(k+1) - 0.6277\hat{x}_1(k)$$
$$- 0.3597\hat{x}_2(k) - 0.025u(k)]$$

while the control law is

$$u(k) = -[1.872 \quad 3.543] \begin{bmatrix} \hat{x}_1 \\ \hat{x}_2 \end{bmatrix}_k$$

This control system was simulated with random inputs for both process noise and measurement noise with initial conditions $x_1(0) = 1$ and $x_2(0) = 0.5$ and the responses of the states and state estimates are shown in Fig. 12.5. If these state histories are compared to those for the open-loop system in Figs. 11.10 and 11.11, we see that the control law is effective in limiting the perturbations due to the disturbance. We evaluate just how effective the use of estimated state feedback has been in minimizing the effects of the random disturbance sequence.

## 12.4 AVERAGE BEHAVIOR OF THE CONTROLLED SYSTEM

In Section 12.3 we considered the optimal control of a randomly disturbed system where the control effort was based on noisy measurements of system output variables. In this section we consider the average performance of this optimally controlled system. From the analysis we can predict the mean-square values of the states and control efforts.

The equation of the plant to be controlled is

$$\mathbf{x}_{k+1} = \mathbf{A}\mathbf{x}_k + \mathbf{B}\mathbf{u}_k + \mathbf{w}_k \tag{12.4.1}$$

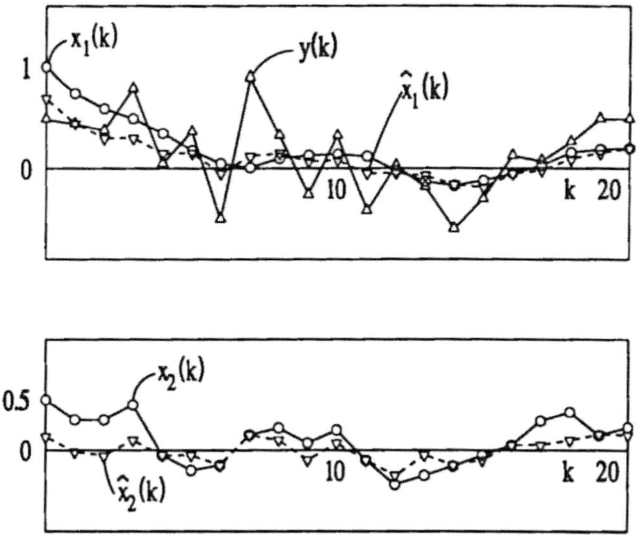


**Figure 12.5.** State, measurement, and state estimate histories for thermal plant.

The two equations (12.3.23) and (12.3.24) can be combined to yield a single estimator equation

$$\bar{\mathbf{x}}_{k+1} = \mathbf{A}[\bar{\mathbf{x}}_k + \mathbf{K}_k(\mathbf{y}_k - \mathbf{C}\bar{\mathbf{x}}_k)] + \mathbf{B}\mathbf{u}_k \tag{12.4.2}$$

Now define the error in the estimate prior to measurement to be

$$\mathbf{e}_k = \bar{\mathbf{x}}_k - \mathbf{x}_k \tag{12.4.3}$$

Subtraction of (12.4.1) from (12.4.2) yields

$$\mathbf{e}_{k+1} = \mathbf{A}(\mathbf{I} - \mathbf{K}_k\mathbf{C})\mathbf{e}_k + \mathbf{A}\mathbf{K}_k\mathbf{v}_k - \mathbf{w}_k \tag{12.4.4}$$

If we note that $\mathbf{u}_k = -\mathbf{F}_k\hat{\mathbf{x}}_k$ and eliminate $\hat{\mathbf{x}}_k$ in terms of $\bar{\mathbf{x}}_k$ from (12.3.24), we can write (12.4.2) as

$$\bar{\mathbf{x}}_{k+1} = [\mathbf{A} - \mathbf{B}\mathbf{F}_k][\bar{\mathbf{x}}_k - \mathbf{K}_k\mathbf{C}\mathbf{e}_k + \mathbf{K}_k\mathbf{v}_k] \tag{12.4.5}$$

We already know that the covariance matrix of the error is

$$\mathbf{M}_k = E[\mathbf{e}_k\mathbf{e}_k^T] \tag{12.4.6}$$

Also, let us define some new covariance matrices as

$$\bar{\mathbf{X}}_k = E[\bar{\mathbf{x}}_k\bar{\mathbf{x}}_k^T] \tag{12.4.7}$$

and as before, the covariance of the state estimate is

$$\mathbf{Z}_k = E[(\hat{\mathbf{x}}_k - \mathbf{x}_k)(\hat{\mathbf{x}}_k - \mathbf{x}_k)^T] \tag{12.4.8}$$

We know from least-squares theory that the error in the estimate is orthogonal to the estimate, so $E[\mathbf{e}_k\bar{\mathbf{x}}_k^T] = \mathbf{0}$, and thus it follows that

$$\mathbf{X}_k = \bar{\mathbf{X}}_k + \mathbf{M}_k \tag{12.4.9}$$

Postmultiplying (12.4.5) by its transpose and taking the expected value yields

$$\bar{\mathbf{X}}_{k+1} = (\mathbf{A} - \mathbf{BF}_k)(\bar{\mathbf{X}}_k + \mathbf{M}_k - \mathbf{Z}_k)(\mathbf{A} - \mathbf{BF}_k)^T \tag{12.4.10}$$

where we have employed the facts that $E[\bar{\mathbf{x}}_k\mathbf{v}_k^T]$ and $E[\mathbf{e}_k\mathbf{v}_k^T]$ are both zero. The stipulation that $\bar{\mathbf{X}}_0$ is zero is a starting condition for this expression [see Eq. (12.3.7)]. Expression (12.4.10) allows us to predict the mean-square histories of the state variables by application of (12.4.9). Let us now examine the control effort, which is given by

$$\mathbf{u}_k = -\mathbf{F}_k\hat{\mathbf{x}}_k \tag{12.4.11}$$

With substitution of (12.3.23) we get

$$\mathbf{u}_k = -\mathbf{F}_k[\bar{\mathbf{x}}_k + \mathbf{K}_k(\mathbf{y}_k - \mathbf{C}\bar{\mathbf{x}}_k)] \tag{12.4.12}$$

Further substitution for $\mathbf{y}_k$ yields

$$\mathbf{u}_k = -\mathbf{F}_k[\bar{\mathbf{x}}_k - \mathbf{K}_k\mathbf{C}\mathbf{e}_k + \mathbf{K}_k\mathbf{v}_k] \tag{12.4.13}$$

Now postmultiply this expression by its transpose and take the expected value to yield the covariance (for zero mean) of the control effort, or

$$E[\mathbf{u}_k\mathbf{u}_k^T] = \mathbf{F}_k[\bar{\mathbf{X}}_k + \mathbf{M}_k - \mathbf{Z}_k]\mathbf{F}_k^T \tag{12.4.14}$$

If the dynamic programming approach of Section 12.3 is continued backward in time to the temporal origin, we find that for optimal control and estimation the total cost for the entire planning horizon is

$$J = \operatorname{Tr}\frac{1}{2}\left\{\mathbf{P}_0\mathbf{X}_0 + \sum_{k=0}^{N-1}\mathbf{P}_{k+1}(\mathbf{W} + \mathbf{BF}_k\mathbf{Z}_k\mathbf{A}^T)\right\} \tag{12.4.15}$$

Some interpretation of the terms of (12.4.15) is now in order. The first term is the total cost associated with the control and is similar to the total cost associated with deterministic control. The first term under the summation is the additional cost associated with uncertainties due to the process noise sequence $\mathbf{w}_k$ as filtered by the plant. The second term in the sum is the cost associated with uncertainties in output measurements contributed by the measurement noise and is essentially the cost of state estimation.

**Example 12.3.** We would like to investigate the average behavior of the stochastic control system designed in Example 12.1. The plant to be controlled is governed by

$$x_{k+1} = 0.8x_k + u_k + w_k$$

with measurements

$$y_k = x_k + v_k$$

where $w_k$ and $v_k$ are zero-mean random sequences with respective mean-square values

$$E[w_k^2] = 0.5 \quad \text{and} \quad E[v_k^2] = 1$$

The optimal stationary values of covariances, feedback gain, and filter gain were found in Example 12.1 to be

$$F = 0.463 \qquad K = 0.438$$

$$Z = 0.438 \qquad M = 0.78$$

The stationary mean-square value of the premeasurement estimate is given by relation (12.4.10) to be

$$\overline{X} = [0.8 - 0.463]^2(\overline{X} + 0.78 - 0.438)$$

Solving for $\overline{X}$, we get

$$\overline{X} = 0.4377$$

Then from relation (12.4.9) the mean-square output response is

$$X = 0.04377 + 0.78 = 0.82337$$

If we compare this value with that given in Example 10.5, which was 1.38, this indicates that the value of the closed-loop control system is the control of fluctuations in output caused by the random disturbance sequence $w_k$.

The stationary mean square value of the control effort is given by relation (12.4.14) to be

$$E[u_k^2] = (0.463)^2(0.04377 + 0.78 - 0.438)$$

and upon evaluation,

$$E[u_k^2] = 0.0827$$

**Example 12.4.** Let us consider the stationary average performance of the control system considered in Example 12.2. This is the thermal control system for which the appropriate matrices are

$$\mathbf{A} = \begin{bmatrix} 0.6277 & 0.3597 \\ 0.0899 & 0.8526 \end{bmatrix} \qquad \mathbf{B} = \begin{bmatrix} 0.0251 \\ 0.1150 \end{bmatrix}$$

$$\mathbf{c}^T = [1 \quad 0], \qquad \mathbf{f}^T = [1.872 \quad 3.543]$$

and

$$\mathbf{k} = \begin{bmatrix} 0.200 \\ 0.248 \end{bmatrix} \qquad V = 0.2 \qquad \mathbf{W} = \begin{bmatrix} 0.0014 & 0.00647 \\ 0.00647 & 0.0298 \end{bmatrix}$$

Solution of expressions (12.3.26), (12.3.27), and (12.4.10) simultaneously on the digital computer yields a covariance matrix for the states of

$$\mathbf{X} = \begin{bmatrix} 0.0649 & 0.0658 \\ 0.0658 & 0.0992 \end{bmatrix}$$

with a mean-square control effort given by (12.4.14) to be

$$E[u_k^2] = 0.5218$$

The rms value of the first state is 0.255 and the result for the uncontrolled system of Example 10.4 is 0.446. For the second state the controlled rms value is 0.395 and the uncontrolled rms value was 0.475. We see immediately that one benefit of feedback is to reduce the output deviation due to disturbances.

## 12.5 STEADY-STATE CONTROL SYSTEM DYNAMICS

We have considered in previous examples the design of time-invariant Kalman filters (observers) and state-feedback gains. The reason for this is pragmatic in that because of their simplicity, these are the types of control systems most often implemented in practice. In this case the Kalman filter has exactly the same form as the "current observer" discussed in Section 8.6. If we neglect the measurement noise sequence $\mathbf{v}_k$ and choose the steady-state feedback gains and Kalman filter gains, the closed-loop control system poles may be shown to be separable and are the roots of

$$\det[z\mathbf{I} - \mathbf{A} + \mathbf{BF}] \det[z\mathbf{I} - \mathbf{A} + \mathbf{KCA}] = 0 \qquad (12.5.1)$$

We have calculated the elements of $\mathbf{F}$ and $\mathbf{K}$ by solving the respective discrete-time Riccati equations, forward in time for Kalman filter gains and backward in time for the estimated state feedback gains. With these gains all known, we may then investigate the resulting closed-loop system dynamics by examining the closed-loop pole locations given by the roots of relation (12.5.1), which is the closed-loop characteristic equation. Two examples that have been pursued throughout this and other chapters will now be explored from a system pole-placement point of view.

**Example 12.5.** Consider the scalar plant of Examples 12.1 and 12.3 governed by the difference equation

$$x_{k+1} = 0.8x_k + u_k + w_k$$

with measurements

$$y_k = x_k + v_k$$

where $W = 0.5$ and $V = 1$ and a performance index of

$$J = E\left\{\frac{1}{2}\sum_{k=0}^{\infty} x_k^2 + u_k^2\right\}$$

The stationary optimal feedback gain was found to be

$$F = 0.463$$

and the optimal estimator gain was

$$K = 0.438$$

The closed-loop roots are given by (12.5.1) to be

$$(z - 0.8 + 0.463)(z - 0.8 + 0.438(0.8)) = 0$$

These poles locations are shown in Fig. 12.6 and will be discussed later in a qualitative fashion.

**Example 12.6.** Consider the thermal control system which was solved in detail in Examples 12.2 and 12.4. The system matrices were

$$\mathbf{A} = \begin{bmatrix} 0.6277 & 0.3597 \\ 0.0899 & 0.8526 \end{bmatrix} \qquad \mathbf{B} = \begin{bmatrix} 0.0251 \\ 0.1150 \end{bmatrix}$$

$$\mathbf{c}^T = [1 \quad 0] \qquad \mathbf{f}^T = [1.872 \quad 3.543]$$

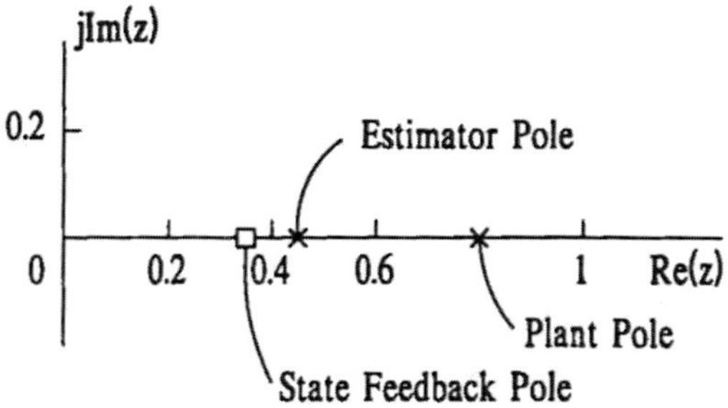


**Figure 12.6.** Poles of scalar stochastic plant.

and

$$\mathbf{k} = \begin{bmatrix} 0.200 \\ 0.248 \end{bmatrix}$$

As in Example 12.5, the roots are given by the roots of (12.5.1), or the first part is

$$\det \begin{bmatrix} z - 0.6277 + 0.0469 & -0.3597 + 0.0889 \\ -0.0899 + 0.215 & z - 0.8526 + 0.407 \end{bmatrix} = 0$$

and the other two roots as roots of

$$\det \begin{bmatrix} z - 0.502 & -0.2877 \\ 0.0657 & z - 0.763 \end{bmatrix} = 0$$

The observer poles are at $z_{1,2} = 0.635 \pm j0.042$ and the state feedback poles are at $z = 0.5135 \pm j0.171$, all of which are shown in Fig. 12.7.

If we examine these figures carefully, we see that the estimator poles are faster (farther to the left) than the open-loop plant poles but are generally slower than the feedback system poles. Recall that in the quadratic sense, these are the optimal pole locations, but the question arises as to why the estimator poles are not also faster than the feedback poles. If we think about the problem from a noise and disturbance point of view, we see that if the estimator poles were faster than the feedback poles, the estimator wold possess excessive bandwidth and hence measurement and process nosie would propagate around the control loop and increase the variance of the output sequence. The reasoning above is a bit crude, but this can be partially verified from the measurement noise point of view since from relation (12.3.31) the Kalman filter gains vary inversely with the covariance of the measurement noise.

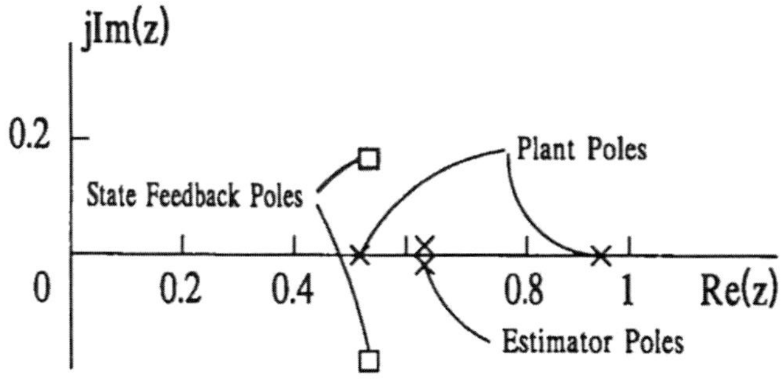


**Figure 12.7.** Poles of stochastic thermal control system.

## 12.6 SUMMARY

In this chapter we have examined the technique of controlling linear discrete-time plants in the presence of random disturbances employing noise-contaminated measurements of variables that are not necessarily the state variables. It is shown that the optimal control policy is to feed back estimates of the state variables. It is also shown that the esimates to be employed should be those from the output of a Kalman filter, which is designed as outlined in Chapter 11. We have shown that the optimal gains employed in feeding back the state estimates are the same as those used in the deterministic regulator problem discussed in Chapter 9. The designs of the estimator and controller are separable and all previous modern design techniques are applicable. We have also derived the expressions necessary to predict the statistical average performance of the resulting control system.

## PROBLEMS

12.1. Let us consider the plant of Problem 10.1, the dynamics of which are governed by $x_{k+1} = 0.6x_k + u_k + w_k$ and measurements governed by $y_k = x_k + v_k$, where $v_k$ and $w_k$ are zero-mean sequences with respective variances $W = 4$, $V = 5$. The performance index is

$$J = E\left\{\frac{1}{2}\sum_{k=0}^{\infty} 2x_k^2 + u_k^2\right\}$$

So $Q = 2$ and $R = 1$.
(a) Find the optimal steady-state feedback gain $F$ to minimize $J$. (b) Find the steady-state Kalman filter gain $K$. (c) Evaluate the rms control effort and the rms system response.

12.2. Consider the liquid-level system, which for a sampling interval of $T = 0.1$ s is represented by the vector difference equation

$$\begin{bmatrix} x_1 \\ x_2 \end{bmatrix}_{k+1} = \begin{bmatrix} 0.8336 & 0.1424 \\ 0.1424 & 0.6200 \end{bmatrix}\begin{bmatrix} x_1 \\ x_2 \end{bmatrix}_k + \begin{bmatrix} 0.0912 \\ 0.0080 \end{bmatrix} u(k) + \begin{bmatrix} 0.0912 \\ 0.0080 \end{bmatrix} w(k)$$

with noise-contaminated measurements of the first tank level, or

$$y(k) = [0 \quad 1]\begin{bmatrix} x_1 \\ x_2 \end{bmatrix}_k + v(k)$$

where $v(k)$ and $w(k)$ are purely random zero-mean sequences with respective variances $E[w^2(k)] = W = 4$ and $E[v^2(k)] = V = 2 \times$

$10^{-3}$. Consider also the scalar performance index

$$J = E\left\{\frac{1}{2}\sum_{k=0}^{\infty} \mathbf{x}^T(k)\mathbf{Q}\mathbf{x}(k) + u^T(k)Ru(k)\right\}$$

where the $Q$ and $R$ matrices are, respectively,

$$\mathbf{Q} = \begin{bmatrix} 100 & 0 \\ 0 & 500 \end{bmatrix} \qquad R = 1$$

Design a time-invariant control system to minimize $J$ which consists of the following operations: (a) Find the stationary Kalman filter gains for the state estimator. (b) Find the stationary feedback gains for feeding back the estimated states.

12.3. Consider the printhead plant considered in Problem 11.3, with respective $A$ and $b$ matrices

$$\mathbf{A} = \begin{bmatrix} 0.569 & 0.0381 \\ -15.26 & 0.4164 \end{bmatrix} \qquad \mathbf{b}_1 = \mathbf{b}_2 = \begin{bmatrix} 0.00216 \\ 0.0763 \end{bmatrix}$$

with state equation $\mathbf{x}(k+1) = \mathbf{A}\mathbf{x}(k) + \mathbf{b}_1 u(k) + \mathbf{b}_2 w(k)$, where $w_k$ is a purely random zero-mean random noise sequence with variance $E[w_k^2] = 500$. State $x_1(k)$ (displacement) is measured but contaminated by purely random, zero-mean measurement noise $v_k$ with variance $E[v_k^2] = 0.02$.

(a) Find the stationary Kalman filter for establishing the estimated states.

(b) Find the feedback gains that minimize

$$J = E\left\{\frac{1}{2}\sum_{k=0}^{\infty} \mathbf{x}^T(k)\mathbf{Q}\mathbf{x}(k) + u^T(k)Ru(k)\right\}$$

where

$$\mathbf{Q} = \begin{bmatrix} 2 & 0 \\ 0 & 400 \end{bmatrix} \qquad R = 1$$

(c) Sketch the complete block diagram to implement this complete control system.

## REFERENCES

Bryson, A. E., Jr., and Y.-C. Ho, 1975. *Applied Optimal Control: Optimization, Estimation and Control*, Halsted, New York.

Goodwin, G. C., and K. S. Sin, 1984. *Adaptive Filtering, Prediction and Control*, Prentice Hall, Englewood Cliffs, NJ.

Gunckel, T. L., and G. F. Franklin, 1968. A general solution for linear sampled-data control, *Trans. ASME: Journal of Basic Engineering*, Ser. D, *85*: 197–203.

Joseph, P. D., and J. Tou, 1961. On linear control theory, *Trans. AIEE*, *80*: 193–196.

Lewis, F. L., 1986. *Optimal Estimation with an Introduction to Stochastic Control Theory*, Wiley, New York.

Stengel, R. F., 1986. *Stochastic Optimal Control: Theory and Applications*, Wiley, New York.

Wonham, W. M., 1968. On the separation theorem of stochastic control, *SIAM Journal of Control*, *6*(2): 312–326.

# 13
# Introduction to System Identification

## 13.1 INTRODUCTION

In the fields of physical science and engineering we are often interested in the development of a mathematical model of some physical phenomenon in order to make analytical predictions about the behavior of the system. In control applications we are often interested in modeling a physical plant, which we wish to control, in order to predict the effect of control efforts and disturbances on that plant.

Often, the system model can be obtained by application of physical principles to the region of space designated as the system and obtaining the governing dynamic equations. Often these equations result from force, mass, energy, or momentum balances or the governing principles of electromechanical systems (i.e., Newton's, Kirchhoff's, Lenz's, and Faraday's laws). Sometimes the model cannot be obtained from physical arguments because of unknown chemical reactions, unknown boundary conditions on the physical processes, or the extreme complexity of the process. In these cases we must resort to the experimental method to develop a system model. This method is applied to the system of Fig. 13.1, wherein measurements of system stimuli and responses are made and the dynamic nature of the system is deduced from the relations between responses and stimuli. This process, known as *system identification* or *characterization*, is the epitome of the scientific method. There are many techniques for accomplishing this process and only two are given here; however, it is interesting to note that the more popular techniques are (1) random or pseudorandom

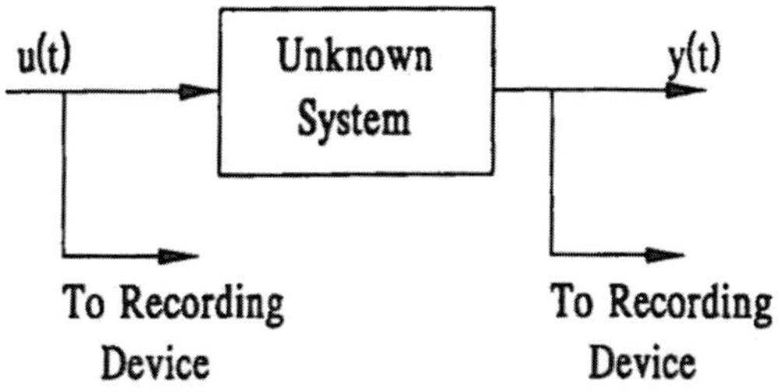


**Figure 13.1.** System to be identified.

excitation; (2) impulse excitation; (3) step excitation; and (4) sinusoidal excitation with a sweep through the frequencies.

A technique is given here that uses an arbitrary deterministic form of input and a system model of fixed structure, leaving only the parameters in that structure to be identified. Before we continue with our discussion of the identification algorithms, we uncover a few of the ideas of the least-squares method of data fitting.

## 13.2 LEAST-SQUARES TECHNIQUE

We will be interested in the identification of systems with constant parameters which will form the parameter vector $\boldsymbol{\theta}$, which will be related to the measurements by the linear vector relation

$$y_i = \mathbf{d}_i^T\boldsymbol{\theta} \qquad i = 1, 2, \ldots, n \tag{13.2.1}$$

where $\mathbf{d}_i$ is a $p$-vector of data corresponding to measurement $y_i$ and $\boldsymbol{\theta}$ is a $p$-vector of parameters which we wish to find based on the data $y_i$ and $\mathbf{d}_i^T$. Let us define our estimate of the parameter vector as $\hat{\boldsymbol{\theta}}$ which may differ from the actual parameter values of $\boldsymbol{\theta}$. The set of equations then are

$$y_i = \mathbf{d}_i^T\hat{\boldsymbol{\theta}} + e_i \qquad i = 1, 2, \ldots, n \tag{13.2.2}$$

where $e_i$ is the error induced because we have employed the estimate of $\boldsymbol{\theta}$. The set of $n$ equations of (13.2.2) can be written as the vector equation

$$\mathbf{y} = \mathbf{D}\hat{\boldsymbol{\theta}} + \mathbf{e} \tag{13.2.3}$$

where $\mathbf{D}$ is the $n \times p$ data matrix with an $i$th row of $\mathbf{d}_i^T$ and $\mathbf{y}$ is commonly called the $n \times 1$ measurement vector and $\mathbf{e}$ is an $n \times 1$ vector of errors.

There are many practical problems that we can put into the form of

(13.2.3), one of which is the curve-fitting problem, which is discussed in Examples 13.1 and 13.2.

**Example 13.1.** An example of least-squares curve fitting involves the set of experimental data shown in Fig. 13.2, which we wish to fit with a straight line, or

$$y_i \cong mx_i + b \qquad i = 1, 2, \ldots, n \tag{13.2.4}$$

or

$$y_i = mx_i + b + e_i \qquad i = 1, 2, \ldots, n \tag{13.2.5}$$

In this case the parameter vector is two-dimensional and contains the slope and $y$ intercept of the straight line, or

$$\boldsymbol{\theta} = \begin{bmatrix} m \\ b \end{bmatrix} \tag{13.2.6}$$

The $n$ relations of (13.2.5) can be written in matrix form as

$$\begin{bmatrix} y_1 \\ y_2 \\ \vdots \\ y_n \end{bmatrix} = \begin{bmatrix} x_1 & 1 \\ x_2 & 1 \\ \vdots & \vdots \\ x_n & 1 \end{bmatrix} \begin{bmatrix} \hat{m} \\ \hat{b} \end{bmatrix} + \begin{bmatrix} e_1 \\ e_2 \\ \vdots \\ e_n \end{bmatrix} \tag{13.2.7}$$

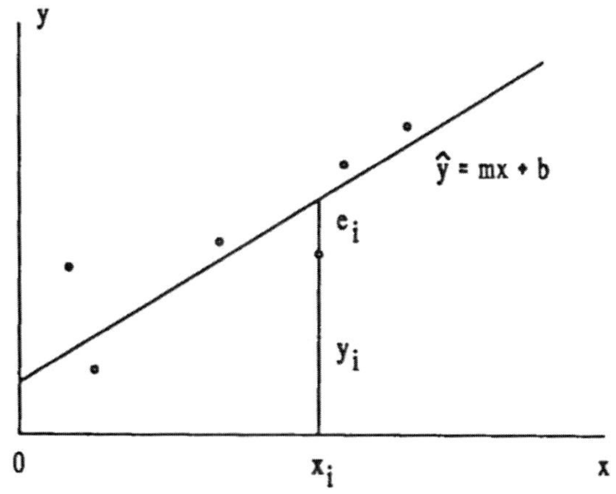


**Figure 13.2.** Data points to be fit with a straight line.

In this example the number of measurements ($n$) must be greater than the number of parameters ($p$), which in this case is two.

The technique of least squares is not restricted to models that are linear in the independent variable $x$, but the models *must* be linear in the parameters, as is this one in $m$ and $b$. Polynomial curve fits are common, but note that these fits are linear in the coefficients which are the parameters. Similarly, Fourier series representations are linear in the parameters, which in that case are the unknown Fourier coefficients.

Let us return now to our general discussion of least squares. The model of the estimator is

$$\mathbf{y} = \mathbf{D}\boldsymbol{\theta} \tag{13.2.8}$$

while in reality

$$\mathbf{y} = \mathbf{D}\hat{\boldsymbol{\theta}} + \mathbf{e} \tag{13.2.9}$$

where the $\mathbf{e}$ vector represents the errors in the estimates of the points $y_i$. We want to minimize the sum of the squares of the errors or minimize the scalar function

$$J = \mathbf{e}^T\mathbf{e} = \sum_{i=1}^{n} e_i^2 \tag{13.2.10}$$

We see that the error vector is the difference of $\mathbf{y}$ and the predicted value of $\mathbf{y}$ (namely $\mathbf{D}\hat{\boldsymbol{\theta}}$), or

$$J = (\mathbf{y} - \mathbf{D}\hat{\boldsymbol{\theta}})^T(\mathbf{y} - \mathbf{D}\hat{\boldsymbol{\theta}}) \tag{13.2.11}$$

Now to minimize $J$ with respect to $\hat{\boldsymbol{\theta}}$ we differentiate the function $J$ with respect to the vector $\hat{\boldsymbol{\theta}}$ and we get

$$\frac{\partial J}{\partial \hat{\boldsymbol{\theta}}} = (\mathbf{y} - \mathbf{D}\hat{\boldsymbol{\theta}})^T(-\mathbf{D}) + (-\mathbf{D}^T)(\mathbf{y} - \mathbf{D}\hat{\boldsymbol{\theta}}) = 0 \tag{13.2.12}$$

These two terms are the same, and hence it is sufficient that one of them be zero and hence

$$\mathbf{D}^T\mathbf{D}\hat{\boldsymbol{\theta}} = \mathbf{D}^T\mathbf{y} \tag{13.2.13}$$

This set of equations is often referred to as the *normal equations*, which must be solved for the parameter estimate vector $\hat{\boldsymbol{\theta}}$. One way would be to invert the matrix $\mathbf{D}^T\mathbf{D}$ to give

$$\hat{\boldsymbol{\theta}} = (\mathbf{D}^T\mathbf{D})^{-1}\mathbf{D}^T\mathbf{y} \tag{13.2.14}$$

The matrix $\mathbf{D}^T\mathbf{D}$ is $p \times p$ and must be inverted. If $p > n$, then the rank of $\mathbf{D}^T\mathbf{D}$ is less than $p$ and it will be singular and hence not invertible.

Clearly, the number of data points ($n$) taken should be larger than the number of parameters to be estimated ($p$). When dealing with a large number of data points it is possible that numerical problems will arise in the inversion required in (13.2.14). It might be easier to solve (13.12.13) by an iterative technique such as relaxation.

**Example 13.2.** Consider the least-squares curve fit of a straight line of the form

$$y = mx + b$$

to the following data and shown in Fig. 13.3.

| $x$ | 0 | 2 | 4 |
|---|---|---|---|
| $y$ | 1 | 2 | 4 |

The parameter vector is

$$\boldsymbol{\theta} = \begin{bmatrix} m \\ b \end{bmatrix}$$

According to relation (13.2.7), the data matrix and measurement vector are

$$\mathbf{D} = \begin{bmatrix} 0 & 1 \\ 2 & 1 \\ 4 & 1 \end{bmatrix} \qquad \mathbf{y} = \begin{bmatrix} 1 \\ 2 \\ 4 \end{bmatrix}$$

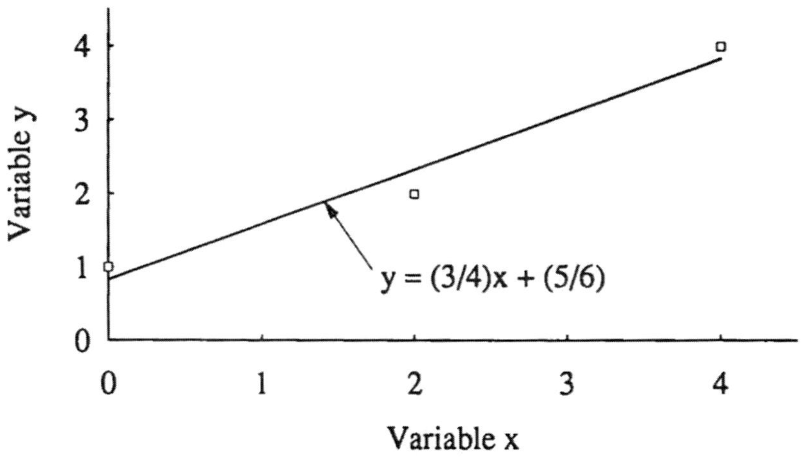


**Figure 13.3.** Data points to be fit by least-squares straight line.

We shall use relation (13.2.14) to solve the problem

$$\mathbf{D}^T\mathbf{D} = \begin{bmatrix} 20 & 6 \\ 6 & 3 \end{bmatrix} \qquad \mathbf{D}^T\mathbf{y} = \begin{bmatrix} 20 \\ 7 \end{bmatrix}$$

Using relation (13.2.14), we get the parameter vector estimate

$$\hat{\boldsymbol{\theta}} = \begin{bmatrix} \frac{3}{4} \\ \frac{5}{6} \end{bmatrix}$$

So the best-fit least-square straight line is

$$y = \frac{3}{4}x + \frac{5}{6}$$

and this is plotted as the straight line in Fig. 13.3.

## 13.3 TRANSFER FUNCTION ESTIMATION USING LEAST SQUARES

We shall assume that we have a discrete-data system and that we have measured stimulus and response sequences $u(k)$ and $y(k)$, respectively. We must assume a priori a form of the $z$-domain transfer function. It is sometimes not an easy task, so one must assume several forms for the transfer function and see which one gives the minimum aggregate error. An example will best illustrate this technique, which then can easily be applied to other transfer functions.

Assume that the process involved is governed by the transfer function

$$G(z) = \frac{Y(z)}{U(z)} = \frac{b_{n-1}z^{n-1} + \cdots + b_1 z + b_0}{z^n - a_{n-1}z^{n-1} - \cdots - a_1 z - a_0} \tag{13.3.1}$$

The equivalent difference equation is

$$\begin{aligned} y(k) = {} & a_{n-1}y(k-1) + \cdots + a_0 y(k-n) \\ & + b_{n-1}u(k-1) + \cdots + b_0 u(k-n) \end{aligned} \tag{13.3.2}$$

We shall assume that $y(k)$ and $u(k)$ are zero for negative indices $k$.

Let us define the $k$th data vector to be

$$\begin{aligned} \mathbf{d}(k) = {} & [y(k-1)\ y(k-2) \cdots \\ & y(k-n)\ u(k-1) \cdots u(k-n)]^T \end{aligned} \tag{13.3.3}$$

and the parameter vector to be

$$\boldsymbol{\theta} = [a_{n-1} \cdots a_0 \quad b_{n-1} \cdots b_0]^T \tag{13.3.4}$$

Then the output at time $k$ can be written as the inner product of the parameter vector and the data vector

$$y(k) = \mathbf{d}^T(k)\boldsymbol{\theta} \tag{13.3.5}$$

If an estimate of $\boldsymbol{\theta}$, say $\hat{\boldsymbol{\theta}}$, is used, there will be an error in the prediction of $y(k)$, or

$$y(k) = \mathbf{d}^T(k)\hat{\boldsymbol{\theta}} + e(k) \tag{13.3.6}$$

If we start $k$ at $n$ and end it at $N$, the resulting set of equations is

$$\begin{aligned} y(n) &= \mathbf{d}^T(n)\hat{\boldsymbol{\theta}} + e(n) \\ &\vdots \\ y(N) &= \mathbf{d}^T(N)\hat{\boldsymbol{\theta}} + e(N) \end{aligned} \tag{13.3.7}$$

Define the vector of output data $\mathbf{y}(N)$ as

$$\mathbf{y}(N) = [y(n) \cdots y(N)]^T \tag{13.3.8}$$

Define the data matrix as

$$\mathbf{D}(N) = [\mathbf{d}(n) \cdots \mathbf{d}(N)]^T = \begin{bmatrix} \mathbf{d}^T(n) \\ \vdots \\ \mathbf{d}^T(N) \end{bmatrix} \tag{13.3.9}$$

Also define the error vector

$$\mathbf{e}(N) = [e(n) \cdots e(N)]^T \tag{13.3.10}$$

If we intend to estimate the parameter vector, the system of equations (13.3.7) can be written as

$$\mathbf{y}(N) = \mathbf{D}(N)\hat{\boldsymbol{\theta}} + \mathbf{e}(N) \tag{13.3.11}$$

If we want to estimate the parameter vector at time $NT$, we want to minimize the sum of the squares of the errors, or

$$J = \sum_{k=n}^{N} e^2(k) = \mathbf{e}^T(N)\mathbf{e}(N) \tag{13.3.12}$$

If we solve relation (13.3.11) for $\mathbf{e}(N)$ and substitute into the relation for $J$,

$$J = (\mathbf{y}(N) - \mathbf{D}(N)\hat{\boldsymbol{\theta}})^T (\mathbf{y}(N) - \mathbf{D}(N)\hat{\boldsymbol{\theta}}) \tag{13.3.13}$$

If we differentiate this product with respect to the vector $\hat{\boldsymbol{\theta}}$, we get

$$\frac{\partial J}{\partial \hat{\boldsymbol{\theta}}} = -\mathbf{y}(N)\mathbf{D}(N) + \hat{\boldsymbol{\theta}}^T\mathbf{D}^T(N)\mathbf{D}(N) = 0 \qquad (13.3.14)$$

or

$$\mathbf{D}^T(N)\mathbf{D}(N)\hat{\boldsymbol{\theta}} = \mathbf{D}^T(N)\mathbf{y}(N) \qquad (13.3.15)$$

This set of equations is called the *normal equations* from least-squares theory, and much attention has been given to the solution of them. We may write the solution symbolically as

$$\hat{\boldsymbol{\theta}} = (\mathbf{D}^T(N)\mathbf{D}(N)^{-1}\mathbf{D}^T(N)\mathbf{y}(N) \qquad (13.3.16)$$

although it is not very likely that we would actually invert the matrix to get the parameter estimate.

**Example 13.3.** We shall consider the identification of the thermal system of Example 3.3. Since the idea would be to control this system with the digital computer in the loop, it stands to reason that we could use the same digital computer to identify a discrete-time model for the plant, the analog-to-digital (A/D) converter, and the digital-to-analog (D/A) converter. The forcing sequence, in this case a unit step, is generated by the computer, and the A/D converter is used to sample the system output. The forcing and response sequences are used in expression (13.3.16) to perform the parameter identification. The physical situation is shown in Fig. 13.4. The driving sequence applied is a unit-step function applied at $k = 0$, so the response sequence $y(k)$ recorded by the A/D converter and the driving sequence are given in Table 13.1. For this problem the data matrix is

$$\mathbf{D} = \begin{bmatrix} 0 & 0 & 1 & 0 \\ 0.01953 & 0 & 1 & 1 \\ \vdots & \vdots & \vdots & \vdots \\ 1.21561 & 1.16435 & 1 & 1 \end{bmatrix}$$

and the measurement matrix is

$$\mathbf{y} = \begin{bmatrix} 0.01953 \\ 0.07567 \\ \vdots \\ 1.26199 \end{bmatrix}$$

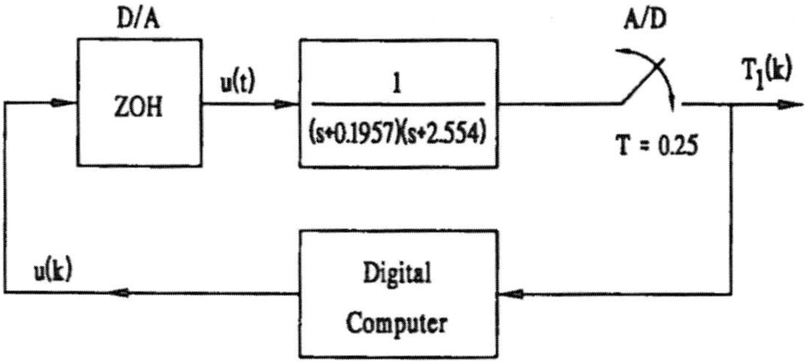


**Figure 13.4.** Digital computer system identification.

Using relation (13.3.16) the estimates of the parameters are

$$\hat{\boldsymbol{\theta}} = \begin{bmatrix} 1.47023 \\ -0.49018 \\ 0.01953 \\ 0.02776 \end{bmatrix}$$

The actual discrete-time parameters found in Example 3.3 are

$$\boldsymbol{\theta} = \begin{bmatrix} 1.48 \\ -0.5026 \\ 0.025 \\ 0.0204 \end{bmatrix}$$

**Table 13.1.** Discrete Data for Thermal System Identification with $T = 0.25$ s

| $k$ | $y(k)$ | $u(k)$ | $k$ | $y(k)$ | $u(k)$ |
|---|---|---|---|---|---|
| 0 | 0 | 1 | 11 | 0.76159 | 1 |
| 1 | 0.01953 | 1 | 12 | 0.82750 | 1 |
| 2 | 0.07567 | 1 | 13 | 0.89096 | 1 |
| 3 | 0.14890 | 1 | 14 | 0.95199 | 1 |
| 4 | 0.22945 | 1 | 15 | 1.00812 | 1 |
| 5 | 0.31001 | 1 | 16 | 1.06427 | 1 |
| 6 | 0.39300 | 1 | 17 | 1.11553 | 1 |
| 7 | 0.47111 | 1 | 18 | 1.16435 | 1 |
| 8 | 0.54922 | 1 | 19 | 1.21561 | 1 |
| 9 | 0.62245 | 1 | 20 | 1.26199 | 1 |
| 10 | 0.69324 | 1 | | | |

The sum of the squares of the errors can be evaluated using relation (13.5.13) to give

$$J = \mathbf{y}^T(N)(\mathbf{y}(N) - \mathbf{D}(N)\hat{\boldsymbol{\theta}}) = 4.63 \times 10^{-5}$$

Using these parameters, a digital simulation was run, and in plots of response data no difference could be detected between the data of Table 13.1 and those given in the digital simulation

The experiment was run several times, and each time data similar to those in Table 13.1, but not exactly the same, were obtained and each time yielded a somewhat different parameter vector estimate. The technique developed in this section is sometimes referred to as a batch mode identification technique because the estimation process used a block of previously collected data.

## 13.4 WEIGHTED LEAST SQUARES

Often it is important to weight the data nearest our current time more heavily than those in the past, so it is sometimes wise to incorporate a weighting matrix into the problem, or

$$J = \mathbf{e}^T(N)\mathbf{W}(N)\mathbf{e}^T(N) \tag{13.4.1}$$

A popular choice for the elements of the weighting matrix is to make it diagonal, such that

$$\mathbf{W}(N) = \begin{bmatrix} a\gamma^{N-n} & & & \bigcirc \\ & \ddots & & \\ & & a\gamma & \\ \bigcirc & & & a \end{bmatrix} \tag{13.4.2}$$

so the $i$th diagonal element is $W_i(N) = a\gamma^{N-n+1-i}$.

Minimizing $J$ yields the following normal equations which involve the weighting matrix:

$$\hat{\boldsymbol{\theta}}(N) = (\mathbf{D}^T(N)\mathbf{W}(N)\mathbf{D}(N)^{-1}\mathbf{D}^T(N)\mathbf{W}(N)\mathbf{y}(N) \tag{13.4.3}$$

If $\mathbf{W}(N) = \mathbf{I}$, we have the ordinary unweighted least squares of relation (13.3.16).

## 13.5 RECURSIVE LEAST SQUARES

In Section 13.3 we explored a batch mode identification technique and in this section we explore a technique with which we may reevaluate the parameter estimate with each newly obtained response sample.

Let us look back at the weighted least-squares technique, which we shall examine for the $(N + 1)$st estimate of the parameter vector

$$\begin{aligned}\hat{\boldsymbol{\theta}}(N + 1) = [&\mathbf{D}^T(N + 1)\mathbf{W}(N + 1)\\ &\times \mathbf{D}(N + 1)]^{-1}\mathbf{D}^T(N + 1)\mathbf{W}(N + 1)\\ &\times \mathbf{y}(N + 1)\end{aligned} \tag{13.5.1}$$

Let us now examine the matrix product in the square brackets, which may be written out as

$$\mathbf{D}^T(N + 1)\mathbf{W}(N + 1)\mathbf{D}(N + 1) = \sum_{k=n}^{N+1} \mathbf{d}^T(k)\mathbf{W}_k(N + 1)\mathbf{d}^T(k)$$

and if we choose exponential weighting,

$$= \sum_{k=n}^{N+1} \mathbf{d}(k)a\gamma^{N+1-k}\mathbf{d}^T(k) \tag{13.5.2}$$

This can be partitioned into two terms:

$$\begin{aligned}&\mathbf{D}^T(N + 1)\mathbf{W}(N + 1)\mathbf{D}(N + 1)\\ &\quad= \sum_{k=n}^{N} \mathbf{d}(k)a\gamma^{N-k}\mathbf{d}^T(k) + \mathbf{d}(N + 1)a\mathbf{d}^T(N + 1)\\ &\quad= \gamma\mathbf{D}^T(N)\mathbf{W}(N)\mathbf{D}(N) + \mathbf{d}(N + 1)a\mathbf{d}^T(N + 1)\end{aligned} \tag{13.5.3}$$

Define a $2n \times 2n$ matrix $\mathbf{P}(N + 1)$ as

$$\mathbf{P}(N + 1) = [\mathbf{D}^T(N + 1)\mathbf{W}(N + 1)\mathbf{D}(N + 1)]^{-1} \tag{13.5.4}$$

From relation (13.5.3) we see that

$$\mathbf{P}(N + 1) = [\gamma\mathbf{P}^{-1}(N) + \mathbf{d}(N + 1)a\mathbf{d}^T(N + 1)]^{-1} \tag{13.5.5}$$

Now employ the matrix inversion lemma, which is proven in Appendix C, to yield

$$\begin{aligned}\mathbf{P}(N + 1) = \frac{\mathbf{P}(N)}{\gamma} - &\frac{\mathbf{P}(N)}{\gamma}\mathbf{d}(N + 1)\\ &\left(\frac{1}{a} + \mathbf{d}^T(N + 1)\frac{\mathbf{P}(N)}{\gamma}\mathbf{d}(N + 1)\right)^{-1}\\ &\mathbf{d}^T(N + 1)\frac{\mathbf{P}(N)}{\gamma}\end{aligned} \tag{13.5.6}$$

We also need to examine the quantity $\mathbf{D}^T(N + 1)\mathbf{W}(N + 1)\mathbf{y}(N + 1)$, which may also be partitioned as

$$\mathbf{D}^T(N+1)\mathbf{W}(N+1)\mathbf{y}(N+1)$$

$$= \sum_{k=n}^{N} \mathbf{d}(k)a\gamma^{N+1-k}\mathbf{y}(N) + \mathbf{d}(N+1)ay(N+1)$$

$$= \gamma\mathbf{D}^T(N)\mathbf{W}(N)\mathbf{y}(N) + \mathbf{d}(N+1)ay(N+1) \qquad (13.5.7)$$

From the weighted least-squares result (13.4.3), we get

$$\hat{\boldsymbol{\theta}}(N+1) = \left[\frac{\mathbf{P}(N)}{\gamma} - \frac{\mathbf{P}(N)}{\gamma}\mathbf{d}(N+1)\left(\frac{1}{a} + \mathbf{d}^T(N+1)\frac{\mathbf{P}(N)}{\gamma}\mathbf{d}(N+1)\right)^{-1}\mathbf{d}^T(N+1)\frac{\mathbf{P}(N)}{\gamma}\right] \times [\gamma\mathbf{D}(N)\mathbf{W}(N)\mathbf{y}(N) + \mathbf{d}(N+1)ay(N+1)] \qquad (13.5.8)$$

When we multiply this out and note that

$$\mathbf{P}(N)\mathbf{D}^T(N)\mathbf{W}(N)\mathbf{y}(N) = \hat{\boldsymbol{\theta}}(N) \qquad (13.5.9)$$

we get

$$\hat{\boldsymbol{\theta}}(N+1)$$

$$= \hat{\boldsymbol{\theta}}(N) + \frac{\mathbf{P}(N)}{\gamma}\mathbf{d}(N+1)ay(N+1) - \frac{\mathbf{P}(N)}{\gamma}\left(\frac{1}{a} + \mathbf{d}(N+1)\frac{\mathbf{P}(N)}{\gamma}\mathbf{d}^T(N+1)\right)^{-1}\mathbf{d}^T(N+1)\hat{\boldsymbol{\theta}}(N)$$

$$- \frac{\mathbf{P}(N)}{\gamma}\mathbf{d}(N+1)\left(\frac{1}{a} + \mathbf{d}^T(N+1)\frac{\mathbf{P}(N)}{\gamma}\mathbf{d}(N+1)\right)^{-1} \times \mathbf{d}^T(N+1)\frac{\mathbf{P}(N)}{\gamma}\mathbf{d}(N+1)ay(N+1) \qquad (13.5.10)$$

Now insert the following indentity between the $\mathbf{d}(N+1)$ and the $a$ in the second term:

$$\left(\frac{1}{a} + \mathbf{d}^T(N+1)\frac{\mathbf{P}(N)}{\gamma}\mathbf{d}(N+1)\right)^{-1}\left(\frac{1}{a} + \mathbf{d}^T(N+1)\frac{\mathbf{P}(N)}{\gamma}\mathbf{d}(N+1)\right) \qquad (13.5.11)$$

After expanding we get the following estimate of the parameter vector at time $(N + 1)T$:

$$\hat{\boldsymbol{\theta}}(N + 1) = \hat{\boldsymbol{\theta}}(N) + \mathbf{k}(N + 1)\big(y(N + 1) - \mathbf{d}^T(N + 1)\hat{\boldsymbol{\theta}}(N)\big) \tag{13.5.12}$$

where the "gain" vector $\mathbf{k}(N + 1)$ is defined as

$$\mathbf{k}(N + 1) = \frac{\mathbf{P}(N)}{\gamma}\,\mathbf{d}(N + 1)\left(\frac{1}{a} + \mathbf{d}^T(N + 1)\,\frac{\mathbf{P}(N)}{\gamma}\,\mathbf{d}(N + 1)\right)^{-1} \tag{13.5.13}$$

One can note the similarity between this and the structure of the Kalman filter from Chapter 11. Note that the parameter estimator given by (13.5.12) is recursive and that it is driven by the error in the predicted system output at time $(N + 1)T$.

This algorithm is useful for real-time identification and hence can be used to monitor systems for which the parameters drift slowly in time. An adaptive control strategy can thus be employed which not only updates the values of the estimates of the system parameters but also updates the control law either in the form of a conventional compensator or a state-space controller composed of a state estimator and estimated state feedback control law. The algorithm can be started in many ways, and hence we make the following suggestions. Let

$$\mathbf{P}(n) = \alpha\mathbf{I}$$

where $\alpha$ is large and positive. If we assume that the first $n$ data points are noiseless, the exact solution for $\hat{\boldsymbol{\theta}}(n)$ would be

$$\hat{\boldsymbol{\theta}}(n) = \begin{bmatrix} \mathbf{d}^T(1) \\ \mathbf{d}^T(2) \\ \vdots \\ \mathbf{d}^T(n) \end{bmatrix}^{1} \begin{bmatrix} y(1) \\ \vdots \\ y(n) \end{bmatrix} \tag{13.5.14}$$

and this would be one way of starting the recursive process.

The algorithm for recursive identification is then summarized as

1. Choose $\mathbf{P}(k)$, $\gamma$, $a$, and $\hat{\boldsymbol{\theta}}(k)$ to start the algorithm.
2. Compute $\mathbf{k}(k + 1)$ from

$$\mathbf{k}(k + 1) = \frac{\mathbf{P}(k)}{\gamma}\,\mathbf{d}(k + 1)\left(\frac{1}{a} + \mathbf{d}^T(k + 1)\,\frac{\mathbf{P}(k)}{\gamma}\,\mathbf{d}(k + 1)\right)^{-1}$$

3. Estimate the parameters at time $k + 1$ or

$$\hat{\boldsymbol{\theta}}(k + 1) = \hat{\boldsymbol{\theta}}(k) + \mathbf{k}(k)\,\big(y(k + 1) - \mathbf{d}^T(k + 1)\hat{\boldsymbol{\theta}}(k)\big)$$

4. Compute $\mathbf{P}(k + 1)$ from

$$\mathbf{P}(k + 1) = \frac{1}{\gamma}[\mathbf{I} - \mathbf{k}(k)\mathbf{d}^T(k + 1)]\mathbf{P}(k)$$

5. Let $k = k + 1$.
6. Go to step 2.

Now we shall illustrate the technique with the data of Example 13.2.

**Example 13.4.** Let us now consider the same data used in the batch mode identification performed in Example 13.3. Let us identify the system recursively and choose an initial estimate of the parameter vector to be

$$\hat{\boldsymbol{\theta}}(0) = \begin{bmatrix} 1 \\ -0.5 \\ 0.02 \\ 0.02 \end{bmatrix}$$

and the initial value of $\mathbf{P}(0)$ as

$$\mathbf{P}(0) = 10^4\mathbf{I}$$

The weighting factors are chosen such the weighting of the past data is exponential, or

$$a = 0.2$$
$$\gamma = (1 - a) = 0.8$$

The algorithm as outlined was programmed and runs very well on a personal computer. In fact, the algorithm is so efficient that it can run in real time in parallel with a control law such that the processor can both control and identify the system at the same time. Plots of the identified parameter histories are given in Fig. 13.5.

## 13.6 EFFECTS OF NOISE

Often in an engineering environment the input and output measurements may be contaminated by random noise. For low levels of noise the methods just presented may produce excellent estimates of the system parameters,

**Figure 13.5.** (a) Parameter estimate history for $\theta_1 = b_1$; (b) parameter estimate history for $\theta_2 = b_0$; (c) parameter estimate history for $\theta_3 = a_1$; (d) parameter estimate history for $\theta_4 = a_0$.

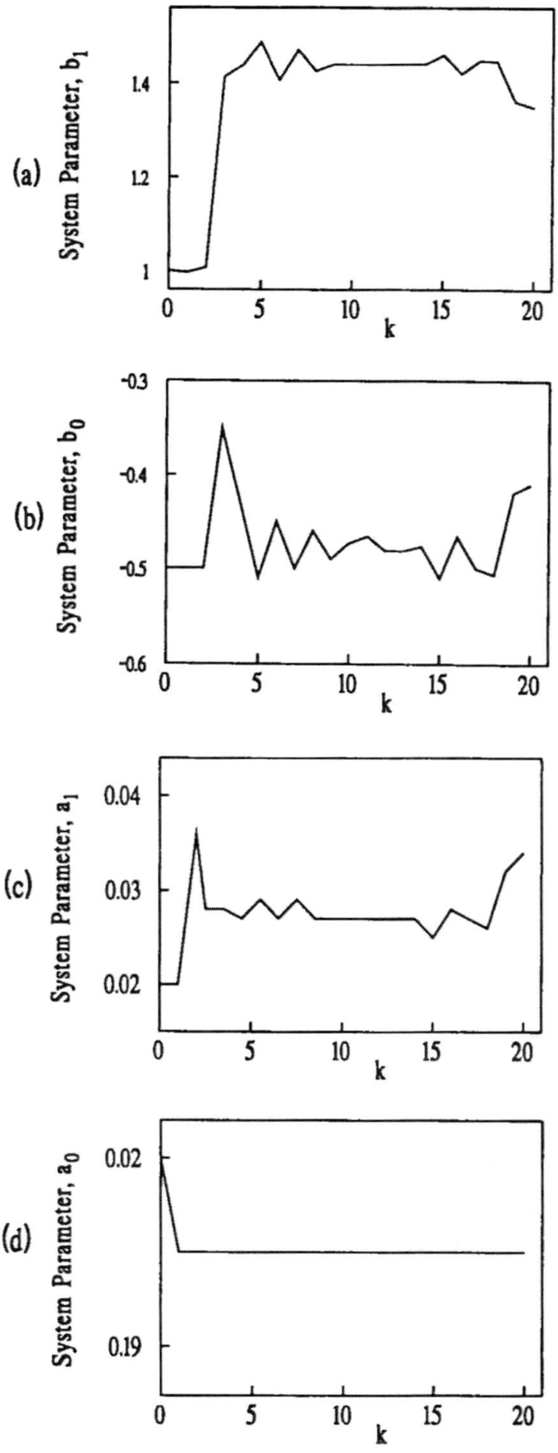

(a)
System Parameter, $b_1$
1.4
1.2
1
0
5
10
15
20
k
(b)
System Parameter, $b_0$
-0.3
-0.4
-0.5
-0.6
0
5
10
15
20
k
(c)
System Parameter, $a_1$
0.04
0.03
0.02
0
5
10
15
20
k
(d)
System Parameter, $a_0$
0.02
0.19
0
5
10
15
20
k

however, with larger levels of noise may require modification of the weighting factors $\gamma$ and $a$. For still larger levels of noise there may be no hope for estimation of the parameters.

Another source of random noise in the measured data is that the system to be identified is also driven by random disturbances at some point. In this case the noise at the output is not purely random, but rather, statistically related (correlated) from sample to sample. There exists a method that will find the least-squares estimate of the parameters, which accounts for the random excitation of the system and thus improved parameter estimates result. This technique is called the *stochastic least-squares method* (Franklin et al., 1990).

One way to handle the random noise in the measurements is to perform the experiment by which the data are gathered a number of times under identical conditions. The data from each of these independent experiments are then averaged at each given time step; thus the random fluctuations in the responses tend to average to zero and the signal-to-noise ratio is greatly increased. In fact, it can be shown that the standard deviation of the noise on the averaged response decreases by a factor of the reciprocal of the square root of the number of records taken.

## 13.7. SUMMARY

In this chapter we have addressed the subject of transfer-function parameter identification. The least-squares batch method was derived and an example given. The recursive least-squares method was also derived. The recursive technique is useful because it can be employed for simultaneous real-time identification and control.

## PROBLEMS

13.1. Given the input–output data given in the table below and a system model of the form

$$y(k+1) = ay(k) + bu(k)$$

identify the parameters $a$ and $b$ using the data given using the least-squares technique in the batch mode. Note that there is a small amount of noise on the measurement data of the output.

| $k$ | $u(k)$ | $y(k)$ | $k$ | $u(k)$ | $y(k)$ |
|---|---|---|---|---|---|
| 0 | 1 | 0 | 6 | 1 | 1.843 |
| 1 | 1 | 0.490 | 7 | 1 | 1.976 |
| 2 | 1 | 0.905 | 8 | 1 | 2.091 |
| 3 | 1 | 1.218 | 9 | 1 | 2.165 |
| 4 | 1 | 1.476 | 10 | 1 | 2.240 |
| 5 | 1 | 1.681 | | | |

13.2. Given the data of Problem 13.1, identify the system employing the recursive least-squares technique. Choose $a = 0.2$ and $\gamma = 0.8$ and $\mathbf{P}(0) = 10^4\mathbf{I}$.

13.3. Consider the sprung-mass positioning problem posed in Problem 3.9. A unit-step force input is employed and the responses are tabulated below.

| $k$ | $y(k)$ | $k$ | $y(k)$ | $k$ | $y(k)$ |
|---|---|---|---|---|---|
| 0 | 0 | 7 | 0.0296 | 14 | 0.0472 |
| 1 | 0.0215 | 8 | 0.0510 | 15 | 0.0584 |
| 2 | 0.0639 | 9 | 0.0686 | 16 | 0.0615 |
| 3 | 0.0877 | 10 | 0.0685 | 17 | 0.0555 |
| 4 | 0.0770 | 11 | 0.0539 | 18 | 0.0469 |
| 5 | 0.0465 | 12 | 0.0387 | 19 | 0.0435 |
| 6 | 0.0254 | 13 | 0.0375 | 20 | 0.0470 |

Identify a second-order model for this system using the least-squares technique.

13.4. For the sprung-mass problem of Problem 13.3., identify the system recursively using $a = 0.2$, $\gamma = 0.8$, and $\mathbf{P}(0) = 10^4\mathbf{I}$.

## REFERENCES

Anderson, B. D. O., and J. B. Moore, 1979. *Optimal Filtering*, Prentice Hall, Englewood Cliffs, NJ.

Franklin, G. F., J. D. Powell, and M. L. Workman, 1990. *Digital Control of Dynamic Systems*, 2nd Ed., Addison-Wesley, Reading, MA.

Goodwin, G. C., and K. S. Sin, 1984. *Adaptive Filtering, Prediction and Control*, Prentice Hall, Englewood Cliffs, NJ.

Johnson, C. R., Jr., 1988. *Lectures on Adaptive Parameter Identification*, Prentice Hall, Englewood Cliffs, NJ.

# Appendix A
# Tables and Properties of z-Transforms

## A.1 PROOF OF THE COMPLEX INVERSION INTEGRAL FOR THE z-TRANSFORM

We have seen that a pragmatic approach to the inversion of the $z$-transform is the use of partial fraction expansions and the $z$-transform table presented in this appendix. As in Chapter 2, the one-sided $z$-transform of a sequence of numbers $f(0), f(1), \ldots$ is defined as

$$F(z) = \mathfrak{Z}[f(k)] = \sum_{k=0}^{\infty} f(k)z^{-k} \tag{A.1}$$

Let us now consider the following integral, which will be denoted as $I$:

$$I = \oint_C F(z)z^{k-1}\,dz \tag{A.2}$$

where the contour C is a circle chosen to contain all the poles of $F(z)$. Let us now substitute (A.1) with (A.2) to yield

$$I = \oint_C \left[\sum_{i=0}^{\infty} f(i)z^{-i}\right] z^{k-1}\,dz \tag{A.3}$$

or

$$I = \oint_C \sum_{i=0}^{\infty} f(i)z^{-i+k-1}\,dz \tag{A.4}$$

The infinite series in (A.4) will converge everywhere on contour $C$ since $C$ has been chosen exterior to all the poles of $F(z)$. With this convergence assumed, the operations of summation and integration in expression (A.4) may be interchanged to give

$$I = \sum_{i=0}^{\infty} f(i) \oint_C z^{-i+k-1}\, dz \tag{A.5}$$

If we employ the Cauchy residue theorem, we can show that

$$\oint_C z^{-m}\, dz = \begin{cases} 0 & m > 1,\ m < 0 \\ 2\pi j & m = 1 \end{cases} \tag{A.6}$$

With this fact the result for $I$ is

$$I = f(k) 2\pi j \tag{A.7}$$

**Table A.1.** Summary of $z$-Transform Operations for the One-Sided Transform

| $f(k)$ | $F(z) = \mathfrak{Z}[f(k)]$ |
|---|---|
| 1. $af(k) + bg(k)$ | $aF(z) + bG(z)$ |
| 2. $f(k+m)$ | $z^m F(z) - \sum_{j=0}^{m-1} f(j) z^{m-j}$ |
| 3. $f(k-m)$ | $z^{-m} F(z)$ |
| 4. $\sum_{j=0}^{k} f(j)$ | $\frac{z}{z-1} F(z)$ |
| 5. $r^k f(k)$ | $F(z/r)$ |
| 6. $\sum_{m=0}^{k} h(k-m) r(m)$ | $H(z)R(z)$ |
| 7. $k^m f(k)$ | $-z^m \frac{d^m F(z)}{dz^m}$ |
| 8. $f(\infty)$ | $\lim_{z \to 1} \frac{z-1}{z} F(z)$ |
| 9. $f(0)$ | $\lim_{z \to \infty} F(z)$ |

Equating (A.7) and (A.2), the result is

$$f(k) = \frac{1}{2\pi j} \oint_C F(z) z^{k-1}\, dz \tag{A.8}$$

where the contour $C$ is chosen to be one that encloses all the poles of $F(k)$.

**Example A.1.** Find the inverse transform of the following function:

$$F(z) = \frac{z - 2}{(z - 0.8)(z - 0.6)}$$

This function has poles at $z = 0.8$ and $z = 0.6$. The inverse transform is given by (A.8) to be

$$f(k) = \frac{1}{2\pi j} \oint_C \frac{(z - 2) z^{k-1}}{(z - 0.8)(z - 0.6)}\, dz$$

where $C$ is a contour containing both the poles. The residue theorem from complex variable theory will give the inverse function to be

$$f(k) = \frac{1}{2\pi j} 2\pi j \text{ (sum of the residues of the integral)}$$

or

$$f(k) = \frac{1}{2\pi j} 2\pi j \left( \frac{0.6 - 2}{0.6 - 0.8} 0.6^{k-1} + \frac{0.8 - 2}{0.8 - 0.6} 0.8^{k-1} \right)$$

If the numerical details are carried out, the resulting sequence is

$$f(k) = 11.66\,(0.6^k) - 7.50\,(0.8^k)$$

**Table A.2.** Table of Laplace and $z$-Transform Pairs

| Transform number | $f(t)$ $t \geq 0$ | $F(s) = \mathcal{L}[f(t)]$ | $f(kT)$ $k \geq 0$ | $F(z) = \mathcal{Z}[f(kT)]$ |
|---|---|---|---|---|
| 1 | — | — | $1, k = 0$<br>$0, k \neq 0$ | 1 |
| 2 | — | — | $1, k = n$<br>$0, k \neq n$ | $z^{n}$ |
| 3 | 1 | $\frac{1}{s}$ | 1 | $\frac{z}{z-1}$ |
| 4 | t | $\frac{1}{s^2}$ | $kT$ | $\frac{Tz}{(z-1)^2}$ |
| 5 | $\frac{t^2}{2}$ | $\frac{1}{s^3}$ | $\frac{(kT)^2}{2}$ | $\frac{T^2z(z+1)}{2(z-1)^3}$ |
| 6 | $e^{at}$ | $\frac{1}{s+a}$ | $e^{aT}$ | $\frac{z}{z-e^{aT}}$ |
| 7 | $te^{at}$ | $\frac{1}{(s+a)^2}$ | $kTe^{akT}$ | $\frac{Te^{aT}z}{(z-e^{aT})^2}$ |

| | | | | |
|---|---|---|---|---|
| 8 | $1 - e^{at}$ | $\dfrac{a}{s(s+a)}$ | $1 - e^{akT}$ | $\dfrac{(1 - e^{-aT})z}{(z-1)(z - e^{aT})}$ |
| 9 | $e^{at} - e^{bt}$ | $\dfrac{b-a}{(s+a)(s+b)}$ | $e^{akT} - e^{bkT}$ | $\dfrac{(e^{-aT} - e^{-bT})z}{(z - e^{aT}(z - b^{bT})}$ |
| 10 | $\sin bt$ | $\dfrac{b}{s^2 + b^2}$ | $\sin bkT$ | $\dfrac{(\sin bT)z}{z^2 \quad 2(\cos bT)z + 1}$ |
| 11 | $\cos bt$ | $\dfrac{s}{s^2 + b^2}$ | $\cos bkT$ | $\dfrac{z^2 - (\cos bT)z}{z^2 - 2(\cos bT)z + 1}$ |
| 12 | $e^{at} \sin bt$ | $\dfrac{b}{(s+a)^2 + b^2}$ | $e^{akT} \sin bkT$ | $\dfrac{(e^{-aT} \sin bT)z}{z^2 - 2(e^{aT} \cos bT)z + e^{2aT}}$ |
| 13 | $e^{at} \cos bt$ | $\dfrac{s+a}{(s+a)^2 + b^2}$ | $e^{akT} \cos bkT$ | $\dfrac{z^2 - (e^{-aT} \cos bT)z}{z^2 - 2(e^{aT} \cos bT)z + e^{2aT}}$ |
| 14 | $\sinh bt$ | $\dfrac{b}{s^2 - b^2}$ | $\sinh bkT$ | $\dfrac{(\sinh bT)z}{z^2 - 2(\cosh bT)z + 1}$ |
| 15 | $\cosh bt$ | $\dfrac{s}{s^2 - b^2}$ | $\cosh bkT$ | $\dfrac{z^2 - (\cosh bT)z}{z^2 - 2(\cosh bT)z + 1}$ |

# Appendix B
# Algebraic Eigenvalue–Eigenvector Problem

## B.1 INTRODUCTION

The eigenvalue–eigenvector problem is a mathematical problem of considerable practical importance. A wide range of problems are studied using this concept from linear algebra. In particular, these are problems of control, vibratory systems, crystal structure, statistics, and quantum mechanics. The approach taken here will then be that with one eye on the modern control systems problem.

## B.2 STATEMENT OF THE PROBLEM

We are interested in the nature of an $n \times n$ matrix $\mathbf{A}$ which can be characterized by the following problem. We wish to find scalars $\lambda$ and vectors $\mathbf{p}$ that satisfy the relation

$$\mathbf{Ap} = \lambda\mathbf{p} \tag{B.1}$$

This is referred to as the *algebraic eigenvalue–eigenvector problem*. The scalars $\lambda$ which satisfy (B.1) are referred to as the *eigenvalues* or *characteristic values* of the matrix $\mathbf{A}$, and the vectors $\mathbf{p}$ that satisfy relation (B.1) are the *eigenvectors* of $\mathbf{A}$. The problem can be rewritten as

$$(\lambda\mathbf{I} - \mathbf{A})\mathbf{p} = 0 \tag{B.2}$$

where the zero represents the null vector. Since we are seeking nonnull vectors **p** which are mapped into null vectors by the linear transformation $(\lambda\mathbf{I} - \mathbf{A})$, the only way nonnull vectors **p** can exist is if the matrix $(\lambda\mathbf{I} - \mathbf{A})$ is singular, or

$$\det(\lambda\mathbf{I} - \mathbf{A}) = 0 \tag{B.3}$$

Expression (B.3) represents an $n$th-degree polynomial in $\lambda$, and this polynomial is said to be the *characteristic polynomial* of matrix **A**. The *fundamental theorem of algebra* states that an $n$th-degree polynomial will have $n$ roots or the matrix **A** has $n$ eigenvalues, which we shall refer to as $\lambda_1, \lambda_2, \ldots, \lambda_n$.

Returning to expression (B.2), we find that it must be satisfied for each of the $\lambda_i$ $(i = 1, \ldots, n)$, and thus there must be a vector **p** for each of the eigenvalues and that associated with $\lambda_i$ will be referred to as $\mathbf{p}_i$ and thus $\mathbf{p}_i$ must satisfy

$$(\lambda_i\mathbf{I} - \mathbf{A})\mathbf{p}_i = 0 \quad i = 1, 2, \ldots, n \tag{B.4}$$

Let us define a diagonal matrix $\mathbf{\Lambda}$ as

$$\mathbf{\Lambda} = \text{diag}(\lambda_1, \lambda_2, \ldots, \lambda_n) \tag{B.5}$$

and a modal matrix **P** as one having the eigenvectors as its $n$ columns

$$\mathbf{P} = \left[\mathbf{p}_1 \,\middle|\, \mathbf{p}_2 \,\middle|\, \cdots \,\middle|\, \mathbf{p}_n\right] \tag{B.6}$$

Let us write the $n$ relations of (E.4) as

$$\left[\mathbf{A}\mathbf{p}_1 \,\middle|\, \mathbf{A}\mathbf{p}_2 \,\middle|\, \cdots \,\middle|\, \mathbf{A}\mathbf{p}_n\right] = \left[\lambda_1\mathbf{p}_1 \,\middle|\, \lambda_2\mathbf{p}_2 \,\middle|\, \cdots \,\middle|\, \lambda_n\mathbf{p}_n\right] \tag{B.7}$$

and now using the definitions of (B.5) and (B.6), we may write (B.7) as

$$\mathbf{AP} = \mathbf{P\Lambda} \tag{B.8}$$

This may readily be verified by expanding (B.8) to yield (B.7). It can be shown (Chen, 1984) that if the eigenvalues are distinct $(\lambda_i \neq \lambda_j,\ i, j = 1, \ldots, n)$, the eigenvectors are linearly independent, and if the columns of **P** are linearly independent, it will be nonsingular and thus possess an inverse $\mathbf{P}^{-1}$.

If we premultiply (B.8) by $\mathbf{P}^{-1}$, we get

$$\mathbf{P}^{-1}\mathbf{AP} = \mathbf{\Lambda} \tag{B.9}$$

In other words, the modal matrix **P** can diagonalize the matrix **A** to give the diagonal matrix of the eigenvalues $\mathbf{\Lambda}$.

## B.3 APPLICATION OF THE EIGENVALUE PROBLEM TO DISCRETE-TIME SYSTEMS

Let us consider the homogeneous linear first-order vector difference equation

$$\mathbf{x}(k+1) = \mathbf{A}\mathbf{x}(k), \qquad \mathbf{x}(0) \text{ known} \tag{B.10}$$

and seek a solution of the form

$$\mathbf{x}(k) = \mathbf{p}\lambda^k \tag{B.11}$$

where $\mathbf{p}$ is an $n \times 1$ vector and $\lambda$ is in general a complex scalar. Substitution of (B.11) into (B.10) yields

$$\mathbf{p}\lambda^{k+1} = \mathbf{A}\mathbf{p}\lambda^k \tag{B.12}$$

or

$$(\lambda\mathbf{I} - \mathbf{A})\,\mathbf{p}\lambda^k = 0 \tag{B.13}$$

Since $\lambda \neq 0$, this reduces to an eigenvalue–eigenvector problem with the eigenvalues given by

$$\det(\lambda\mathbf{I} - \mathbf{A}) = 0 \tag{B.14}$$

which is the characteristic polynomial for the difference equation. The roots of this polynomial are the system eigenvalues denoted by $\lambda_1, \ldots, \lambda_n$. Associated with each eigenvalue $\lambda_i$ will be an eigenvector $\mathbf{p}_i$ satisfying

$$(\lambda_i\mathbf{I} - \mathbf{A})\,\mathbf{p}_i = 0 \qquad i = 1, 2, \ldots, n \tag{B.15}$$

Once the $\lambda_i$ and $\mathbf{p}_i$ are determined, we know that the total solution to the difference equation will be a linear combination of these solutions, or

$$\mathbf{x}(k) = \sum_{i=1}^{n} c_i\mathbf{p}_i\lambda_i^k \tag{B.16}$$

and the unknown $c_i$ are given by the initial conditions to be

$$\mathbf{x}(0) = \sum_{i=1}^{n} c_i\mathbf{p}_i \tag{B.17}$$

which can be written in matrix form as

$$\mathbf{x}(0) = \mathbf{P}\mathbf{c} \tag{B.18}$$

where

$$\mathbf{c} = \begin{bmatrix} c_1 \\ \vdots \\ c_n \end{bmatrix} \tag{B.19}$$

or the vector **c** is given by

$$\mathbf{c} = \mathbf{P}^{-1}\mathbf{x}(0) \tag{B.20}$$

The solution (B.16) can thus be written as

$$\mathbf{x}(k) = \mathbf{P}\mathbf{\Lambda}^k\mathbf{P}^{-1}\mathbf{x}(0) \tag{B.21}$$

We know that the solution to (B.10) is

$$\mathbf{x}(k) = \mathbf{A}^k\mathbf{x}(0) \tag{B.22}$$

Examination of expressions (B.21) and (B.22) yields the following:

$$\mathbf{A}^k = \mathbf{P}\mathbf{\Lambda}^k\mathbf{P}^{-1} \tag{B.23}$$

This may also be proven in a different fashion since from equation (B.9) we may write

$$\mathbf{A} = \mathbf{P}\mathbf{\Lambda}\mathbf{P}^{-1} \tag{B.24}$$

and if we raise this relation to the $k$th power,

$$\mathbf{A}^k = \mathbf{P}\mathbf{\Lambda}\mathbf{P}^{-1}\mathbf{P}\mathbf{\Lambda}\mathbf{P}^{-1} \cdots \mathbf{P}\mathbf{\Lambda}\mathbf{P}^{-1} \tag{B.25}$$

but the $\mathbf{P}^{-1}\mathbf{P}$ terms yield identity matrices and the relation of (B.23) is proven.

**Example B.1.** Find the eigenvalues and eigenvectors associated with the following matrix:

$$\mathbf{A} = \begin{bmatrix} 1 & -0.2 \\ 0.4 & 0.4 \end{bmatrix}$$

Let us first find the eigenvalues

$$\det\,[\lambda\mathbf{I} - \mathbf{A}] = \det \begin{bmatrix} \lambda - 1 & 0.2 \\ -0.4 & \lambda - 0.4 \end{bmatrix} = 0$$

and upon expanding,

$$\lambda^2 - 1.4\lambda + 0.48 = 0$$

The roots of this polynomial are given by the quadratic formula to be

$$\lambda_{1,2} = 0.8 \text{ and } 0.6$$

The first eigenvector is given by (B.4) to be

$$\begin{bmatrix} 0.8 - 1 & 0.2 \\ -0.4 & 0.8 - 0.4 \end{bmatrix} \begin{bmatrix} p_{11} \\ p_{21} \end{bmatrix} = \begin{bmatrix} 0 \\ 0 \end{bmatrix}$$

The first equation is

$$-0.2p_{11} + 0.2p_{21} = 0$$

or

$$p_{21} = p_{11}$$

So the first eigenvector is

$$\mathbf{p}_1 = \begin{bmatrix} 1 \\ 1 \end{bmatrix}$$

and the second eigenvector is given by

$$\begin{bmatrix} 0.6 - 1 & 0.2 \\ -0.4 & 0.6 - 0.4 \end{bmatrix} \begin{bmatrix} p_{12} \\ p_{22} \end{bmatrix} = \begin{bmatrix} 0 \\ 0 \end{bmatrix}$$

or

$$p_{22} = 2p_{12}$$

So

$$\mathbf{p}_2 = \begin{bmatrix} 1 \\ 2 \end{bmatrix}$$

Thus the modal matrix **P** is

$$\mathbf{P} = \begin{bmatrix} 1 & 1 \\ 1 & 2 \end{bmatrix}$$

Inversion of this matrix yields

$$\mathbf{P}^{-1} = \begin{bmatrix} 2 & -1 \\ -1 & 1 \end{bmatrix}$$

We can verify that indeed $\mathbf{P}^{-1}\mathbf{A}\mathbf{P}$ gives a matrix of the eigenvalues.

## Decoupling of the State Equation

Let us consider the system again described by the linear vector difference equation

$$\mathbf{x}(k + 1) = \mathbf{A}\mathbf{x}(k) \tag{B.26}$$

and let us define a new set of state variables $q_i(k)$ linearly related to the states $x_j(k)$ by

$$\mathbf{x}(k) = \mathbf{P}\mathbf{q}(k) \tag{B.27}$$

Substitution of (B.27) into (B.26) gives

$$\mathbf{Pq}(k + 1) = \mathbf{APq}(k) \tag{B.28}$$

and solving for $\mathbf{q}(k + 1)$ gives

$$\mathbf{q}(k + 1) = \mathbf{P}^{-1}\mathbf{APq}(k) \tag{B.29}$$

and from relation (B.9) we see that

$$\mathbf{q}(k + 1) = \mathbf{\Lambda q}(k) \tag{B.30}$$

where $\mathbf{\Lambda}$ is a diagonal matrix of the eigenvalues. This relation can be written as a set of first-order difference equations

$$q_i(k + 1) = \lambda_i q_i(k) \quad i = 1, 2, \ldots, n \tag{B.31}$$

which very simply has a solution

$$q_i(k) = q_i(0)\lambda_i^k \qquad i = 1, 2, \ldots, n \tag{B.32}$$

where the initial condition is yet unspecified. The initial value in the new coordinates is related to the initial condition in the original coordinates by (B.27) or its inverse relation

$$\mathbf{q}(0) = \mathbf{P}^{-1}\mathbf{x}(0) \tag{B.33}$$

and the solution (B.32) can be written as

$$\mathbf{q}(k) = \mathbf{\Lambda}^k\mathbf{P}^{-1}\mathbf{x}(0) \tag{B.34}$$

The mapping of this back to the original coordinates using (B.27) yields

$$\mathbf{x}(k) = \mathbf{P\Lambda}^k\mathbf{P}^{-1}\mathbf{x}(0) \tag{B.35}$$

where is the same as given in expression (B.21).

## The Forced Problem

Let us now consider the inhomogeneous problem where

$$\mathbf{x}(k + 1) = \mathbf{Ax}(k) + \mathbf{Bu}(k) \tag{B.36}$$

In this case the solution can be given directly by $z$-transformation, but this involves the symbolic inversion of the $(z\mathbf{I} - \mathbf{A})$ matrix, so let us seek an alternative method. As in Section B.2, let us define a new set of states as given by relation (B.27), which upon substitution into (B.36) yields)

$$\mathbf{Pq}(k + 1) = \mathbf{APq}(k) + \mathbf{Bu}(k) \tag{B.37}$$

Premultiplication by $\mathbf{P}^{-1}$ yields, after noting relation (B.9),

$$\mathbf{q}(k + 1) = \mathbf{\Lambda q}(k) + \mathbf{P}^{-1}\mathbf{Bu}(k) \tag{B.38}$$

If we let $\mathbf{P}^{-1}\mathbf{B}$ equal another matrix $\mathbf{Q}$, this relation becomes

$$\mathbf{q}(k + 1) = \mathbf{\Lambda q}(k) + \mathbf{Qu}(k) \tag{B.39}$$

The output equation is given by

$$\mathbf{y}(k) = \mathbf{Cx}(k) \tag{B.40}$$

and if relation (B.27) is substituted,

$$\mathbf{y}(k) = \mathbf{CPq}(k) \tag{B.41}$$

We shall now investigate the controllability and observability of this system in the new coordinates. The set of coordinates that cause the system to be decoupled are commonly called the *normal coordinates*.

## B.4 CONTROLLABILITY AND OBSERVABILITY

In Chapter 6 we examined the conditions for linear discrete-time dynamic systems to be controllable or observable. We found that the necessary condition could be expressed as a rank requirement on the controllability and observability matrices respectively.

The condition for controllability was given by expression (6.7.6) to be

$$\text{rank}\,[\mathbf{A}^{n-1}\mathbf{B} \quad \mathbf{A}^{n-2}\mathbf{B} \quad \cdots \quad \mathbf{AB} \quad \mathbf{B}] = n \tag{B.42}$$

while the condition for observability was given in expression (6.8.10) as

$$\text{rank}\begin{bmatrix} \mathbf{C} \\ \mathbf{CA} \\ \vdots \\ \mathbf{CA}^{n-1} \end{bmatrix} = n \tag{B.43}$$

In this section the goal will be to present alternative conditions to these in order that these concepts might be more easily understood.

### Controllability

If we examine relation (B.39) and find that the $i$th row of the $\mathbf{Q}$ matrix is composed entirely of zeros, the elements of the control vector sequence $\mathbf{u}(k)$ will not affect the $i$th normal coordinate sequence $q_i(k)$. The $i$th normal coordinate will then be said to be uncontrollable. More than one of the normal coordinates may be uncontrollable. If the mode that is uncontrollable has an unstable eigenvalue (exterior to the unit circle), feedback control may *not* be used to stabilize this mode. On the other hand, if the normal coordinate which is uncontrollable is associated with a stable eigenvalue, then under the action of feedback control, the response in that coordinate will occur in an open-loop fashion and decay in an exponential fashion.

**Example B.2.** Consider the matrix-vector difference system governed by the **A** matrix of Examples 6.10 and B.1:

$$\begin{bmatrix} x_1 \\ x_2 \end{bmatrix}_{k+1} = \begin{bmatrix} 1 & -0.2 \\ 0.4 & 0.4 \end{bmatrix} \begin{bmatrix} x_1 \\ x_2 \end{bmatrix}_k + \begin{bmatrix} 1 \\ 1 \end{bmatrix} u(k)$$

If we decouple this set of difference equations, we get

$$\begin{bmatrix} q_1 \\ q_2 \end{bmatrix}_{k+1} = \begin{bmatrix} 0.8 & 0 \\ 0 & 0.6 \end{bmatrix} \begin{bmatrix} q_1 \\ q_2 \end{bmatrix}_k + \begin{bmatrix} 2 & -1 \\ -1 & 1 \end{bmatrix} \begin{bmatrix} 1 \\ 1 \end{bmatrix} u(k)$$

Upon carrying out the multiplication, we get

$$\begin{bmatrix} q_1 \\ q_2 \end{bmatrix}_{k+1} = \begin{bmatrix} 0.8 & 0 \\ 0 & 0.6 \end{bmatrix} \begin{bmatrix} q_1 \\ q_2 \end{bmatrix}_k + \begin{bmatrix} 1 \\ 0 \end{bmatrix} u(k)$$

Note now that the second difference equation is

$$q_2(k + 1) = 0.6q_2(k)$$

and it is apparent that the control effort $u(k)$ cannot influence the system dynamics associated with the second eigenvalue ($\lambda_2 = 0.6$). The conclusion here is the same as that reached in Example 6.10, where the system was found to be uncontrollable. In the case of this system there is no state feedback control scheme that will control this second mode.

There exist physical problems where the uncontrollable modes are those which are unstable, and in this case the closed-loop system will be unstable because no feedback control efforts can be used to stabilize the unstable mode.

## Observability

We have already discussed the condition for observability developed in Chapter 6. Let us now examine the output relation given by relation (B.41). If the **CP** matrix in that relation has a $j$th column that is zero, the normal coordinate $q_j(k)$ will not appear in the output sequence $\mathbf{y}(k)$. No continued measurement of $\mathbf{y}(k)$ will allow us to infer anything about $q_j(k)$. The $j$th mode, in this case, is said to be unobservable. That portion of the state vector sequence due to $q_j(k)$ can never be deduced because the information simply is not in the measurements.

**Example B.3.** Consider the system of Example B.2 except now let us specify a scalar output given by

$$y(k) = [1 \quad -1] \begin{bmatrix} x_1 \\ x_2 \end{bmatrix}_k$$

Investigate from a modal coordinate point of view the observability of this system.

The decoupled equations from Example B.2 are

$$\begin{bmatrix} q_1 \\ q_2 \end{bmatrix}_{k+1} = \begin{bmatrix} 0.8 & 0 \\ 0 & 0.6 \end{bmatrix} \begin{bmatrix} q_1 \\ q_2 \end{bmatrix}_k + \begin{bmatrix} 1 \\ 0 \end{bmatrix} u(k)$$

If we now apply expression (B.41), the output equation is

$$y(k) = [1 \quad -1] \begin{bmatrix} 1 & 1 \\ 1 & 2 \end{bmatrix} \begin{bmatrix} q_1 \\ q_2 \end{bmatrix}_k$$

and upon carrying out the indicated multiplication, we get

$$y(k) = [0 \quad -1] \begin{bmatrix} q_1 \\ q_2 \end{bmatrix}_k$$

This expression tells us that the dynamics of the first mode ($\lambda_1 = 0.8$) are not available in the output measurements. In other words, it will be impossible to reconstruct state histories regardless of how many measurements of $y(k)$ are taken.

If the transfer function between input $u(k)$ and output $y(k)$ were found, we would find it to be zero since the first mode is unobservable and the second is uncontrollable.

## B.5 CAYLEY–HAMILTON THEOREM

The Cayley–Hamilton theorem is perhaps one of the most important theorems from linear algebra which is directly applicable to modern control systems problems. The theorem states that every matrix satisfies its own characteristic equation. In mathematical terms, if an $n \times n$ matrix $\mathbf{A}$ has a characteristic equation given by

$$\lambda^n + d_1\lambda^{n-1} + \cdots + d_{n-1}\lambda + d_n = 0 \tag{B.44}$$

then

$$\mathbf{A}^n + d_1\mathbf{A}^{n-1} + \cdots + d_{n-1}\mathbf{A} + d_n\mathbf{I} = 0 \tag{B.45}$$

There are $n$ eigenvalues which satisfy (B.45), and thus

$$\lambda_i^n + d_1\lambda_i^{n-1} + \cdots d_{n-1}\lambda_i + d_n = 0 \qquad i = 1, \ldots, n \tag{B.46}$$

These $n$ relations may be written in matrix form by noting that a diagonal matrix raised to a power is also a diagonal matrix with each of the original matrix elements raised to that power, or

$$\mathbf{\Lambda}^n + d_1\mathbf{\Lambda}^{n-1} + \cdots + d_{n-1}\mathbf{\Lambda} + d_n\mathbf{I} = \mathbf{0} \tag{B.47}$$

Recall from the inverse of relation (B.24) that $\mathbf{\Lambda}^k = \mathbf{P}^{-1}\mathbf{A}^k\mathbf{P}$, so we may rewrite (B.47) as

$$\mathbf{P}^{-1}\mathbf{A}^n\mathbf{P} + d_1\mathbf{P}^{-1}\mathbf{A}^{n-1}\mathbf{P} + \cdots + d_{n-1}\mathbf{P}^{-1}\mathbf{A}\mathbf{P} + d_n\mathbf{P}^{-1}\mathbf{P} = \mathbf{0} \tag{B.48}$$

Factoring $\mathbf{P}^{-1}$ to the left and $\mathbf{P}$ to the right, the result is

$$\mathbf{P}^{-1}[\mathbf{A}^n + d_1\mathbf{A}^{n-1} + \cdots + d_{n-1}\mathbf{A} + d_n\mathbf{I}]\mathbf{P} = \mathbf{0} \tag{B.49}$$

If we premultiply both sides by $\mathbf{P}$ and postmultiply by $\mathbf{P}^{-1}$, the result is

$$\mathbf{A}^n + d_1\mathbf{A}^{n-1} + \cdots + d_{n-1}\mathbf{A} + d_n\mathbf{I} = \mathbf{0} \tag{B.50}$$

and the theorem is proven. Let us now solve for $\mathbf{A}^n$ to yield

$$\mathbf{A}^n = -d_1\mathbf{A}^{n-1} - \cdots - d_{n-1}\mathbf{A} - d_n\mathbf{I} \tag{B.51}$$

We may premultiply by $\mathbf{A}$ to yield

$$\mathbf{A}^{n+1} = -d_1\mathbf{A}^n - \cdots - d_n\mathbf{A}^2 - d_n\mathbf{A} \tag{B.52}$$

We may substitute from (B.51) for $\mathbf{A}^n$ to give

$$\begin{aligned}\mathbf{A}^n = {} & -d_1(-d_1\mathbf{A}^{n-1} - \cdots - d_{n-1}\mathbf{A} \\ & - d_n\mathbf{I}) - d_2\mathbf{A}^{n-1} - \cdots - d_n\mathbf{A}^2 - d_n\mathbf{A}\end{aligned} \tag{B.53}$$

From relations (B.51) and (B.53) it is clear for any $n \times n$ matrix $\mathbf{A}$ that powers of $\mathbf{A}$ greater than or equal to $n$ may be represented by a linear combination of the powers of $\mathbf{A}$ less than $n$. This theorem will be useful in proof of the conditions for controllability and observability in Chapter 6.

## REFERENCES


Chen, C. T., 1984. *Linear System Theory and Design*, Holt, Rinehart and Winston, New York.

Brogan, W. L., 1985. *Modern Control Theory*, Prentice Hall, Englewood Cliffs, NJ.

Kailath, T., 1980. *Linear Systems*, Prentice Hall, Englewood Cliffs, NJ.

# Appendix C
# Proof of the Matrix Inversion Lemma

In a number of problems of optimal control, state estimation, and parameter estimation, one is faced with the task of computing the inverse of a composite matrix to give its inverse **M**. The matrix **M** to be calculated is of a particular form, which is

$$\mathbf{M} = (\mathbf{P}^{-1} + \mathbf{B}\mathbf{R}^{-1}\mathbf{B}^T)^{-1} \tag{C.1}$$

where **P** and **R** are such that **P** is $n \times n$, **R** is $m \times m$, and **B** is $n \times m$. In most problems of interest $m$ is less than $n$. Let us evaluate $\mathbf{M}^{-1}$ or

$$\mathbf{M}^{-1} = \mathbf{P}^{-1} + \mathbf{B}\mathbf{R}^{-1}\mathbf{B}^T \tag{C.2}$$

Now premultiply by **M** to give

$$\mathbf{I} = \mathbf{M}\mathbf{P}^{-1} + \mathbf{M}\mathbf{B}\mathbf{R}^{-1}\mathbf{B}^T \tag{C.3}$$

Now postmultiply by **P** to give

$$\mathbf{P} = \mathbf{M} + \mathbf{M}\mathbf{B}\mathbf{R}^{-1}\mathbf{B}^T\mathbf{P} \tag{C.4}$$

Now postmultiply by **B** and factor the product **MB** out to the left to give

$$\mathbf{P}\mathbf{B} = \mathbf{M}\mathbf{B}(\mathbf{I} + \mathbf{R}^{-1}\mathbf{B}^T\mathbf{P}\mathbf{B}) \tag{C.5}$$

Factor $\mathbf{R}^{-1}$ noting that $\mathbf{I} = \mathbf{R}^{-1}\mathbf{R}$; then the result is

$$\mathbf{P}\mathbf{B} = \mathbf{M}\mathbf{B}\mathbf{R}^{-1}(\mathbf{R} + \mathbf{B}^T\mathbf{P}\mathbf{B}) \tag{C.6}$$

Postmultiply by the inverse of the matrix in parentheses:

$$\mathbf{P}\mathbf{B}(\mathbf{R} + \mathbf{B}^T\mathbf{P}\mathbf{B})^{-1} = \mathbf{M}\mathbf{B}\mathbf{R}^{-1} \tag{C.7}$$

Now postmultiply by $\mathbf{B}^T$ to give

$$\mathbf{PB}(\mathbf{R} + \mathbf{B}^T\mathbf{PB})^{-1}\mathbf{B}^T = \mathbf{MBR}^{-1}\mathbf{B}^T \tag{C.8}$$

The matrix $\mathbf{BR}^{-1}\mathbf{B}^T$ may be replaced by $\mathbf{M}^{-1} - \mathbf{P}^{-1}$ from relation (C.2) to yield

$$\mathbf{PB}(\mathbf{R} + \mathbf{B}^T\mathbf{PB})^{-1}\mathbf{B}^T = \mathbf{M}(\mathbf{M}^{-1} - \mathbf{P}^{-1}) \tag{C.9}$$

Recognizing that $\mathbf{MM}^{-1}$ is the identity matrix gives

$$\mathbf{PB}(\mathbf{R} + \mathbf{B}^T\mathbf{PB})^{-1}\mathbf{B}^T = \mathbf{I} - \mathbf{MP}^{-1} \tag{C.10}$$

Now postmultiply by $\mathbf{P}$ to give

$$\mathbf{PB}(\mathbf{R} + \mathbf{B}^T\mathbf{PB})^{-1}\mathbf{B}^T\mathbf{P} = \mathbf{P} - \mathbf{M} \tag{C.11}$$

The solution for $\mathbf{M}$ is thus

$$\mathbf{M} = \mathbf{P} - \mathbf{PB}(\mathbf{R} + \mathbf{B}^T\mathbf{PB})^{-1}\mathbf{B}^T\mathbf{P} \tag{C.12}$$

Now we see that the only matrix which must be inverted is $(\mathbf{R} + \mathbf{B}^T\mathbf{PB})$, which is $m \times m$, and hence smaller than the original $n \times n$ matrix. In the case where $m = 1$, which is not uncommon, $\mathbf{R}$ is scalar and hence the inverse term is given by a simple division.

# Appendix D
# Digital Control Designer

The software that accompanies this book provides a computational environment to support the learning associated with Chapters 2 through 9. It runs on any IBM-compatible personal computer with VGA graphics. A popular alternative would be to use the Control Systems Toolbox associated with MATLAB (The MathWorks, 1992), which gives a more flexible environment at the expense of more programming. An attractive simulation environment for digital control systems is that of VisSim (Visual Solutions, 1993).

The structure of the program, which is dictated by the menu structure, is outlined briefly below. The program consists of three modules which address transfer function design, state variable design, and a set of utility routines, respectively. The structure of the program is outlined below.

## Transfer Function Design

Transfer function–based design as addressed in Chapters 3 and 4 is addressed by this module. The following tasks can be accomplished.

1. The pulse transfer function for the zero-order hold, plant transfer function, and sampler can be calculated.
2. The pulse transfer function for the plant can be entered directly.
3. Data for the cascade compensator/controller are entered directly.
4. The root locus for the closed-loop system as a function of a gain parameter is calculated.
5. The open-loop frequency response for the controller-plant transfer function is calculated for Nyquist, Bode, or Nichols chart design.

6. Closed-loop step or ramp response is calculated.
7. Closed-loop frequency response can be calculated for purposes of checking specifications.

### State-Space Design

State-space-based design as examined in Chapters 6, 8, and 9 is addressed by this module. The following tasks are performed:

1. Calculate the discrete-time state-variable model for the zero-order hold, continuous-time plant, and samplers along with open-loop pole locations.
2. Direct input of the discrete-time plant model.
3. Calculate state feedback gains for arbitrary pole placement and calculate observer gains for both prediction and current observers.
4. Calculate state feedback gains using linear-quadratic optimal control and calculation of the resulting closed-loop pole locations.
5. Simulate the time response for the open-loop plant, state feedback, or estimated state feedback for arbitrary initial conditions and constant reference inputs.

### Utility Routines

This module addresses a group of computationally tedious tasks involved in digital control system design. These are useful in other contexts as well. The tasks performed are:

1. Multiply two polynomials to obtain the product polynomial.
2. Find the roots and factors of a polynomial of arbitrary degree.
3. Find the inverse $z$-transform of a $z$-domain function in equation form and provide a graph of the inverse sequence.
4. Solve a set of real linear equations of arbitrary order.
5. Invert a real matrix of arbitrary order.
6. Calculate and plot the frequency response of an arbitrary discrete-time transfer function.
7. Calculate the eigenvalues of a real matrix of arbitrary order.

## REFERENCES

The MathWorks, Inc., 1992. *Matlab Reference Guide*, Natick, MA.
Visual Solutions, Inc., 1993. *VisSim User's Guide—Version 1.2*, Westford, MA.

# Index

* 9 7 8 0 8 2 4 7 8 9 1 4 5 *

T - #0077 - 101024 - C0 - 234/156/23 [25] - CB - 9780824789145 - Gloss Lamination